Trends in Counterfeit Drugs

Counterfeit drugs continue to infiltrate the drug market in the United States, causing illness and death. This book addresses this issue and examines the recent trends in drug counterfeiting over the past 5–10 years. The text shows perspectives from federal lab, state and city crime lab and toxicology lab personnel and academic researchers, and includes topics such as a history of cases and issues with counterfeit drugs, trends observed in forensic labs, instrumental methods and approaches used in detecting counterfeit medicines, and policy approaches for controlling counterfeit drugs. There is a focus on ways to reduce counterfeit drugs in the market, to help improve the health and safety of people all over the world.

Features

- Focuses on the recent (5–10 years) trends in counterfeit drugs and analysis.
- Shows perspectives from crime lab and toxicology lab personnel and academic researchers.
- Focuses on drugs seized by law enforcement and approaches to reducing counterfeit medicine in the market.
- Discusses the detection and analysis of counterfeit drugs, and appropriate tools for combating this issue.
- Emphasizes the global impact of illegal medicines.

Counterfeit Drugs Series

Series Editor
Kelly M. Elkins

Trends in Counterfeit Drugs
Edited by Kelly M. Elkins

Trends in Counterfeit Drugs

Edited By
Kelly M. Elkins
Chemistry Department and Forensic Science Program,
Towson University

CRC Press is an imprint of the
Taylor & Francis Group, an **informa** business

Front cover image: Andre Boukreev/Shutterstock

First edition published 2024
by CRC Press
6000 Broken Sound Parkway NW, Suite 300, Boca Raton, FL 33487-2742

and by CRC Press
4 Park Square, Milton Park, Abingdon, Oxon, OX14 4RN

CRC Press is an imprint of Taylor & Francis Group, LLC

ISBN: 978-1-032-02427-1 (HB)
ISBN: 978-1-032-02428-8 (PB)
ISBN: 978-1-003-18332-7 (EB)

DOI: 10.1201/9781003183327

Typeset in Times
by codeMantra

Dedication

To my parents, Dale and Rachel, and all people who rely on authentic medicines for their health.

Contents

Foreword

I am delighted to have been invited by my colleague, Kelly Elkins, to write a foreword for this book. The subject of this work represents an important and timely topic, counterfeit drugs, that is having a detrimental impact on our society. The editor has assembled an outstanding list of authors representing different areas of academia, research, and government agencies. I know many of the authors personally and can attest to their expertise. These are experts in this field, and they have contributed detailed examples and technical information in chapters on a variety of topics related to counterfeit drugs.

For many years, counterfeit drugs have been distributed to pharmacies and consequently ended up in people's medicine cabinets. This occurrence is on the rise and has become a worldwide problem due to increased internet sales and new methods of manufacturing and distribution. The expanding prevalence of counterfeit medications is not only a concern for patients but also for pharmacies and pharmaceutical companies. Counterfeit drugs are not only illegal but can also have severe detrimental health consequences if the user consumes the wrong medication. High-demand medicines such as chemotherapeutic agents, antibiotics, and others are targets for this kind of illegal activity. Many who are deceived by counterfeit drugs are those using these drugs illegally or people seeking a cheaper alternative to the expensive medicines. The counterfeit medications generally resemble legitimate commercially manufactured products such as packaging labels, pills, and tablets. Often, they may have much more of the active ingredient than prescribed, no active ingredient at all, or may contain harmful substances. They often are overlooked by pharmacies and consumers, hence, leading to dangerous and harmful results, including sometimes death.

The counterfeit drug issue is not only a societal problem but is also a challenge for analytical, pharmaceutical, and forensic scientists. The topics covered in this text are of particular interest to these professions. An introduction and history of counterfeit drug cases is presented as well as a survey of trends in seized counterfeit drugs and toxicological analysis. Opiate overdoses have become a critical problem over the recent years, and a thorough coverage of fentanyl adulteration of drugs is discussed. The detection and analysis of counterfeit drugs are described, and tools for combating counterfeit drugs are presented in a separate chapter. The reader should find the chapter with examples of poisoning cases caused by fake Xanax® tablets of interest. Finally, approaches to countering the problem of counterfeit drugs are offered.

Written for readers of various levels of experience in this field, the information does not presume prior knowledge of the subject. The book will be most useful to the practicing scientists in the pharmaceutical field who analyze counterfeit drug formulations and to the forensic scientists involved in the analysis of drugs. It will also be a valuable resource to students and instructors of higher learning, specifically students and college professors in pharmaceutical and forensic science programmes. Others, however, will find this book of interest and useful. Attorneys, judges, and law enforcement should find the information of use. The public, and perhaps mystery writers, will find the information interesting, particularly the case examples and applications presented. It is hoped that the information here will help promote public awareness and educate the public about the counterfeit drug problem leading to improved global health.

Thomas A. Brettell, Ph.D., *ABC-GKE*
Chair, Professor of Chemistry
Department of Chemical, Physical, &
Forensic Sciences
Cedar Crest College
Allentown, PA 18104

Preface

Americans and foreign nationals living in the United States rely upon authentic pharmaceuticals to treat a variety of medical issues. According to the United States (U.S.) Centers for Disease Control (CDC) National Center for Health Statistics (NCHS), from 2015 to 2018, 48.6% of people in the U.S. took at least one prescription drug in the past 30 days. A reported 2.9 billion drugs were ordered or provided by physicians according to 2016 national data.

The world's supply of medicines is subject to regulation, monitoring, and testing. Fake medicines pose significant dangers to consumers but are lucrative for counterfeiters. In the U.S., pharmaceutical drugs are regulated by the Food and Drug Administration (FDA)—a Department of Health and Human Services (HHS) agency. According to a 2015 report, the World Health Organization (WHO) estimates that the world's market of fake medicines is worth $200 billion a year and costs the pharmaceutical industry $83 billion a year in lost revenue. It is estimated that 1% of pharmaceuticals in the U.S. markets are fake, although the percentage is increasing. Insufficient surveillance and monitoring enable counterfeit drugs to penetrate the marketplace and pose risks to consumers. The FDA reported that 85% of drugs sold from online pharmacies claiming to be Canadian pharmacies are delivered from other countries.

U.S. border monitoring and commerce are regulated by the Customs and Border Protection (CBP) law enforcement agency and the U.S. Coast Guard (USCG), both agencies of the Department of Homeland Security (DHS). Together CBP and USCG monitor 95,439 miles of coastline and 3,796,742 square miles of land. Federal, state, and local law enforcement officers and forensic lab personnel report that the number of counterfeits identified by their labs is increasing. Consumers can report suspected counterfeit drugs to the FDA's MedWatch office. The FDA Forensic Chemistry Center (FCC) applies its array of analytical techniques and has reported on the identification of counterfeit medicines. Numerous state and regional labs, private testing labs, academic labs, and public–private entities have reported the same—counterfeit drugs are a threat to Americans and all people living in the country's borders. Scientists continue to work to improve the analytical testing methodologies and instruments to identify counterfeit drug products. The goal of this book is to elucidate the trends in counterfeit drugs in pharmaceutical commerce and methods for identification and approaches to consumers, and reducing the threat of fake medicines. A team of experts have collaborated to write the chapters contained herein and present their most current data. We anticipate that the information will be especially valuable to citizens, doctors and their patients, law enforcement personnel, lawmakers, and students of the forensic sciences.

Kelly M. Elkins
July 2021

Acknowledgements

Hilary LaFoe, CRC Press acquisitions editor, invited me to serve as editor of their new series on Counterfeit Drugs and enthusiastically supported me in producing this first volume. Her team including Jessica Poile, Danielle Zarfati, and Sukirti Singh were terrific to work with. I want to thank Dr. David Newman for a thoughtful review of my proposal for this volume, and Erin Artigiani, Tom Carr and Jerry Daley for helpful conversations, and discussing the goals of HIDTA and their work. This project and the data contained herein is the result of years of work and extensive experience by numerous scientists and law enforcement personnel; it would not have been possible without them. I look forward to hearing from our readers and continuing the conversation and adding additional approaches to further reduce and ultimately eradicate the problem of counterfeit drugs in our markets in years to come.

List of Abbreviations

Term	Abbreviation
4-Aminophenyl-1-phenethylpiperidine	**4-ANPP**
American Academy of Forensic Sciences	**AAFS**
Acquired immune deficiency syndrome	**AIDS**
Alternate light source	**ALS**
Amplitude modulated	**AM**
Abbreviated New Drug Application	**ANDA**
Atmospheric pressure chemical ionization	**APCI**
Asia-Pacific Economic Cooperation	**APEC**
Active pharmaceutical ingredients	**API**
Atmospheric solids analysis probe	**ASAP**
American Society of Crime Laboratory Directors	**ASCLD**
Attenuated total reflectance Fourier transform infrared spectroscopy	**ATR FT-IR**
Area under the curve	**AUC**
Computer aided diagrams	**CAD**
Cannabidiol	**CBD**
Customs and Border Protection	**CBP**
Controlled dangerous substances	**CDS**
Centers for Disease Control	**CDC**
Collision-induced dissociation	**CID**
Continuous inject printing	**CIJ**
Counterfeit drug identifier	**CODI**
Controlled Substances Act	**CSA**
Chemical warfare agents	**CWAs**
Code of Federal Regulations	**FR**
Direct analysis in real time	**DART**
Drug Enforcement Administration	**DEA**
Desorption electrospray ionization	**DESI**
Department of Homeland Security	**DHS**
Department of Health and Human Services	**DHHS**
Digital light microscope	**DLM**
Difference measurements	**DM**
Deoxyribonucleic acid	**DNA**
Drop on demand	**DOD**
Express courier hubs	**ECHs**

Environmental Protection Agency	**EPA**
Federal Bureau of Investigation	**FBI**
Forensic Chemistry Center	**FCC**
Food and Drug Administration	**FDA**
Federal Food, Drug, and Cosmetic Act	**FD&C**
Fast Fourier transform	**FFT**
Frequency modulated	**FM**
Fourier transform infrared spectroscopy	**FT-IR**
Gas chromatography–flame ionization detection/mass spectrometry	**GC-FID/MS**
Gas chromatography-infrared spectroscopy	**GC-IR**
Gas chromatography–mass spectrometry	**GC-MS**
Grey-Level Co-Occurrence Matrix	**GLCM**
Global Pharma Health Fund	**GPHF**
Global positioning system	**GPS**
High-Intensity Drug Trafficking Area	**HIDTA**
High-pressure (or performance) liquid chromatography	**HPLC**
High-performance liquid chromatography with ultraviolet detection	**HPLC-UV**
High-pressure mass spectrometry	**HPMS**
Headspace gas chromatography–mass spectrometry	**HS-GC-MS**
Homeland Security Investigations	**HSI**
2D image analysis	**2DIA**
3D image analysis	**3DIA**
International Council of Harmonization of Technical Requirements for Pharmaceuticals for Human Use	**ICH**
Inductively coupled plasma-mass spectrometry	**ICP-MS**
International mail facilities	**IMF**
Ion mobility spectrometry	**IMS**
International Criminal Police Organization	**INTERPOL**
Intellectual property	**IP**
Institutional Review Board	**IRB**
Internal Revenue Service	**IRS**
International unit of activity	**IU**
Infrared	**IR**
Laser ablation multicollector inductively coupled plasma-mass spectrometry	**LA-MC-ICP-MS**
Liquid chromatography	**LC**
Liquid chromatography–mass spectrometry	**LC-MS**
Liquid chromatography–mass spectrometry/mass spectrometry	**LC-MS/MS**
Light-emitting diodes	**LEDs**
Low-and middle-income countries	**LMIC**

Life Sciences Innovation Forum	**LSIF**
3,4-Methylenedioxymethamphetamine	**MDMA**
Magnetic resonance imaging	**MRI**
Mass spectrometry	**MS**
Neutron activation analysis	**NAA**
National Agency for Food and Drug Administration and Control	**NAFDAC**
New drug application	**NDA**
National Institute for the Control of Pharmaceutical and Biological Products of China	**NICPBP**
National Institutes of Health	**NIH**
National Institute of Standards and Technology	**NIST**
Nuclear magnetic resonance spectroscopy	**NMR**
New psychoactive substance	**NPS**
Office of Criminal Investigations	**OCI**
Organization for Economic Cooperation and Development	**OECD**
Paper analytical devices	**PADs**
para-fluorobutyryl fentanyl	**para-FBF**
Principal components analysis	**PCA**
Pharmacists Council of Nigeria	**PCN**
Polymerase chain reaction	**PCR**
Phosphodiesterase type-5	**PDE-5**
Particle-induced X-ray emission	**PIXE**
Polarized light microscopy	**PLM**
Partial least squares	**PLS**
Partial least-squares regression	**PLSR**
Profile measurement	**PM**
Performance management process	**PMP**
Port of entry	**POE**
Personal protective equipment	**PPE**
Pharmaceutical Security Institute	**PSI**
Quality Assurance	**QA**
Quality Control	**QC**
Quick response	**QR**
Regulatory Harmonization Steering Committee	**RHSC**
Radio frequency	**RF**
Radio frequency identification device	**RFID**
Receiver Operator Curves	**ROC**
Region of interest	**ROI**
Sodium dodecyl sulfate polyacrylamide gel electrophoresis	**SDS-PAGE**
Scanning electron microscopy with energy dispersive X-ray detection	**SEM-EDX**

Surface-enhanced Raman spectroscopy	**SERS**
Stereo light microscopy	**SLM**
Single lens reflex	**SLR**
Signal-to-noise ratio	**SNR**
Spatially Offset Raman Spectroscopy	**SORS**
Solid-Phase Microextraction	**SPME**
Surface plasmon resonance	**SPR**
Size, Weight, and Power	**SWaP**
Severe acute respiratory syndrome coronavirus 2 2019	**COVID-19**
Tuberculosis	**TB**
Time domain-nuclear magnetic resonance spectroscopy	**TD-NMR**
Triboelectric nanogenerator	**TENG**
Total ion chromatogram	**TIC**
Toxic industrial chemicals	**TICs**
Thin layer chromatography	**TLC**
Trimethyl silyl	**TMS**
Time of flight	**TOF**
United Kingdom	**UK**
United States	**US**
U.S. Coast Guard	**USCG**
U.S. dollars	**USD**
U.S. Postal Inspection Service	**USPIS**
Ultraviolet	**UV**
Verified internet pharmacy practice sites	**VIPPS**
Visible	**VIS**
Visible-near infrared	**VNIR**
Video spectral comparator	**VSC**
World Health Assembly	**WHA**
World Health Organization	**WHO**
X-ray powder diffraction	**XRD**
X-ray fluorescence	**XRF**

About the Editor

Dr. Kelly Elkins is the author and editor of several books including *Introduction to Forensic Chemistry, Forensic DNA Biology: A Laboratory Manual, Next Generation Sequencing in Forensic Science: A Primer,* and *International Ethics in Chemistry: Developing Common Values across Cultures.* Dr. Elkins is a Professor of Chemistry at Towson University and the author of more than 50 papers and invited book chapters on her research. Her work has been published in the *Journal of Forensic Sciences, Analytical Biochemistry, Drug Testing and Analysis,* and *Medicine, Science and the Law.* She is also one of the three founding editors of the *Journal of Forensic Science Education.* She has taught courses in forensic chemistry and forensic biology under various course numbers at four colleges and universities since 2006. She is a Fellow of the American Academy of Forensic Sciences, a member of the American Chemical Society, and a member of the Council of Forensic Science Educators and served as its President in 2012. She consults on forensic cases.

List of Contributors

Douglas Albright
Forensic Chemistry Center
U.S. Food & Drug Administration
Cincinnati, Ohio

Richard A. Crocombe
Crocombe Spectroscopic Consulting, LLC
Winchester, Massachusetts

Theresa DeAngelo
Quality Assurance Manager
Maryland State Police Forensic Sciences Division
Pikesville, Maryland

Kirby B. Drake
Kirby Drake Law
PLLC
Dallas, Texas

Kelly M. Elkins
Chemistry Department and Forensic Science Program
Towson University
Towson, Maryland

Haley Fallang
Montana Department of Justice Forensic Science Division
Missoula, Montana
Chemistry Department and Forensic Science Program
Towson University
Towson, Maryland

Cheryl L. Flurer
Forensic Chemistry Center
U.S. Food & Drug Administration
Cincinnati, Ohio

Lianji Jin
Forensic Chemistry Center
U.S. Food & Drug Administration
Cincinnati, Ohio

Brooke W. Kammrath
Henry C. Lee College of Criminal Justice and Forensic Sciences
Henry C. Lee Institute of Forensic Science
University of New Haven
West Haven, Connecticut

Martin Kimani
Forensic Chemistry Center
U.S. Food & Drug Administration
Cincinnati, Ohio

Adam Lanzarotta
Forensic Chemistry Center
U.S. Food & Drug Administration
Cincinnati, Ohio

Pauline E. Leary
Noble Worldwide Headquarters
Boston, Massachusetts

Nikolas P. Lemos
Cameron Forensic Medical Sciences, Barts
The London School of Medicine and Dentistry
Queen Mary University of London
London, England

Nicola Ranieri
Forensic Chemistry Center
U.S. Food & Drug Administration
Cincinnati, Ohio

Mark R. Witkowski
Forensic Chemistry Center
U.S. Food & Drug Administration
Cincinnati, Ohio

1 Introduction to Counterfeit Drugs

Mark R. Witkowski, Adam Lanzarotta, and Cheryl L. Flurer
U.S. Food & Drug Administration

CONTENTS

Disclaimer: The contents of this introduction are the authors' opinions and should not be considered as opinions or policy of the U.S. FDA. The mention of trade names and manufacturers is for technical accuracy and should not be considered as endorsement of a specific product or manufacturer.

1.1 INTRODUCTION

In 1967, Margaret Kreig wrote the book *Black Market Medicine* (Kreig, 1967). In the book, she provided a firsthand account of the world of drug counterfeiting by working with and interviewing Food and Drug Administration (FDA) investigators along with her own research using resources available to her at the

DOI 10.1201/9781003183327-1

time. Published over 50 years ago, some may consider Kreig's book to be dated; however, much of her work concerning drug counterfeiting still holds true today. The types of drug products counterfeited today may be different from those in the 1960s, but the world of counterfeit drugs continues to be a dangerous place to the consumer and financially lucrative for the counterfeiter.

The FDA's Forensic Chemistry Center (FCC) has been involved in the analysis of suspect counterfeit active pharmaceutical ingredients (APIs), pharmaceutical products, and packaging for more than 25 years. In the middle to late 1990s, the focus was on the analysis of suspect counterfeit APIs. The first examples of counterfeit finished dosage products were not received and analyzed until the early 2000s (e.g., Viagra tablets and injectable products such as Neupogen and Procrit) (Kaine et al., 1994; U.S. Congress, 2000; Erikson, 2003; Rudolf et al., 2004). This early work was key in establishing testing protocols and developing methods for analyzing suspect counterfeit drugs, which continue to be used today. Since the early 2000s, the number of suspect counterfeit drug products submitted to the FCC for analysis has increased and has coincided with the expansion of the internet and increased use of online pharmacies. The number and types of products found to be counterfeit have expanded from lifestyle drugs (e.g., Viagra, Alli) to medically necessary drugs (e.g., Lipitor, Plavix), biologics (e.g., Avastin), and opioids (e.g., oxycodone) (Rudolf et al., 2004; The Partnership for Safe Medicines, 2010; Flaherty and Gaul, 2003; Department of Justice, 2013; The Partnership for Safe Medicines, 2014; and U.S. FDA, 2018 and 2019).

Based on its past and current experiences, the FCC uses a multidisciplinary approach to analyze suspect counterfeit drug products. This approach incorporates a broad array of laboratory instrumentation, methods, and expertise that are especially necessary when analyzing sophisticated suspect counterfeit drugs where a single technique is not adequate to determine if the product is counterfeit or authentic. Depending on the type of product (e.g., tablet, injectable), a variety of physical and instrumental techniques may be required to distinguish between a suspect counterfeit and the authentic product (Lanzarotta et al., 2011; Degardin et al., 2013, Chapter 1; Rebiere et al., 2017; Bakker-'t Hart et al., 2021). Many of the laboratory-based techniques and methods that have been developed by the FCC are now being implemented to analyze suspect counterfeit drug products both inside and outside of the laboratory setting (Ranieri et al., 2014; Lanzarotta et al., 2015, 2020, 2021; Vickers et al., 2018).

A Google search of the term "counterfeit drugs" produces approximately 11 million search results. The literature and information available on counterfeit drugs, methods and detection technology can be overwhelming. The intent of this introduction chapter is to provide the reader with a starting point of basic information related to the subject of counterfeit drugs. It combines general information with examples based on FCC's experiences and information gathered through the years related to the topic of counterfeit drugs. This provides the reader with a real-world firsthand perspective of the information presented in this chapter. However, this chapter is not meant to be a complete comprehensive review of all things related to counterfeit drugs. To the reader who is experienced in this area, this information may be very familiar. However, for the reader who is new to the area or has an interest in the analysis of counterfeit drugs, this chapter will provide a starting point.

1.1.1 Counterfeit Drugs

The sentence "Counterfeit drug products pose a major public health threat" can be found in some form or another in most articles, books or presentations written on the subject of counterfeit drugs. It is certainly not an overstatement, and it captures the single biggest problem with the manufacture, sale, and consumption of counterfeit drugs: they are illegally manufactured products, which have been and continue to be distributed widely throughout the world (Koczwara, 2017; Blackstone et al., 2014; Finlay, 2011). Counterfeit drug products are unregulated, meaning that they are produced by uncontrolled processes. Therefore, the lack of manufacturing control and batch record information leads to untraceability of events that may be associated with a suspect counterfeit product. The counterfeit product ingredients can be changed at any time, may be harmful, and/or contain dangerous impurities. The bioavailability and environmental stability of these products are unknown, and furthermore, there is no assurance that these products contain the declared active ingredient at the right dosage level, if at all (Degardin et al., 2014). In some cases, adulteration can occur where a different drug substance or drug analog may be substituted for the declared active ingredient (Gratz et al., 2006; Nickum and Flurer, 2015; Patel et al. 2014; Jannetto et al., 2019). Estimates vary, but the highest percentages of counterfeit products are often found in low- to middle-income countries (Mackey et al., 2015; Bakker-'t Hart et al., 2021). Examples of counterfeit drugs include brand name and generic pharmaceutical products covering a wide variety of therapeutic groups such as antibiotics, antihistamines, antimalarial drugs, lifestyle drugs, oncological drugs, steroids, opioids, and other products (Koczwara, 2017; Finlay, 2021; Venhuis et al., 2018).

Patients with prolonged or chronic illnesses may experience exacerbated symptoms or death as a result of taking a counterfeit medication (Koczwara, 2017; Blackstone et al., 2014; Cockburn et al., 2005). For example, if counterfeit versions of a preventative medication such as antimalarial drugs make their way into the legal distribution chain or if a consumer purchases these drugs outside the legal distribution chain, the unsuspecting individual or patient may unknowingly consume a counterfeit product. In the case of malaria, this may leave the patient untreated and those around the patient may still be at risk of developing and transmitting the disease. The purchase of pharmaceutical drugs via the legal supply chain reduces the risk of individuals or patients being exposed to counterfeit drugs. This is extremely important when treating chronically or seriously ill patients (e.g., HIV, cancer). In these instances, treatment protocols are necessary to control the disease or increase survival of the patient. The use or consumption of a counterfeit drug may affect treatment protocols that may be harmful and lessen the prospect of a positive outcome for the patient (Venhuis et al., 2018; WHO, 2019). It is the legal distribution chain that provides security and minimizes the threat of counterfeit drugs. However, when an individual intentionally goes outside the legal distribution chain to purchase pharmaceutical drugs (e.g., oxycodone), the individual has a high risk of encountering extremely dangerous counterfeit drugs. In this case, the individual may unknowingly consume a counterfeit drug containing a highly potent API (e.g., fentanyl) causing extreme harm or death (Mohr et al., 2016; Armenian et al., 2018; Daniulaityte et al., 2019; CDC, 2021).

With access to the internet, the consumer now has the ability to purchase pharmaceutical drug products online. This direct-to-consumer selling of prescription pharmaceuticals weakens the distribution chain by removing safeguards and potentially exposes the consumer to dangerous counterfeit products. This is further exacerbated by countries that have less developed regulatory systems, weak drug regulations, and weak enforcement, which allows for easier importation and exportation of counterfeit drugs (Blackstone et al., 2014; Finlay, 2011; Degardin et al., 2014). Weak distribution chains, laws and regulations give the counterfeiter the perfect environment to move counterfeit products in and out of a country with relative ease. Alternatively, more complex supply chains can provide opportunities for counterfeiters to infiltrate the legitimate supply chain. For example, the increase in outsourcing of drug product manufacturing (i.e., a manufacturer using other companies or contractors to perform portions of the manufacturing process) can offer the counterfeiter an opportunity to exploit and introduce counterfeit products into the legitimate supply chain. All these factors create an environment that favors the counterfeiter, hinders enforcement, and endangers the consumer.

Given the significant danger that they pose to public health, combating counterfeit drugs requires ongoing cooperation and communication among numerous national and global stakeholders. This includes sharing information on counterfeit drugs with national and global regulatory authorities, customs authorities, law enforcement agencies, pharmaceutical manufacturers, pharmaceutical repackagers/distributors, healthcare providers, and consumers/patients. Many national and global regulatory agencies provide information on counterfeit drugs or counterfeit product alerts that is accessible to the public (U.S. FDA, 2022; EMA, 2022; WHO, 2022). In addition to sharing information, it is also useful and necessary to share methods of analysis that will allow others to easily detect and identify counterfeit drugs in their supply and distribution chains. Further, as these groups begin to use new technologies to detect counterfeit drugs and protect the supply chain, it is important that these new users understand the capabilities and limitations of these authentication technologies.

Over the years, the FCC and others have identified and adopted key areas that are important for establishing a robust counterfeit drug analysis program. These include having a clear understanding of what criteria or thresholds are needed to establish that a product is counterfeit, unapproved, or authentic, establishing an authentic product library for use in direct comparison analysis of suspect samples, understanding a product's manufacturing process, understanding the factors that contribute to drug counterfeiting and establishing multidisciplinary analysis techniques and methods to both identify and provide a comprehensive analysis of suspect counterfeit products.

1.1.2 Defining a Counterfeit Drug

It is important to define what criteria or thresholds need to be met to determine if a submitted sample is authentic, unapproved, or counterfeit (Finlay, 2011; Degardin et al., 2014). The term counterfeit is a legal term defined by government laws and regulations being applied to the sample once the analysis is complete. It should be noted that drug manufacturers may also apply intellectual property law to their product providing a broader set of criteria to determine a suspect product as counterfeit (e.g., product shape, color). The FDA relies on the definition of what constitutes a counterfeit, as stated in the Federal Food, Drug, and Cosmetic (FD&C) Act, which is provided in Table 1.1 (21 U.S. Code § 321). For example, the FCC uses analytical instrumental techniques and methods to definitively

TABLE 1.1
Definitions of Authentic, Counterfeit, Diverted, Unapproved Generic, and Unapproved Drug Products Used in This Chapter

Authentic	A product that is produced and approved for distribution in the U.S. markets or other world markets.
Counterfeit	The term "counterfeit drug" means a drug, container, or labeling without authorization, bears the trademark, trade name, or other identifying mark, imprint, or device, or any likeness thereof, of a drug manufacturer, which thereby falsely purports or is represented to be the product of, or to have been packed or distributed by such other drug manufacturer….[a]
Diverted	An authentic product, produced by the recognized/approved manufacturer that is designated for distribution in non-U.S. markets.
Unapproved generic	A term used to describe a product that contains or purports to contain an API, which is still under patent and for which no US-approved generics exist.
Unapproved	A product that contains API(s) not approved for sale in the U.S. market.[b]

[a] See U.S. Code of Federal Regulations, under the Federal Food, Drug, and Cosmetic Act, Sec. 201 [21 U.S.C. 321] for complete definition.

[b] U.S. FDA Orange Book: Approved Drug Products with Therapeutic Equivalence Evaluations.

show that a suspect counterfeit sample is different from the authentic product, which provides the evidence to support the FD&C act definition of a counterfeit drug. Samples collected for testing as part of a market survey, criminal investigation or import seizure are considered a suspect counterfeit product, and once the analysis is complete, the product is reported as either consistent or not consistent with the authentic product. The results of the analysis are then used to support the legal definition during legal proceedings, which then determine if the drug is a counterfeit or not.

While analyzing a suspect counterfeit product, the methods and instrumentation must be robust and reliable to avoid the misidentification of a suspect counterfeit drug. Samples submitted as suspect counterfeit drugs may not be suspect counterfeit products at all but may fall in other categories, which include diverted, unapproved generic, and unapproved drug products (Table 1.1). When developing methods or choosing instrumentation to be used to differentiate suspect counterfeit products from authentic products, it is important to ensure that these methods and instrumentation can differentiate among these other types of drug products.

Additionally, these terms and definitions allow for developing a flow or sample triage chart, which helps guide the analyst through the examination of the suspect product (Figure 1.1). This guide helps ensure that the proper analysis is applied to a given sample and allows for a streamlined process so that only necessary tests are performed.

1.1.2.1 Establishment of an Authentic Drug Product Reference Library

If a suspect counterfeit drug does not contain the correct API, or contains a different API, the determination of product authenticity is straightforward. However, if the correct API is present, then a direct comparison between the

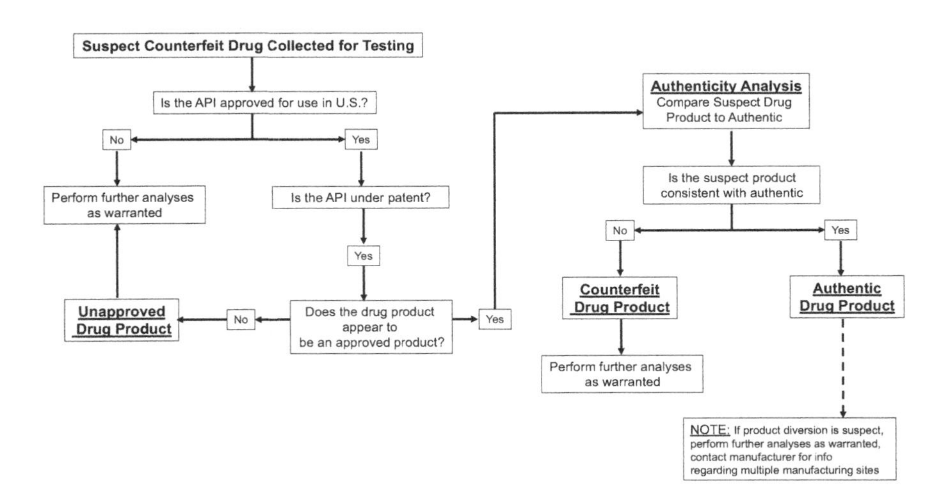

FIGURE 1.1 Example of sample flow chart for the analysis of suspect counterfeit products used at the FDA Forensic Chemistry Center.

suspect and the authentic product is necessary. It is essential to have the correct authentic product to perform the direct comparison, which is not always straightforward. Care must be taken when creating a reference library of authentic drug products to ensure that the tablet shape, color, and debossing are the same as the suspect products collected and submitted for testing (Hefferren, 1962; National Drug Code Directory, n.d.; Drugs.com Pill Identifier, n.d.). An authentic drug product reference library should contain physical examples of the drug product and physical examples of packaging components if possible. In the absence of physical examples, information from the manufacturer about the packaging (e.g., materials used, printing processes) can be used for the comparison by an expert with knowledge in printing processes.

In addition to examples of the authentic product itself, other information that may be helpful include physical and chemical characteristics of the API (e.g., polymorphs, impurities, residual solvents) (Bugay, 2001), the number of manufacturing sites used to produce the product, product formulation and impurity profiles. In the case of a poor counterfeit, all this information may not be necessary to determine authenticity, but this information may be necessary for sophisticated counterfeit products. The authentic product and product information combined with the physical and chemical information obtained from the suspect counterfeit drug will provide sufficient points of comparison to determine the authenticity of a suspect counterfeit product.

Obtaining physical examples and information of the authentic product is a straightforward process for the authentic drug manufacturer. Since they manufacture and package the product, it is easy for them to access retained samples or current lots of an authentic product to compare to suspect counterfeit drug submitted to them for analysis. However, for regulatory agencies such as the FDA or other large law enforcement/government agencies, establishing a comprehensive authentic reference library can be a challenge. Therefore, it is important to establish contacts with the different authentic product manufacturers who can assist with providing examples of authentic products, packaging, and information. Obtaining the authentic drug product directly from the manufacturer ensures a provenance or chain of custody. It not only provides a high degree of confidence in the results obtained, but it also reduces the burden of proving that the authentic product used in the comparison is the correct authentic product. One may look to a local pharmacy to obtain examples of authentic drug products; however, this may raise questions of the origin of the authentic product, especially in regions of the world where the supply chain is not as secure.

For a regulatory agency such as FDA, the number of approved pharmaceutical products for use in the U.S. is very large and to try to establish a comprehensive reference library of all products is impossible. In some cases, a straightforward approach to establish an authentic product library would be to focus the scope or group of drug products of interest (e.g., opioid-containing products). Other approaches can be taken to build an authentic product reference library, such as reactive and proactive approaches. The reactive approach consists of requesting authentic products as needed based on the suspect counterfeit samples received by the laboratory for testing. The proactive approach involves reviewing the scientific literature, medical journals, or news media and looking for trends in what products may have the highest potential to be counterfeited in the future.

1.1.2.2 The Manufacturing Process

Understanding how the authentic drug product is manufactured and packaged can be helpful when trying to decide the best techniques or methods to use to analyze a suspect counterfeit drug. A basic understanding of the manufacturing process of a product helps one understand which differences observed between a suspect counterfeit drug and an authentic are significant and which may be indicative of manufacturing variability.

The science behind the formulation and manufacture of drug products can be sophisticated, complex, and specialized depending on the type of product. There are many different types of pharmaceutical drug preparations including but not limited to solid dosage forms (e.g., tablets, capsules), liquids, lyophilized powders, injectables and inhalers (Allen Jr., 2018). For each different type, a different manufacturing process is required to produce the finished drug product. Additionally, each different product type requires certain types of packaging materials and containers. The product formulation, the manufacturing process, and the packaging all impart unique characteristics on the authentic product. Since the manufacturing process of a pharmaceutical product is regulated, the first dosage unit manufactured for a given batch or lot will be consistent with the last. The unique manufacturing characteristics and consistency of the process can be used to differentiate an authentic from a suspect counterfeit product.

The drug product itself is a drug delivery system, and the most common type of drug delivery system produced is solid dosage forms (e.g., tablets, capsules). They are also the easiest type of drug product to manufacture. Figure 1.2 is a general flow chart of the manufacturing process of a solid dosage form. Although the equipment used has modernized, the general processes used to manufacture solid dosage forms have not changed significantly over the last several decades. The starting point is the raw materials or ingredients. The wet process includes an additional two steps, granulation and drying/milling, followed by blending, tablet compression/capsule filling and tablet coating/capsule printing. Once the product has been manufactured, it is ready for packaging.

The ingredients and amounts used to manufacture a tablet or capsule are commonly known as the formulation. The formulation or list of ingredients on a product carton, bottle label or patient insert may seem unimportant or arbitrary, but to the experienced pharmaceutical formulator or drug chemist, each ingredient has an important role in the stability and function of the authentic product. The raw materials or ingredients used in the product formulation can be separated into two groups: the API and excipients. The API is the pharmacologically active compound that generates the therapeutic effect on the body. The excipients or inactive ingredients have a variety of uses in the drug product formulation (Bugay, 2001; Rowe et al., 2009). Excipients can be used as fillers, binders, and lubricants, also provide physical stability (hardness) and chemical stability, and help the dosage

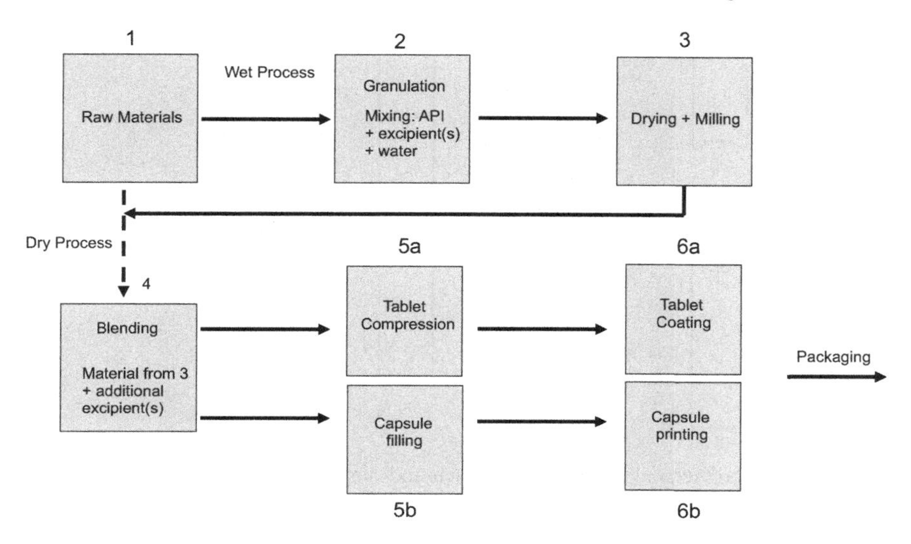

FIGURE 1.2 General flow chart for the manufacture of a solid dosage from pharmaceutical product.

form to release the API into the body at a given rate. All ingredients, both API and excipients, used to manufacture the drug product are tested to meet required specifications prior to being used to manufacture the finished drug product.

Once all the formulation ingredients have been granulated, milled and blended, they are ready to either be pressed into tablets or encapsulated in capsule shells. Compressed tablets use a punch-and-die combination to compress the powder into a tablet core. The tablet compression step is where the tablet core takes form. Figure 1.3 shows a general example of a punch-and-die set with a top and bottom punch, and a die. The blended ingredients fill the die, and the powder is compressed by the top and bottom punches. The compression force used to shape the tablet core is determined by the product manufacturer and is product specific. Once the powder is compressed, the tablet core is ejected and the process starts again.

In this example, the tablet produced is a beveled, elliptical, ½ scored tablet debossed "A 1" on one side. In practice, a tablet press used by a manufacturer will have numerous punch-and-die sets, compressing tablets at very high rates and making thousands of tablets in a single batch.

Once the tablet cores are prepared, they can be coated and polished. Like the tablet core, the ingredients used in the coating formulation used by the manufacturer provide product stability (e.g., environmental, chemical), taste and unique color for branding purposes. Once coated, the finished tablets are packaged, labeled and prepared for shipment. The finished product is tested by the manufacturer to ensure that a particular batch of product meets their approved product specifications prior to being released for sale to the consumer.

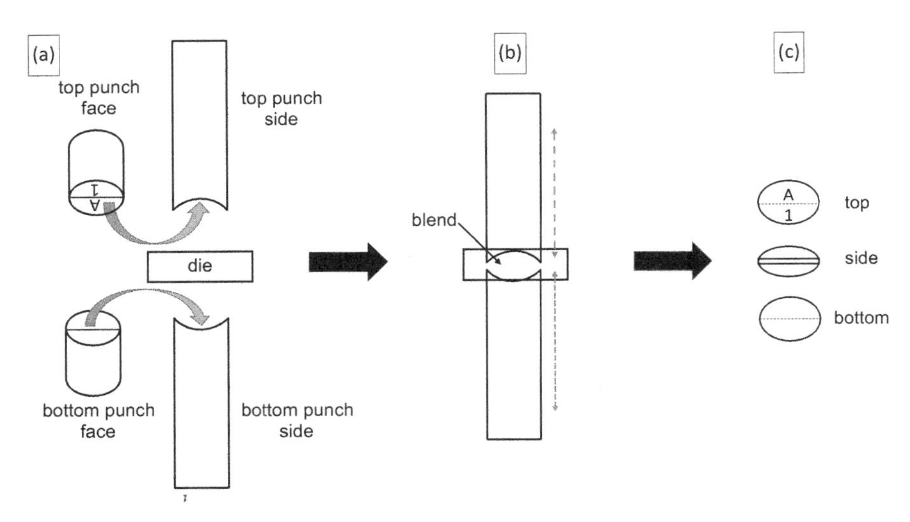

FIGURE 1.3 (a) General example of a punch-and-die set, (b) position of punches, die, and blend at the point of compression, and (c) the top, side, and bottom views of the table produced.

The pharmaceutical drug manufacturing process is a validated process that has been inspected and approved by regulatory authorities; therefore, products manufactured using this process will be of consistent quality. To achieve this, the manufacturer will test raw materials to make certain that they meet required specifications, perform in-process testing (e.g., in-line or off-line) as the product is being manufactured, and use tooling (e.g., punches and dies) machined to tolerances that ensure that the physical (e.g., weight, shape, debossing), chemical and performance characteristics of the first tablet of a batch are statistically similar to the last tablet of the batch.

Unlike the authentic drug product manufacturer, the goal of the drug counterfeiter's manufacturing process is to impart enough visible authentic product characteristics to deceive the importer, distributer and patient. Compared to the authentic manufacturer, a counterfeiter's process may be a simple garage operation using a kitchen blender and a single-stage, hand press or as complex as a using commercial-grade pharmaceutical manufacturing equipment. Regardless of the type of process used, the drug counterfeiter will have a difficult time producing an exact copy of the authentic product.

1.1.2.3 Types of Counterfeits

Counterfeit drugs can fall into one of three different classes: poor, intermediate, and sophisticated (Figure 1.4) (Rebiere et al., 2014; Koczwara, 2017; Degardin et al., 2014). It is important not to use the term "quality" when classifying different counterfeits. The term "quality" implies that the counterfeit was manufactured using a set of standards recognized by a regulatory authority. However, due to the clandestine nature of the manufacture of counterfeit drugs, the importer,

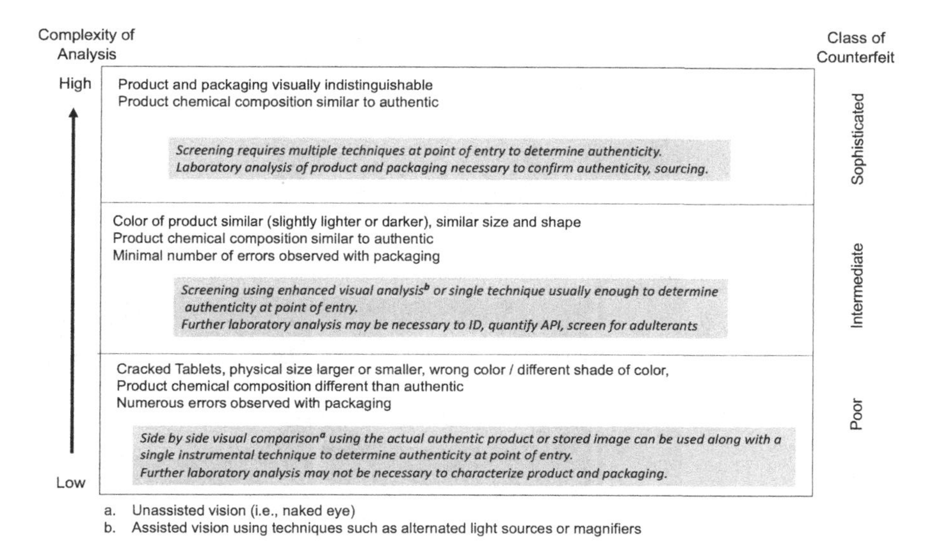

FIGURE 1.4 Characteristics of the different classes of counterfeit products that may be encountered and increasing complexity of analysis.

distributer, patient, investigator or chemist has zero knowledge or confidence in how the suspect counterfeit drugs are manufactured and what they contain.

Using a classification system allows one to develop specific testing protocols to determine if a suspect product is counterfeit or not. In this way, the types of analyses used will produce the information needed to determine if the product is counterfeit.

At the bottom of the scale are poor counterfeits. The visual differences between the poor counterfeit and the authentic can be so extreme, and a minimal amount of analysis is needed to determine if these products are counterfeit. They may contain one or two excipients; the correct API may not be present or may be substituted by a different API or analog. The substituted API present in a poor counterfeit may provide some therapeutic relief or in other examples may be life threating. For example, acetaminophen found in counterfeit antimalarial drugs itself is not an antimalarial drug, but it may provide the user some symptomatic relief. However, counterfeit oxycodone hydrochloride tablets containing fentanyl presents a life-threating situation to the unsuspecting individual. In most cases, they are meant to fill high demand for a given drug. These types of counterfeits can be easily passed off because they appear authentic to the untrained eye.

Examples of poor counterfeit drug products are shown in Figure 1.5a–c. Figure 1.5a is a side-by-side comparison of an authentic vial and a counterfeit vial of testosterone. The authentic vial in Figure 1.5a (left) is smaller in diameter, and the liquid in the vial is darker yellow in color compared to the counterfeit vial shown in Figure 1.5a (right). Figure 1.5b is an example of a poor counterfeit version of a hydrocodone bitartrate/acetaminophen (10/325 mg) tablet. The yellow coating

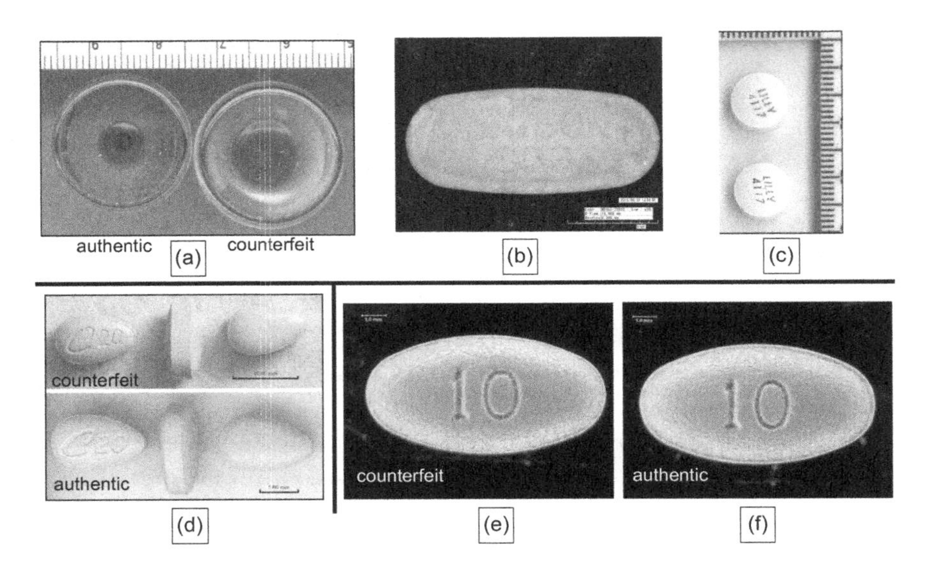

FIGURE 1.5 Examples of the different classes of counterfeit products. Poor counterfeits: (a) counterfeit testosterone injection compared to authentic, (b) counterfeit hydrocodone bitartrate/acetaminophen (10/325 mg) tablet, and (c) counterfeit Zyprexa 10 mg tablets; intermediate counterfeits: (d) counterfeit Cialis 20 mg tablets compared to authentic; sophisticated counterfeit: (e) counterfeit Lipitor 10 mg tablet and (f) authentic Lipitor 10 mg tablet.

is chipped and cracked, and the authentic manufacturer of this product no longer manufactures or markets a yellow-coated version of this product. Chemical analysis revealed that the counterfeit contained acetaminophen and diphenhydramine, but no hydrocodone bitartrate. The last example (Figure 1.5c) shows the type of variability one may encounter in a single package of a counterfeit product. In this case, the two Zyprexa 10 mg tablets are from the same blister package; however, the color and placement of the printing on the tablets is different. Chemical analysis and comparison to the authentic product confirmed that these tablets were counterfeit.

Intermediate-type counterfeits can appear very similar in size and shape to the authentic product. The chemical composition of the intermediate counterfeit may be similar to the authentic product, mainly the highest-concentration excipients. For example, the most abundant excipient found in the intermediate counterfeit formulation may be the same as that used in the authentic; however, the other minor excipients may be missing or different than the authentic. In many cases, the correct API may be present or an API with the same therapeutic response. For example, a counterfeit Cialis tablet may contain sildenafil citrate (API in Viagra) and not tadalafil, but both APIs are PDE5 inhibitors and are used to treat erectile dysfunction. Figure 1.5d is an example of an intermediate counterfeit. The Cialis 20 mg counterfeit tablets (Figure 1.5d top) are thicker and a darker yellow than the authentic (Figure 1.5d bottom).

The final class is the sophisticated counterfeit. This type of counterfeit product can be very difficult to differentiate visually from the authentic product, and, therefore, often requires the use of multiple physical and chemical analysis techniques. In this case, a side-by-side comparison may not be enough to differentiate the suspect counterfeit from the authentic, and more information from the manufacturer may be needed. Figure 1.5e is an example of a sophisticated counterfeit compared to the authentic product Figure 1.5f. Visually, the counterfeit Lipitor 10 mg tablets look nearly identical in shape, color, and debossing as the authentic. In this case, further analysis (2D Image Analysis (IA)) and chemical analysis of the suspect 10 mg Lipitor tablet was necessary to determine that it was counterfeit.

1.1.3 Factors Fostering Drug Counterfeiting

In principle, drug counterfeiting is a financially driven crime (Finlay, 2011; Degardin et al., 2014). Counterfeiters use low-cost or substandard materials and manufacturing processes to make counterfeit drugs; this, in turn, allows the counterfeited product to be sold at the same or lower cost than the authentic product. Even if the counterfeiter chooses to sell his product at or slightly below the authentic product price, the profit margin is large enough for the counterfeiter to turn a profit. The counterfeiter can ship their products globally, directly to the patient/consumer, and by using a complex distribution network that allows them to evade law enforcement and regulatory authorities. There are several major factors that drive drug counterfeiting, which include but are not limited to the internet and demand, health crises (e.g., epidemics, pandemics), and expensive products (e.g., oncology, biopharmaceuticals).

1.1.3.1 Internet and Demand

The internet has helped open the global prescription drug market where once prescription products were only available via brick-and-mortar pharmacies. Counterfeiters have exploited the internet to meet the demand or popularity of a drug product. If the demand for a drug is high or greater than what can be produced by the legitimate manufacturer, a counterfeiter will fill the void with a counterfeit version of the authentic product.

When the popular lifestyle drugs Viagra and Cialis were approved and entered the U.S. market in the early 2000s, it was not long before unapproved generic and counterfeit versions of the drugs began to surface. Unapproved generic versions and suspect counterfeit versions of Viagra and Cialis were some of the first counterfeit drugs analyzed at the FCC. In 2003, counterfeit versions of the popular medically necessary statin drug Lipitor, used to control cholesterol, were analyzed at the FCC followed by additional types of counterfeit life-saving drugs in 2006 (Flaherty and Gaul, 2003; DOJ, 2003; The Partnership for Safe Medicines, 2014). Most recently, the demand for opioid-containing drugs reached crisis levels in 2017–2018, and these drugs continue to be counterfeited and analyzed at the FCC to this day (Daniulaityte et al., 2019; CDC, 2021).

1.1.3.2 Health Crises

Counterfeit life-saving drugs that find their way into areas experiencing severe health crises can increase the amount of harm to a population in a country or region. Epidemics and pandemics can fuel counterfeit drug manufacturing and distribution. If these counterfeit drugs find their way into the normal distribution chain, it can have devastating effects on the patients who are dependent on these drugs. A counterfeit version of a drug meant to slow or stop the spread of a disease may have a reverse effect by prolonging illness and increasing the spread of the disease. For example, in areas where malaria, a parasitic blood disease transmitted by mosquitos, is prevalent, there is a constant high demand for antimalarial drug therapies. Because of the continuous high demand for these products, large amounts of counterfeit antimalarial drugs can be found in these areas. In many cases, the counterfeit antimalarial drugs do not contain the correct API or contain no API at all (Newton et al., 2011; CDC, 2018). Therefore, unsuspecting individuals taking counterfeit medications are still at risk of developing and transmitting the disease.

In 2004 and 2005, a strain of H5N1 avian flu severely affected poultry farms in Asia (WHO, 2014; CIDRAP, 2005; CIDRAP, 2006; Kaye and Pringle, 2005). As the number of infected birds increased, there was fear that this strain of H5N1 would pass from infected birds to humans and potentially result in a global pandemic. At the time, the antiviral Tamiflu (oseltamivir phosphate) was suggested as a possible treatment for H5N1 infections in humans (Liem et al., 2009; NBC News, 2006). With the potential threat of a global pandemic, many individuals and governments sought out Tamiflu to stockpile.

In December 2005, FCC received the first samples of "Tamiflu" (oseltamivir phosphate) capsules being sold on the internet, and imported into the U.S. for personal use and appeared to be an unapproved generic form of Tamiflu (Figure 1.6) (Kaye and Pringle, 2005). These unapproved generic "Tamiflu" capsules were found to contain ascorbic acid (vitamin C) and lactose (Cheng, 2006). No oseltamivir phosphate was detected; however, trace amounts of other APIs were detected in the capsule contents of these samples (Table 1.2). In early 2006, the first counterfeit Tamiflu capsules were received and analyzed at FCC. These counterfeit Tamiflu capsules (Type 1) contained flour and no oseltamivir phosphate. Then, in June of 2006, additional counterfeit Tamiflu capsules were received and analyzed. This second group of counterfeit Tamiflu capsules (Type 2) did contain oseltamivir phosphate, but the capsule formulation and the printing on the outside of the capsule shells were not consistent with the authentic product (Table 1.2).

From the end of 2005 through 2006, multiple samples of the unapproved generic Tamiflu, the counterfeit Type 1 and the counterfeit Type 2, were collected and sent to FCC for analysis. However, once the potential of a H5N1 global pandemic had passed, the number of suspect counterfeit Tamiflu collected and sent to FCC for analysis decreased dramatically. This case study is an example of how counterfeiters took advantage of people's fears of a possible pandemic to profit off the sale of dangerous unapproved and counterfeit Tamiflu products. Except for the Type 2 counterfeits, the other samples received did not contain any API,

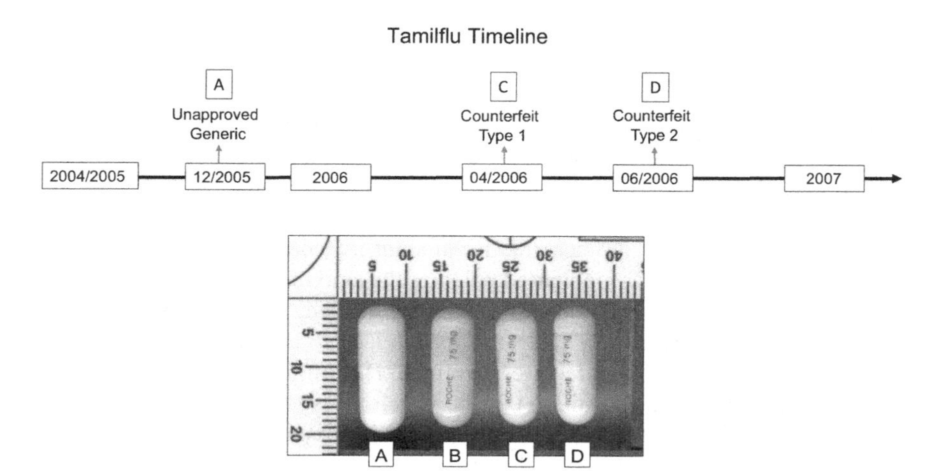

FIGURE 1.6 Tamiflu timeline showing examples of (a) unapproved generic Tamiflu, (b) authentic Tamiflu 75 mg capsule (used for comparison), (c) counterfeit (Type 1) Tamiflu 75 mg capsule, and (d) counterfeit (Type 2) Tamiflu 75 mg capsule.

TABLE 1.2
Summary of the Analysis of the Different Types of Suspect Tamiflu Products Analyzed at FCC between 2005 and 2007

Sample Type	Capsule Contents Found to Contain...	API Present[a]
Unapproved generic	Ascorbic acid (Vitamin C), lactose, trace amounts of other APIs[b]	No
Counterfeit Type 1	Flour	No
Counterfeit Type 2	The capsule formulation was not consistent with authentic[c]	Yes

[a] Oseltamivir phosphate is the API used in authentic Tamiflu.
[b] Trace amounts of the APIs sibutramine and celecoxib were detected in the capsule contents using LC–MS.
[c] Individual components of the capsule content formulation were not determined.

and these products would have exacerbated the spread of the virus due to lack of efficacy. Even though the Type 2 counterfeit contained the correct API, it also contained other APIs, making this product a threat to the public health.

Another example of counterfeiters capitalizing on a major health crisis is the ongoing opioid epidemic (Jannetto et al., 2019; Armenian et al., 2018; Daniulaityte et al., 2019; CDC, 2021; Wolf, 2021). Although the opioid health crisis has primarily affected the U.S., it is also having an adverse effect in other parts of the world.

This has resulted in many who are addicted to opioids to seek out opioid-containing products from unscrupulous and illegal sources. This high demand for opioids has created a large and lucrative illegal market for counterfeit opioid tablets where these drugs are illegally manufactured and distributed to unsuspecting individuals. Recently, some of the most encountered opioid-containing counterfeit drug products are those labeled to contain oxycodone but are actually found to contain the much more potent and deadly opioid, fentanyl. Due to fentanyl having a higher potency than oxycodone, a much smaller amount is needed to induce the same desired effect. Because of its increased potency, the amount of fentanyl needed per tablet is much less than the amount of oxycodone needed, which means the counterfeiter can manufacture more tablets. This can lead to a significant increase in profits for the counterfeit manufacturer (Wolf, 2021). Ultimately, it's the unsuspecting individual who purchases these tablets, which leads to deadly consequences. They consume a tablet(s) thinking that they are taking oxycodone, but, in fact, they have consumed a drug more potent than oxycodone.

1.1.3.3 Expensive Drug Products

As medicine and science increase their understanding of diseases and how to treat them, new therapies become available to treat patients. Many of these new and innovative life-saving therapies are expensive, which makes them an appealing target for counterfeiters (Mackey et al., 2015; Venhuis et al., 2018). Unlike other drugs that have a low-cost relative to these new therapies, the counterfeiter does not have to produce millions of tablets to generate a profit. For example, if an oral drug costs $3 per tablet and the counterfeiter wants to generate $100,000, the counterfeiter would have to manufacture and sell over 33,000 tablets. However, for a novel biotherapeutic which costs $1,000 per vial, the counterfeiter would only need to manufacture and sell 100 vials to make the $100,000.

For example, in 2012–2013, a counterfeit version of injectable oncology drug Altuzan/Avastin (bevacizumab), which costs around $800 (25 mg/mL), found its way into the U.S. supply chain via importation (U.S. FDA, 2018). The samples sent to the laboratory for analysis were found to contain no active ingredient, only water. Another example is Iculsig (ponatinib hydrochloride), an oral chemotherapy drug approved for the treatment of chronic myelogenous leukemia and certain types of acute lymphoblastic leukemia. For a supply of 30 tablets, the cost for Iclusig (10 mg) is around $18,000, which is around $600 per tablet. A counterfeit version of Iclusig 15 mg tablets was found by health authorities in Switzerland in 2019 and found only to contain paracetamol (acetaminophen) (WHO, 2019). In both cases, the counterfeiters were potentially able to make a large amount of money with limited legal exposure, since the amount of counterfeit product required to generate a sizeable profit would be small. A larger concern is that the counterfeiters did not need a large operation to generate counterfeit versions of either drugs. In the case of Avastin, the counterfeiters used discarded authentic Avastin vials, counterfeit labels, and water to make their products, making it more difficult to detect the counterfeit.

1.1.4 Analysis of Counterfeit Drugs

The type of counterfeit drug products encountered will determine the types of analysis required. In most cases, it is not enough to analyze a suspect counterfeit drug product for the presence/absence of the correct API(s), because APIs are often readily available via the internet. While it is important to examine counterfeit products for the presence of the correct API(s) as part of a thorough analysis, the laboratory instrumentation and methods must also be able to differentiate the suspect product formulation (e.g., coating, core, excipients, API) and packaging from those of the authentic product. The laboratory instrumentation and methods employed by the FCC and others are capable of this type of differentiation, and can be broken down into four general categories: physical (P), authenticity (A), chemical assay—impurity (C), and field-deployable rapid screening technologies (FA) (Table 1.3).

In general, the suspect counterfeit drugs collected and sent to the FCC for analysis are first screened to determine whether they are counterfeit or authentic (Figure 1.7). Once the determination of authenticity has been completed, more comprehensive analysis may be performed to identify specific attributes of the suspect product that would allow product sourcing (Figure 1.7). This library of counterfeit chemical profiles or "fingerprints" may then potentially be used to link counterfeit samples from different criminal investigations. The FCC is currently working on field-deployable technologies that can be used to screen suspect counterfeit drugs (Ranieri et al., 2014; Lanzarotta et al., 2015, 2020, 2021). Table 1.3 provides an overview of the instrumentation or techniques used for a particular method in the analysis of a suspect counterfeit drug.

1.1.4.1 Physical Methods (P)

Several physical methods can be used to grossly differentiate counterfeit from authentic products. These same types of methods can also be used on the packaging materials. Images of all suspect samples are collected. The suspect product weight, size and color are compared to those of the authentic. The package (e.g., bottle, carton and blister package) dimensions are compared to those of the authentic packaging. Additionally, a gross visual exam can be conducted on the product packaging to determine the type of printing processes used to manufacture the suspect packaging, which can then be compared to the authentic.

1.1.4.2 Authenticity Methods (A)

Authenticity methods are designed to differentiate even the most sophisticated counterfeit product and packaging from authentic products. The instrumentation used in these methods has been chosen for their discriminating power, when comparing suspect counterfeit to authentic products. The information obtained by these methods can also be used to establish a "fingerprint" of the counterfeit, which can be used for sourcing and connecting counterfeit investigations.

TABLE 1.3
Summary of Methods, Instruments/Techniques Useful in Counterfeit Drug Analysis

Application	Instrument / Technique	Categories[b]				
		P	A	C	S	FA
Physical/Unassisted Visual Examination: weight and physical dimensions of finished dosage form	Balance, Digital micrometers	X	X[a]			X
Assisted Visual examination of dosage forms and packaging materials (bottle labeling, cartons, blister package, printing inks)	Alternate Light Sources (**ALS**); Crime Scope; Video Spectral Comparator (**VSC**), magnifiers	X	X			X
Physical examination (tablet dimensions), tablet debossing, capsule shell printing, tool mark analysis (e.g., tablet punches), packaging material analysis, determination of printing processes, printing defect analysis	Stereo Light Microscope (**SLM**); Digital Light Microscope (**DLM**), Video Spectral Comparator (**VSC**); 2D Image Analysis (**2DIA**); 3D Image Analysis (**3DIA**)	X	X		X	
Formulation Analysis; excipient/inactive ingredients ID; ID of pharmaceutical drugs, illicit drugs, adulterants; analogs; unknowns; Determination of crystalline form of drugs (polymorphs); Trace materials; Packaging materials (adhesives)	Polarized Light Microscopy (**PLM**); Fourier transform infrared spectroscopy (**FT-IR**); FT-IR microspectroscopy; FT-IR imaging; Raman spectroscopy; X-Ray powder diffraction (**XRD**); Nuclear Magnetic Resonance (**NMR**)		X	X[a]	X[a]	X
Residual solvents analysis and profiles	Headspace Gas Chromatography with Mass Spectrometric Detection (**HS-GC-MS**)		X[a]	X	X[a]	
ID of pharmaceutical drugs, illicit drugs, drug analogs, adulterants, unknowns; Impurity profiles	Gas Chromatography with Mass Spectrometric Detection (**GC – MS**); Liquid Chromatography with Mass Spectrometric Detection (**LC–MS**); Direct Analysis in Real Time with Mass Spectrometric Detection (**DART-MS**)		X[a]	X		X
Quantitation of drugs and impurities, Impurity profiles	High Performance Liquid Chromatography with Ultraviolet Detection (**HPLC/UV**)			X		
Elemental analysis, elemental profiles, Packaging materials (foils)	Inductively Coupled Plasma with Mass Spectrometric Detection (**ICP-MS**); X-Ray Fluorescence (**XRF**); Scanning Electron Microscopy with Energy Dispersive X-Ray detection (**SEM-EDX**)		X	X	X	X

[a] Can be used on a limited basis
[b] P, Physical; A, Authenticity; S, Sourcing/Link Analysis; C, Chemical-Assay-Impurity Analysis; FA, Field Adaptable.

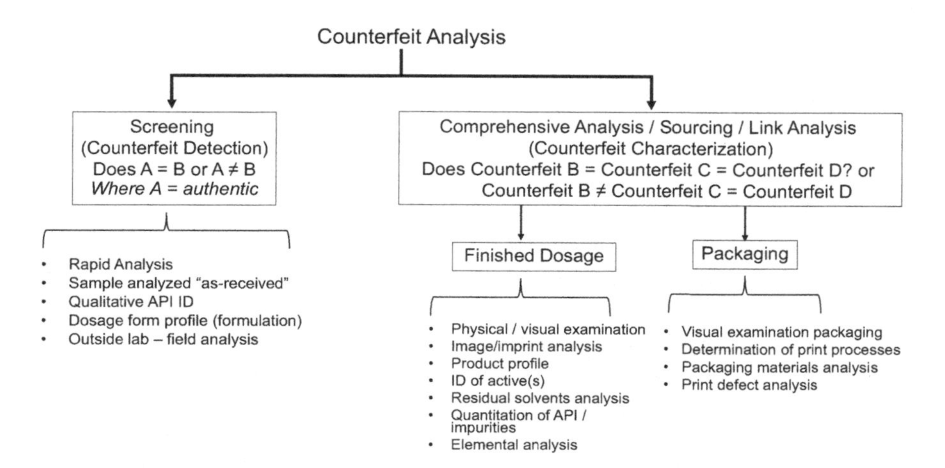

FIGURE 1.7 Analysis of counterfeit drugs—screening vs. comprehensive analysis describing the types of analysis performed for each.

1.1.4.3 Chemical-Assay-Impurity Methods (C)

The chemical-assay-impurity methods can be used to differentiate counterfeit from authentic products on a limited basis. The focus of these methods is more about the API and less on the excipient (inactive/filler) materials found in a product. These methods are helpful for identifying APIs that are present, analyzing for trace impurities, and determining the amount of APIs and impurities present in a counterfeit product. These methods can determine whether the correct or wrong API is present in the counterfeit product or whether other APIs have been substituted. Additionally, these methods can determine if other potentially harmful chemical compounds are present in the counterfeit formulation. The strength in these methods is their ability to detect a wide variety of chemical compounds at extremely low levels. These methods can also be used to determine what type of synthetic process was used to manufacture the API present. For example, the types of residual solvent(s) found and the impurities that are present in the counterfeit can be used to determine if the API was manufactured using a process similar to that used in the manufacture of the authentic product.

1.1.4.4 Field-Deployable Rapid Screening Technologies (FA)

These include either miniaturized versions of laboratory instrumentation (e.g., Raman), new technology, or small, compact laboratory instrumentation that can be used in an environment outside the laboratory. They are relatively simple to use and provide the user with the ability to rapidly screen large numbers of samples. This intelligent sampling allows the user to make sound sampling decisions when faced with examining large quantities of products entering through a port of entry or large group of samples collected as part of a search warrant. Methods that are developed on the laboratory versions of these devices can be modified and transferred to the field devices for use outside the laboratory. These devices also give customs and regulatory authorities the ability to quickly determine if a product is counterfeit or not, thus keeping it from entering the legal distribution chain.

1.2 CONCLUSIONS

Given the dangers that counterfeit drugs pose, combating counterfeit drugs is of major importance to public health. The multidisciplinary analysis approach is very effective at detecting poor and sophisticated counterfeit drug products. The physical and screening methods allow for a quick assessment of a suspect product either in the laboratory or in the field. Comprehensive analysis methods provide details about a suspect counterfeit, and they can be used to source or link known counterfeit products together. As drug counterfeiting grows as a clandestine industry, it is not only important to detect and intercept counterfeit drugs entering the legal supply chain but also to find and shut down the source of these dangerous products. Over the last 20 years, the growth in sophisticated copies of counterfeit products and packaging has increased and so too has the importance of implementing this type of approach to analyze counterfeit drugs.

As the type and sophistication of counterfeit products increase, new and more sophisticated counterfeit detection methods and instrumentation will be needed to counter the threat of these counterfeits. As medical science and the pharmaceutical industry develop novel drug therapies to treat existing and new diseases, these too will become a target for the counterfeiter. In many of these new products, the API will consist of large biomolecules that will require sophisticated instrumentation and methods to detect the presence of the active molecule. Understanding the factors that contribute to drug counterfeiting will help the regulatory and law enforcement agencies keep ahead of the drug counterfeiter by anticipating or quickly reacting to a new counterfeit threat.

Given the significant danger that counterfeit drugs potentially pose to public health, combating counterfeit drugs will require ongoing and increased cooperation among regulatory authorities, law enforcement officials, pharmaceutical companies, and healthcare providers. With the increase of the global market and medicines being available and shipped all over the world, more and more field analysis screening techniques will be needed to screen shipments of products. The deployment and use of field analysis devices globally will help to secure not only the U.S. supply chain but the global supply chain, too. Finally, it is important to continue to increase public awareness to the likely hazards and dangers of counterfeit drugs, which will help protect the public from these products.

REFERENCES

Allen Jr., L.V. *ANSEL'S* Pharmaceutical *Dosage Forms and Drug Delivery Systems*; Wolters Kluwer, 11th edition, 2018.

Armenian, P.; Vo, K.T.; Barr-Walker, J.; Lynch, K.L. *Neuropharmacology.* **134**, 121 (2018).

Bakker-'t Hart, I.M.E.; Ohana, D.; Venhuis, B.J.; *J. Pharm. Biomed. Anal.* **197**, 113948 (2021).

Blackstone, E.A.; Fuhr Jr., J.P.; Pociask *Am. Health Drug Benefits.* **7(4)**, 216 (2014).

Bugay, D.E. *Adv. Drug Deliv.Rev.* **48**, 43 (2001).

CDC, Centers for Disease Control and Prevention, Counterfeit and Substandard Antimalarial Drugs, 2018, https://www.cdc.gov/malaria/malaria_worldwide/reduction/counterfeit.html (last accessed 11/01/2021).

CDC, Centers for Disease Control and Prevention, 2021, Synthetic Opioid Overdose Data, https://www.cdc.gov/drugoverdose/deaths/synthetic/index.html (last accessed 11/01/2021).

Cheng, M.H. *Lancet Infect. Dis.* **6(2)**, 79 (2006).

CIDRAP, Center for Infectious Diseases Research and Policy, US agents seize shipments of fake Tamiflu, 12/20/2005, https://www.cidrap.umn.edu/news-perspective/2005/12/us-agents-seize-shipments-fake-tamiflu (last accessed 11/17/2021).

CIDRAP, Center for Infectious Diseases Research and Policy, YEAR-END REVIEW: Avian flu emerges as high-profile issue in 2005, 01/05/2006, https://www.cidrap.umn.edu/news-perspective/2006/01/year-end-review-avian-flu-emerged-high-profile-issue-2005 (last accessed 11/17/2021).

Cockburn, R.; Newton, P.N.; Agyarko, E.K.; Akunyili, D.; White, N.J. *PLoS Med.* **2(4)**, e100 (2005).

Daniulaityte, R.; Juhascik, M.P.; Strayer, K.E.; Sizemore, I.E.; Zatreh, M.; Nahhas, R.W.; Harshbarger, K.E.; Antonides, H.M.; Martins, S.S.; Carlson, R.G. *Drug Alcohol Depend.* **198**, (2019).

Degardin, K.; Roggo, Y.; Fisher, A.; Margot, P. In *Counterfeit Medicines Volume II: Detection, Identification and Analysis*, Wang, P.E.; Wertheimer, A.I., Ed.: ILM Publications, 2013; Ch. 1.

Degardin, K.; Roggo, Y.; Margot, P. *J. Pharm. Biomed. Anal.* **87**, 167 (2014).

DOJ, Department of Justice, 2013, Canadian Citizen Sentenced in Scheme to Defraud Consumers Purchasing Pharmaceuticals Online, https://www.justice.gov/opa/pr/canadian-citizen-sentenced-scheme-defraud-consumers-purchasing-pharmaceuticals-online (last accessed 10/29/2021).

Drugs.com Pill Identifier, Copyright 2000–2023, https://www.drugs.com/pill_identification.html (last accessed 11/01/2021).

Erikson, J. *Oncology Times* **25(8)**, 5 (2003).

European Medicines Agency (EMA) website, Counterfeit Medicine, https://www.ema.europa.eu/en/glossary/counterfeit-medicine (last accessed 1/24/2022).

Finlay, B.D. Counterfeit Drugs and National Security Stimson Center, (2011), https://www.stimson.org/2011/counterfeit-drugs-and-national-security/ (last accessed 11/01/2021).

Flaherty, M.P.; Gaul, G.M. 2003, Man Charged with Selling Counterfeit Lipitor, https://www.washingtonpost.com/archive/business/2003/12/06/miami-man-charged-with-selling-counterfeit-lipitor/3c6033b8-91e5-4811-855f-def3ca7a1b57/ (last accessed 10/29/2021).

Gratz, S.R.; Gamble, B.M.; Flurer, R.A. *Rapid Commun. Mass Spectrom.* **20**, 2317, (2006).

Hefferren, J.J. *JAMA* **182(12)**, 1145 (1962).

Jannetto, P.J.; Helander, A.; Garg, U.; Janis, G.C.; Goldberger, B.; Ketha, H. *Clin. Chem.* **65(2)**, 242 (2019).

Kaine, L.A.; Heitkemper, D.T.; Jackson, D.S.; Wolnik, K.A. *J. Chromatogr. A* **671**, 303 (1994).

Kaye, D.; Pringle, C.R. *Clin. Infect. Dis.* **40**, 108 (2005).

Koczwara, A. Dressman, *J. Pharm. Sci.* **106**, 2921 (2017).

Kreig, M. *Black Market Medicine*; Prentice-Hall: New Jersey, 1967.
Lanzarotta, A.; Lakes, K.; Marcott, C.A.; Witkowski, M.R.; Sommer, A.J. *Anal. Chem.* **83**, 5972 (2011).
Lanzarotta A.; Ranieri N.; Albright D.A.; Witkowski, M.R.; Batson J.S.; Fulcher, M. *Am. Pharm. Rev.* **18(3)**, 24 (2015).
Lanzarotta, A.; Kimani, M.M.; Thatcher, M.D.; Lynch, J.; Fulcher, M.; Witkowski, M.R.; Batson, J.S. *J. Forensic Sci.* **65**, 1274 (2020).
Lanzarotta, A.; Kern, S.; Batson, J.; Falconer, T.M.; Fulcher, M.; Gaston, K.W.; Kimani, M.M.; Lorenz, L.; Morales-Garcia, F.; Ranieri, N.,; Skelton, D.; Thatcher, M.D.; Toomey, V.M.; Voelker, S.; Witkowski, M.R. *J. Pharm. Biomed. Anal.* **203**, 114183 (2021).
Liem, N.T.; Tung, C.V.; Hien, N.D.; Hien, T.T.; Chau, N.Q.; Long, H.T.; Hien, N.T.; Mai, L.Q.; Taylor, W.R.J.; Wertheim, H.; Farrar, J.; Khang, D.D.; Horby, P. *Clin. Infect Dis.* **48**, 1639 (2009)
Mackey, T.K.; Liang, B.A.; York, P.; Kubic, T. *Am. J. Trop. Med. Hyg.* **92(suppl 6)**, 59 (2015)
Mackey, T.K.; Cuomo, R.; Guerra, C.; Liang, B.A. *Nat. Rev. Clin. Oncol.* **12(5)**, 302 (2015)
Mohr, A.L.A.; Friscia, M.; Papsun, D.; Kacinko, S.L.; Buzby, D.; Logan, B.K. *J. Anal. Toxicol.* **40**, 709 (2016).
NBC News, 2006, Tamiflu works against bird flu, study shows https://www.nbcnews.com/id/wbna10939656 (last accessed 11/18/2021).
National Drug Code Directory (NDC), U.S Department of Health and Human Services, Public Health Service, Food and Drug Administration, Center for Drug Evaluation and Research, Division of Data Management and Services,https://www.accessdata.fda.gov/scripts/cder/ndc/index.cfm(last accessed 11/01/2021).
Newton, P.N.; Green, M.D.; Mildenhall, D.C.; Plancon, A.; Nettey, H.; Nyadong, L.; Hostetler, D.M.; Swamidoss, I.; Harris, G.A.; Powell, K.; Timmermans, A.E.; Amin, A.A.; Opuni, S.K.; Barereau, S.; Faurant, C.; Soong, R.CW., Faure, K.; Thevanayagam, J.; Ferandes, P.; Kaur, H.; Angus, B.; Stepniewska, K.; Guerin, P.J.; Fernandez, F.M.et al. *Malar. J.* **10**, 352 (2011).
Nickum, E.A.; Flurer, C.L. *J. Chromatogr. Sci.* **53**, 38 (2015).
Patel, D.N.; Li, L.; Kee, C.L.; Ge, X.; Low, M.Y.; Koh, H.L. *J. Pharm. Biomed. Anal.* **87**, 176 (2014).
Ranieri, N.; Tabernero, P.; Green, M.D.; Verbois, L.; Herrington, J.; Sampson, E.; Satzger, R.D.; Phonlavong, C.; Thao, K.; Newton, P.N.; Witkowski, M.R. *Am. J. Trop. Med. Hyg.* **91(5)**, 920 (2014).
Rebiere, H.; Guinot, P.; Chauvey, D.; Brenier, C. *J. Pharm. Biomed. Anal.***142**, 286 (2017).
Rowe, R.C.; Shesky, P.J.; Quinn, M.E., Ed., *Handbook of Pharmaceutical Excipients*, Pharmaceutical Press, 6th Edition, 2009.
Rudolf, P.M.; Bernstein, I.B.G. *N. Engl. J. Med.* **350**, 1384 (2004).
The Partnership for Safe Medicines,01/26/ 2010, FDA Alert Regarding Counterfeit Alli, https://www.safemedicines.org/2010/01/subject-fda-alert-regarding-counterfeit-alli.html (last accessed 10/29/20210).
The Partnership for Safe Medicines, 08/06/2014, Bahamian Pharmacist Who Shipped Drugs for RXNorth Ordered to Face Retrial on Counterfeit Drug Shipping Charges, https://www.safemedicines.org/2014/08/bahamian-pharmacist-who-shipped-drugs-for-rxnorth-ordered-to-face-retrial-on-counterfeit-drug-shippi-8-5-14.html (last accessed 10/29/2021).
U.S. Congress, House of Representatives, Sub Committee on Oversight and Investigations of the Committee on Commerce, Counterfeit Bulk Drugs; 106th Congress, 2d Sess, Serial No. 106–164 U.S. Government Printing Office: Washington DC, 2000.

U.S. FDA, Health Care Provider Alert: Another Counterfeit Cancer Medicine Found in United States, 02/05/2013, https://www.fda.gov/drugs/counterfeit-medicine/health-care-provider-alert-another-counterfeit-cancer-medicine-found-united-states (last accessed 11/18/2021).

U.S. FDA, Canadian Drug Firm Admits Selling Counterfeit and Misbranded Prescription Drugs Throughout the United States, 04/13/2018 https://www.fda.gov/inspections-compliance-enforcement-and-criminal-investigations/criminal-investigations/april-13-2018-canadian-drug-firm-admits-selling-counterfeit-and-misbranded-prescription-drugs (last accessed 10/29/2021).

U.S. FDA, Jury Convicts Shamo of Leading Drug Trafficking Network, 08/09/2019, https://www.fda.gov/inspections-compliance-enforcement-and-criminal-investigations/press-releases/jury-convicts-shamo-leading-drug-trafficking-network (last accessed 10/29/2021).

U.S. FDA website, Counterfeit Medicines, https://www.fda.gov/drugs/buying-using-medicine-safely/counterfeit-medicine (last accessed 01/24/2022).

Venhuis, B.J.; Oostlander, A.E.; Di Giorgio, D.; du Plessis, I. *Lancet Oncol.* **19**, e209 (2018).

Vickers, S.; Bernier, M.; Zambrzycki, S.; Fernandez, F.M.; Newton P.N.; Caillet, C. *BMJ Glob. Health* **3(4)**, 1 (2018).

World Health Organization (WHO), Influenza A(H5N1) highly pathogenic avian influenza: timeline of major events, 12/04/2014, https://www.who.int/publications/m/item/influenza-a(h5n1)-highly-pathogenic-avian-influenza-timeline-of-major-events (last accessed 11/17/2021).

World Health Organization (WHO), Medical Product Alert N°2/2019: Falsified ICLUSIG traded globally, 02/01/2019, https://www.who.int/news/item/01-02-2019-medical-product-alert-n-2-2019-(english-version) (last accessed 11/19/2021).

World Health Organization (WHO),, Medical Product Alert N° 3/2019: Falsified ICLUSIG (Asia and globally), 02/21/2019, https://www.who.int/news/item/21-02-2019-medical-product-alert-n-3-2019-(english-version) (last accessed 11/18/2021).

World Health Organization (WHO),, Substandard and falsified medical products, https://www.who.int/health-topics/substandard-and-falsified-medical-products#tab=tab_1 (last accessed 01/24/2022).

Wolf, Z.B. Analysis, CNN Americans are overdosing on a drug they don't know they're taking, 11/18/2021, https://www.cnn.com/2021/11/17/politics/fentanyl-overdose-deaths-what-matters/index.html (last accessed 11/19/2021).

2 History and Overview of Counterfeit Drugs Cases

Haley Fallang
Montana Department of Justice Forensic Science Division

CONTENTS

2.1 THE HISTORY OF THE COUNTERFEIT DRUG MARKETPLACE

Counterfeit drugs have been found in global markets for ages. Dating as far back as the 1600s, counterfeit versions have been found in the supply chain of antimicrobial medicines, such as counterfeit cinchona bark (Newton et al., 2009). In the 1800s, counterfeit quinine was discovered in various pharmaceutical markets (Newton et al., 2009). Although counterfeit drugs and medicines have been found all over the world for centuries, it was not until the Tylenol poisoning deaths of 1982 that the U.S. general public became aware of the threat of such drugs and the dangers that may ensue from consuming them.

DOI 10.1201/9781003183327-2

In the 1980s, Tylenol was the highest-selling nonprescription pain reliever available on the market in the United States and could be found on the shelves of pharmacies everywhere; interestingly, acetaminophen/Tylenol remains one of the top medications taken by Americans (Markel, 2014) (NIH, 2020). From when the U.S. Food and Drugs Act was signed in 1906 up until 1982, American consumers generally did not worry about whether the drug they had purchased was an authentic product, nor did they fear that any harm might come to them from taking the drug as instructed (Elkins, 2019). That all changed on September 29, 1982, when Mary Kellerman, a 12-year-old child in Illinois, took one extra-strength capsule of Tylenol and died within hours (Markel, 2014). In a shocking event, Mary had unknowingly consumed a Tylenol capsule that had been laced with the extremely poisonous chemical potassium cyanide. That day also saw the loss of three family members, all of whom had consumed poisoned Tylenol capsules (Markel, 2014). Within days, three more individuals from different regions of Illinois died after taking the tainted Tylenol.

After the connection was made between the Tylenol capsules and the poisoning deaths, Tylenol manufacturer McNeil Consumer Products (a subsidiary of Johnson & Johnson) took swift action to recall more than 31 million bottles of Tylenol. Johnson & Johnson was able to quickly determine that the cyanide had been introduced after the cases of Tylenol had been shipped out of the factory. The most popular hypothesis was that someone had taken bottles of Tylenol from the shelves of grocery stores laced the Tylenol capsules with potassium cyanide and returned the packages to the store shelves (Markel, 2014).

This hypothesis led Johnson & Johnson to create and integrate new product protections, such as tamper-proof packaging and new versions of Tylenol pills called caplets, which could not be easily opened or contaminated (Markel, 2014). These measures reassured consumers that they would be able to immediately recognize whether the packaging of their Tylenol product had been tampered with and prevented them from consuming potentially poisoned Tylenol pills (Markel, 2014). In 1983, the U.S. Congress passed "the Tylenol bill," which made tampering with consumer products a federal offense. In 1989, the U.S. Food and Drug Administration (FDA) created federal guidelines for drug manufacturers that ensured all products would be tamper-proof (Markel, 2014).

The Tylenol poisonings shattered the perception of drug safety and awakened Americans to the threat of counterfeit drugs. This tragedy claimed the lives of unsuspecting Americans and inspired the integration of new safety measures to protect consumers. Many measures have been put in place to make drugs safer for consumers. The Tylenol drug poisoning caused an awakening of global concern regarding drug regulations, drug interference, and packaging tampering (Markel, 2014).

While public concern for proper drug packaging and possible tampering was instigated by the Tylenol poisonings, counterfeit medicines have since taken over as the prime concern of drug consumers. Counterfeit drugs that have invaded modern markets range from fentanyl tablets masquerading as oxycodone and new psychoactive substances parading as synthetic cannabinoids to lifestyle drugs and even life-saving medicines. Increased access to the Internet has allowed people to gain access to global markets in which these counterfeit pharmaceuticals flow freely (Wertheimer, 2008). The increased access to counterfeit drugs has created a significant cause for concern, as it has become more and more difficult to discern fake medicines from authentic medicines, and to track drugs from manufacturing through the supply chain to the consumer. In consequence, counterfeit drugs are a major public health concern throughout the globe.

2.1.1 Definition of Counterfeit Drugs

Counterfeit drugs are known by various names throughout the world, including fake drugs, falsified drugs, and spurious drugs (Mages and Kubic, 2016). Regardless of the moniker, however, the definition remains essentially the same. Counterfeit drugs are drugs that have been purposefully and deceitfully produced and/or improperly labeled so that the drug will appear as an authentic product (Mages and Kubic, 2016). The Pharmaceutical Security Institute defines counterfeit medicines as "products deliberately and fraudulently produced and/or mislabeled with respect to identity and/or source to make it appear to be a genuine product" (PSI, 2021). Drugs are considered counterfeit if they have been fraudulently advertised, manufactured, or packaged (Mohamed Ibrahim, 2019).

The individuals most often affected by the counterfeit drug trafficking trade are those of low income and those residing in developing countries (Mhando et al., 2016). However, the counterfeit drug trade is not exclusively an issue of poverty, nor is it only found in developing countries. The counterfeit drug trade is a threat to all nations and concerns the public health of every individual, regardless of income level. There are many severe consequences for those who consume counterfeit versions of drugs, including therapeutic failure, adverse physical effects, economic concerns, drug resistance, severe illness, and death, as well as a loss of confidence in medicine and healthcare (Mhando et al., 2016). Because of the severity of its consequences, this public health issue should be known to all drug consumers.

The majority of counterfeiting methods involve the manipulation of active pharmaceutical ingredients (APIs) and/or packaging. Some counterfeit drugs will have no APIs and simply contain placebos. Others will contain a small quantity of API in comparison to the authentic drug. Alternatively, some counterfeit drugs will contain a completely different active or inactive ingredient that is not found in the authentic drug (Obi-Eyisi et al., 2012). The counterfeiting of medicines may affect both brand-name products and generic products. The quality of counterfeit drugs can be difficult to determine, as they may contain the correct ingredients in improper ratios, contain incorrect ingredients, lack the active ingredient(s), contain a small quantity of the correct ingredients, and/or have fraudulent packaging. The quality of counterfeit drugs can also be affected by the manufacturing location, as the drugs may have been created in a typical facility, such as a laboratory, or in a "backyard" location that is unknown and unregulated. Ascertaining the quality of the manufacturing practice and the expertise and qualifications of those who are manufacturing the drugs is next to impossible, as the manufacturing of many counterfeit pharmaceutical products takes place outside of patent laws, contracts, and certified factories (Obi-Eyisi et al., 2012).

The level of sophistication concerning counterfeit drugs often depends on the location of manufacturing. Some outfits produce drug products that may be easily identified as counterfeit due to their physical appearances and qualities. Other counterfeiting outfits have rather advanced manufacturing practices that create high-quality products that are difficult to discern from authentic products without employing sophisticated chemical analysis methods. In fact, some manufacturers produce counterfeit products during the off-hours in facilities that manufacture the authentic drugs during normal work hours to limit detection (Obi-Eyisi et al., 2012).

Technological advancements have aided counterfeiters in producing difficult-to-detect counterfeit drugs. By using advanced computers and other technologies, counterfeiters can easily manipulate items so that they closely mimic authentic products. For example, counterfeiters are now able to produce counterfeit drug packaging that so closely resembles authentic packaging that it is nearly impossible to distinguish the two, even down to identification codes and holographic marks (Obi-Eyisi et al., 2012).

A huge concern for patients and consumers is the inability to quickly and easily determine whether they have consumed fake medicines. Detecting counterfeit drugs is not always as simple as visually examining pills or their packaging, as the packaging is often thrown away after use and the pills ingested. For instance, a patient's failure to respond well to a medicine may not be attributed to the consumption of counterfeit drugs but rather the variation often seen in patient results (Blackstone et al., 2014). By the time counterfeit drugs may actually be suspected by a medical professional or a patient, the evidence is often long gone, and no testing may be performed. This phenomenon can result in deadly effects if a patient has not been consuming the authentic life-saving drug she requires but rather a bogus drug without any of the active ingredients. Another concern remains for those who purchase counterfeit recreational or dangerous drugs, as these individuals are less likely to come forward and report the counterfeited drugs for fear of their own guilty association.

Overall, counterfeit drugs have a huge impact on the economy and can be very difficult to detect from their morphology and packaging to the challenging nature of identifying drugs as counterfeit postconsumption. Consequently, counterfeiters have joined the counterfeit drug trafficking trade, causing dramatic increases in fake drugs detected by the Drug Enforcement Administration (DEA) and policing groups. Barring commerce policies directed toward the fight against counterfeit medicines and increased high-quality jobs for educated and skilled individuals in developing countries, such as the chemists employed in manufacturing fake medicines, the counterfeit drug production, and trade, will continue to explode.

2.1.2 Methods of Counterfeit Drug Identification

Counterfeit drug manufacturing processes are rather simple, depending upon the facility, making it easy for skilled and less-skilled individuals to produce mass amounts of fake drugs at an affordable cost (Obi-Eyisi et al., 2012). A counterfeiter simply needs to produce a pill or tablet that has approximately the same shape, size, and overall color as the authentic drug. With advancements in technology, packaging no longer presents an issue either. Labeling and packaging pills are made simple through the use of easy-to-access digital printing instruments and scanners. As a result, counterfeiters can produce a product that is nearly, if not entirely, indistinguishable from the authentic pharmaceutical. Counterfeit drugs are also small and durable enough to be easily concealed and shipped anywhere in the world (Obi-Eyisi et al., 2012).

Even though counterfeits are getting more difficult to detect, labs and trained individuals are still able to identify counterfeits. Methods for identifying counterfeit medicines are both diverse and complex, each with certain abilities and limitations. The first approach utilized in identifying a counterfeit medicine is to visually compare the suspected counterfeit medicine to its authentic version, looking for differences in physical appearance, text, and coloration (Martino et al., 2010), as exemplified in Figure 2.1. A counterfeit drug may show differences in dosage between the packaging and the number of drugs inside the packaging, differences in the described and actual pill structure (for example, a drug labeled as a capsule when the pill itself is a tablet), and differences in expiration dates and batch numbers between the packaging and the pills (Mhando et al., 2016). However, it is important to note that quality assurance issues, such as incorrect expiration dates and batch numbers, do not necessarily constitute a counterfeit drug.

A visual inspection offers a starting point for analysis; however, most counterfeiters have sophisticated manufacturing methods that require more complex testing (Martino et al., 2010). Most suspected counterfeit drug products will be examined using chemical analyses, in-field assays, or advanced laboratory methods (Martino et al., 2010; Elkins, 2019). Table 2.1 provides a brief list with descriptions of some of the methodologies available to those tasked with identifying fake medicines.

Valisure is an online pharmacy that performs drug testing in addition to visual inspections (Valisure, 2021). Valisure takes samples from each batch of medicines it dispenses and runs chemical tests to analyze the drugs for carcinogens,

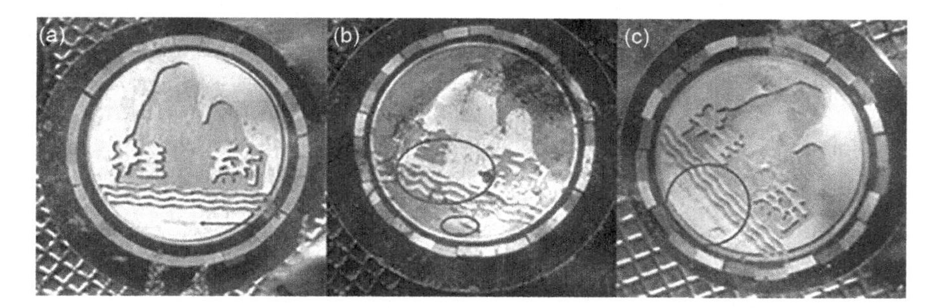

FIGURE 2.1 Authentic and bogus versions of an artesunate blister pack detected through the use of packaging holograms. (a) is the authentic medicine. (b) and (c) are counterfeit artesunate with fake holograms (Cockburn et al., 2005).

TABLE 2.1
Counterfeit Drug Detection Methods

Drug Identification Method	Description of Analysis
Visual Examination	Visually examine the packaging, tamper-proof packaging, text, lot number, expiration date, pill morphology, pill color, and pill markings for any discrepancies.
Colorimetry	These tests may provide a semi-quantitative measure of drug ingredients from a visual examination of the depth of the color present. They may also provide a quantitative measure of the color absorbance by using a portable photometer (Martino et al., 2010) or color analysis app (Elkins et al., 2017).
Thin-layer chromatography	These tests provide ingredient separation followed by colorimetric or fluorescent detection, and yield semi-quantitative results using TLC plates (Martino et al., 2010).
Gas chromatography	This method separates the volatile chemical ingredients present in the drug to determine the identity and quantity of labeled ingredients (Martino et al., 2010). It is easily coupled to mass spectrometry and referred to as GC–MS.
High-performance liquid chromatography	This method is the most commonly used technique for drug analysis, as it provides a quantitative analysis of the components present in a drug. An advantage of using this technique is its ability to collaborate with other methods, such as electrochemical techniques, evaporative light scattering, UV spectroscopy mass spectroscopy, and photodiode arrays and its application to larger nonvolatile molecules (Martino et al., 2010).
Mass spectrometry	This overall method presents a variety of techniques, such as GC–MS, direct analysis in real time (DART), desorption electrospray ionization (DESI), and HPLC–MS. These technologies have complex quantitative procedures; however, they are sensitive and provide specific chemical ingredient information. The DESI method can provide the surface spatial distribution of drug components (Martino et al., 2010).
FTIR spectroscopy	Fourier transform infrared spectroscopy (FTIR) can identify the functional groups of chemical ingredients with absorption frequencies in the infrared region (Martino et al., 2010).
NIR spectroscopy	This technique can characterize a drug through its absorption in the near infrared (NIR) 700–2,500 nm wavelength region. This method is particularly useful in identifying a drug through its packaging materials, as the spectra are collected in reflectance mode (Martino et al., 2010).
Raman spectroscopy	This technique collects scattering data, rather than absorption. Raman techniques illuminate a sample with a laser, which a sample scatters at particular wavelengths, allowing drug components to be identified (Martino et al., 2010, Frosch et al., 2019).

(*Continued*)

TABLE 2.1 (*Continued*)
Counterfeit Drug Detection Methods

Drug Identification Method	Description of Analysis
NMR spectroscopy	This method utilizes changes in active nuclear magnetic resonance (NMR) nuclei orientation using a particular magnetic field strength to detect all of the components and estimate proportions present in a sample in a single run, and can be used for structure determination of new compounds (Martino et al., 2010; Krummel et al., 2015).
X-ray diffractometry	This technique analyzes X-ray diffraction (XRD) patterns, essentially the "fingerprint" of a sample to identify differences in tablet coatings and drug composition after coating removal (Martino et al., 2010).
X-ray fluorescence spectroscopy	XRF is a nondestructive technique that can be used to perform elemental analysis.
SEM–EDX	Scanning electron microscopy (SEM) with energy dispersive X-ray analysis (EDX) can be used to analyze pigment and ink morphology, and elemental composition.
ICP–MS	Inductively Coupled Plasma-Mass Spectrometry (ICP–MS) can be used to identify and profile the elemental composition of a sample.

elemental impurities, active ingredient identification, dosage, dissolution, and/or other ingredients. By performing these analyses, Valisure ensures that the medicines it dispenses are not counterfeit (Valisure, 2021).

2.1.3 Internet Pharmacies and Counterfeit Drugs

Many counterfeit drugs found in the United States are manufactured overseas and brought into the homes of American consumers when they purchase these drugs while overseas or from Internet pharmacies (Wertheimer, 2008). The online drug market, known as the gray market, has stemmed a massive growth in counterfeit drug sales (Pascu et al., 2020). There are different categories of consumers that seek out Internet pharmacies to purchase pharmaceuticals, many of whom fall victim to drug counterfeiters. Internet pharmacy customers often do not have the means to pay for the authentic drugs they need, or they may be unable to afford the doctor's visit necessary to obtain a prescription (Obi-Eyisi et al., 2012). Consequently, they seek out alternate avenues, such as websites offering medicines for significantly lower prices without a prescription requirement (Obi-Eyisi et al., 2012). One of the primary factors contributing to the expansion of the counterfeit drug trade is the consistent increase of pharmaceutical costs and the inability of many consumers to afford genuine medications (Obi-Eyisi et al., 2012). Other factors include the availability of drugs online that are not approved in their home markets and discreet purchase capability for drugs than physical stores.

The Internet supplies the largest unregulated global market (Sanofi, 2013). Online pharmacies present a challenge to consumers, because there are both legal and fraudulent pharmacies available, both of which may offer authentic and/or counterfeit drugs, making it difficult to choose a valid online pharmacy and find genuine medicines (Mohamed Ibrahim, 2019). Average consumers fall victim to fraudulent pharmacies found online (Sanofi, 2013). The nature of the Internet has allowed thousands of fraudulent online pharmacies to create websites, often advertising medications at attractive prices without requiring a valid prescription (Obi-Eyisi et al., 2012).

Differentiating legal pharmacies from illegal pharmacies and genuine pharmaceuticals from counterfeit drugs on the Internet may be all but impossible to the average consumer, particularly because the websites often conceal or bury their true physical addresses while producing a site that looks genuine and professional (Obi-Eyisi et al., 2012). Online pharmacies acting fraudulently may also make appealing offers to consumers, such as by providing access to controlled medicines, unapproved medicines, or drugs currently unavailable elsewhere without requiring any prescriptions or personal identification (Mohamed Ibrahim, 2019). This book is aimed at increasing consumer awareness to the threat of counterfeit drugs.

One of the most alarming concerns regarding Internet pharmacies and their propagation of counterfeit drugs is that consumers are often completely unaware of the dangers that go hand-in-hand with purchasing drugs off of the Internet (Blackstone et al., 2014). The sheer number of counterfeit drugs on the market would cause a consumer to pause before purchasing any kind of pharmaceutical product on the Internet if they were aware of the extent of the problem (Blackstone et al., 2014). Unfortunately, the offers propagated by Internet pharmacies often outweigh the concern an average consumer has regarding counterfeit drugs (Blackstone et al., 2014).

Some consumers seek out online pharmacies to avoid any embarrassment or shame that they may feel when filling prescriptions for certain conditions, such as erectile dysfunction, anxiety, or depression in a traditional pharmacy (Pitts, 2020). Other consumers seek online pharmacies to purchase dangerous or recreational drugs that may not be legal or easily obtained (Pitts, 2020). Whatever the reason for using online pharmacies, the results are often the same – counterfeiters target drugs sold on the Internet, and consumers can end up with bogus drugs, wasted money, sickness, and/or death.

2.1.3.1 The Problem of Internet "Canadian" Pharmacies

Internet pharmacies and social media have greatly contributed to the increase in counterfeit drug trafficking, especially in the United States. In 2014, the FDA found that approximately 25% of Internet users have purchased some sort of pharmaceutical from an online pharmacy (Blackstone et al., 2014). Online pharmacies peddling fake drugs try to reassure customers by conveying that they are based out of Canada. Some online pharmacies claim to sell authentic medicines that were manufactured in Canada; typically, the drugs were not manufactured in Canada, and they were not approved by the FDA or the Canadian government. The drugs are often counterfeit and potentially dangerous to consumers (Blackstone et al., 2014).

A 2005 study determined that of the 11,000 online pharmacies claiming Canadian origin at that time, only 214 of the online pharmacies were actually registered in Canada (Blackstone et al., 2014). The danger to consumers of an online pharmacy falsely claiming Canadian origin is that drugs that are not sold or consumed in Canada are not subject to Canadian laws or regulations (Blackstone et al., 2014). As long as the counterfeit drug sales are made in countries other than Canada, online pharmacies operate under false pretenses of Canadian affiliation or ownership but operate from anywhere in the world (Blackstone et al., 2014). Simply displaying the Canadian flag on the pharmacy's website is enough for consumers to assume the pharmacy is located in Canada and immediately feel a certain level of safety and comfort in the transaction (Wertheimer, 2008). As a result, the comfort that consumers feel in purchasing a drug from a Canadian online pharmacy is completely ill-founded, as the main supplier of counterfeit drugs in the United States is online pharmacies claiming to be based in Canada (Blackstone et al., 2014).

2.1.3.2 Illegal Counterfeit Markets and the Dark Web

A significant factor in the increased presence of illegitimate online pharmacies on the Internet is the ease and simplicity with which an Internet presence can be made and the strong consumer market already in place (Mohamed Ibrahim, 2019). According to a report by the European Alliance for Access to Safe Medicines, 94.3% of the monitored websites that sell health-related products have sold illegal pharmaceutical products (EAASM, 2015). A 2018 study identified 30 marketplaces that were active on the dark web (DEA, 2020). This study found several illicit drugs and prescription medications available in these marketplaces, some of which have been connected to overdose deaths (DEA, 2020). Dream Market, one of the most popular dark web marketplaces available, listed more than 3,500 prescription drugs and approximately 5,000 opioids such as heroin, fentanyl, and prescription opioids for sale (DEA, 2020). After the arrest of some of Dream Market's vendors, the marketplace announced that it would cease its dark web operations (DEA, 2020).

In 2011, the National Association of Boards of Pharmacy discovered that 96% of the 8,000 online pharmacies they examined were not in compliance with United States laws. They also found that 85% of the pharmacies did not require prescriptions, a huge tactic used to lure consumers (Pitts, 2020). In 2017, the International Criminal Police Organization (INTERPOL) successfully removed 3,584 websites and suspended over 3,000 advertisements for the online sale of illicit medicines (INTERPOL, 2017). It seems that any country with Internet access is subject to counterfeit drug and pharmaceutical product traffickers who intend to sell illegal and/or counterfeit drugs to consumers, leading to a largely unregulated market and a worldwide public health concern (Mohamed Ibrahim, 2019).

2.1.3.3 Global Counterfeit Drug Trafficking

In 2015, the Pharmaceutical Security Institute (PSI) analyzed global trends in counterfeit drug incidents (Mages and Kubic, 2016). PSI documented a total of 3,002 pharmaceutical crime incidents in 2015, an increase of 51% from 2011. In 2019, PSI documented a total of 5,081 pharmaceutical crime incidents, an all-time high and an increase of 69% from just five years prior (PSI, 2021). Of course, this data relies on incidents that have been documented; the true number of counterfeit drug incidents is unknown but likely to be significantly higher. Incident numbers are not only affected by the level of counterfeit drug sales in a country but also by the nation's effectiveness in identifying the incidents (PSI, 2021).

In 2019, an increase in counterfeit drug incidents was seen in nearly every region of the world (PSI, 2021). A 15% increase in incidents was observed worldwide, and 150 countries were affected. North America topped the list of regions affected by counterfeit drug incidents with a total of 2,091 incidents in 2019. Next on the list was Asia, with 1,509 incidents in 2019. Africa ranked lowest on the list, with 161 recorded incidents in 2019. Additionally, the number of arrests made for pharmaceutical crime incidents increased globally by 12% from 2018 to 2019. Asia saw the most arrests at 40%, followed by North America with 20% (PSI, 2021).

The World Health Organization (WHO) estimates that around 10%–15% of all drugs sold worldwide are counterfeit (Wertheimer, 2008). In Asian and Latin American nations, however, the percentage of fake drugs likely increases to anywhere from 30% to 60% (Mhando et al., 2016). Interestingly, the WHO estimates that counterfeit drugs may account for over half of the drug market worldwide (Glass, 2014).

As of 2008, India led the world in the manufacturing and export of counterfeit drugs (Wertheimer, 2008). The European Commission reported that up to 75% of counterfeit drugs found around the globe were manufactured in India before being exported elsewhere (Gautam et al., 2009). Currently, the primary economies producing counterfeit drugs are India and China (OECD/EUIPO, 2020). Other nations, including Nigeria, Mexico, and Russia, contribute significantly to the trade as well (Obi-Eyisi et al., 2012). Figure 2.2 shows counterfeit products seized in Nigeria. Counterfeit drug exports from these nations are of particular concern for consumers in the United States, where approximately 40% of available drugs are manufactured abroad, and almost 80% of the active ingredients of drugs are imported (Blackstone et al., 2014).

The primary producers of fentanyl, a drug commonly found in counterfeit pharmaceutical products in the United States, are China and Mexico (DEA, 2020). Fentanyl is typically smuggled across the Southwest border from Mexico or shipped into the United States via mail, primarily in counterfeit pill form or powder (DEA, 2020). Data from the DEA and U.S. Customs and Border Protection (CBP) indicate that fentanyl from China tends to be shipped in smaller quantities and has a purity of over 90% (DEA, 2020). In contrast, the fentanyl trafficked across the Southwest border from Mexico tends to be transported in larger quantities with low levels of purity (less than 10% on average) (DEA, 2020). While

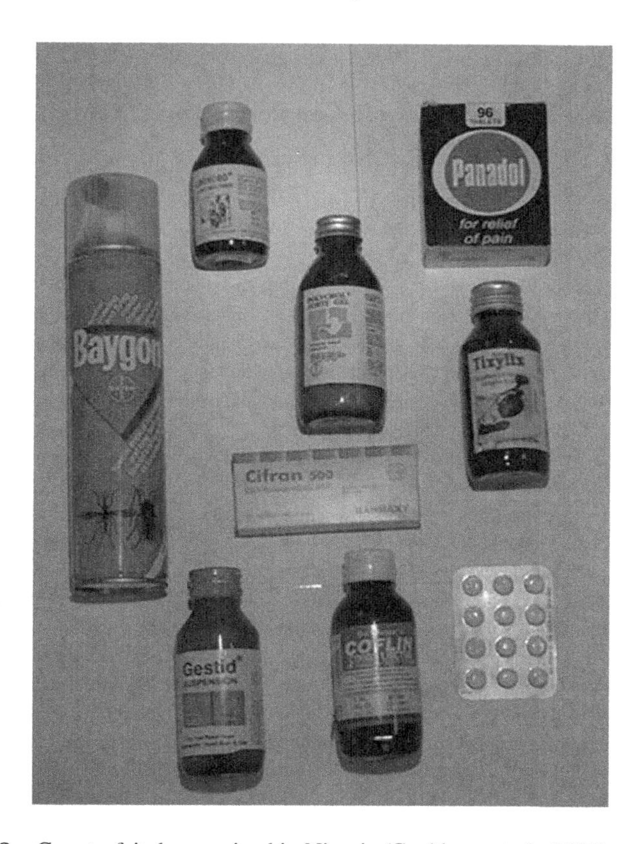

FIGURE 2.2 Counterfeit drugs seized in Nigeria (Cockburn et al., 2005).

most fentanyl is sourced from China or Mexico, India has recently emerged as a source of fentanyl (DEA, 2020). In 2018, DEA and Indian officials conducted two operations in India that resulted in a seizure of 110 kilograms of fentanyl or fentanyl-related chemicals set for distribution to Mexico (DEA, 2020). Analyzing more of the material at the border using portable instruments could curb this flow.

2.1.4 Counterfeit Drug Seizures

This section explores the trends in counterfeit drug seizures over the last 20 years as well as some of the efforts made to inhibit the growth of counterfeit drug trafficking. Table 2.2 presents a survey that is exemplary of the global counterfeit drug trade, although not all-inclusive.

2.1.5 Response to the Counterfeit Drug Trade

Table 2.3 lists recent laws and policies enacted to reduce the counterfeit drug economy.

TABLE 2.2
Chronology of Selected Worldwide Counterfeit Drug Events: From 2000 to Present

Year	Case
2000	Counterfeit malaria medicines found in Cambodia resulted in at least 30 deaths (Mohamed Ibrahim, 2019).
2003	The United States FDA formed a Counterfeit Drug Task Force (Pitts, 2020).
2008	Counterfeit heparin, a blood thinner, was found in the United States. The active ingredient had been replaced with a less costly ingredient that caused adverse side effects (Blackstone et al., 2014).
2009	Egyptian law enforcement and WHO seized counterfeit drugs whose authentic versions were intended for cancer, diabetes, epilepsy, schizophrenia, heart disease, and organ-transplant patients (Mohamed Ibrahim, 2019).
2011	An INTERPOL operation seized 2.4 million illegal and counterfeit drugs worth approximately $6.3 million (Mohamed Ibrahim, 2019).
2011	Two Chinese manufacturing sites and 26 storage areas were raided. Approximately 200 million doses of counterfeit and unregulated generic drugs were seized from at least five pharmaceutical companies. Over 50 instruments and 1,200 molds, tools, and dyes used to manufacture counterfeit medicines were also seized. Chinese law enforcement made 26 arrests in relation to this drug counterfeiting operation (Mages and Kubic, 2016).
2012	Over 80 million doses of illegal drugs were seized across the coasts of Africa. Officials found counterfeit antibiotics, contraceptives, antiparasitic drugs, antimalarial drugs, and cough syrup, all estimated to be worth over $40 million (Mohamed Ibrahim, 2019).
2013	Indian law enforcement linked 8,000 patient deaths in a Himalayan hospital to a counterfeit antibiotic that lacked the active ingredient (Ossola, 2015).
2013	Approximately, 1.2 million counterfeit Viagra and Cialis tablets were seized in Saudi Arabia (Mages and Kubic, et al., 2016).
2016	Over 1,000 packages of counterfeit and illicit drugs worth over $4 million were seized. Thousands of counterfeit pills were discovered among 64 Indonesian factories and drug facilities (Mohamed Ibrahim, 2019).
2017	Counterfeit Hepatitis C medicines were discovered in drug markets throughout Germany (Mohamed Ibrahim, 2019).
2018	Three Florida residents pleaded guilty to charges related to counterfeit drug trafficking. Large amounts of steroid and counterfeit medicine ingredients had been shipped from China to Alabama and Florida (PSM, 2018).
2020	INTERPOL reported that law enforcement had seized approximately 4.4 million units of illegal medicines globally. These units included vitamins, nervous system medicines, steroids, cancer medicines, erectile dysfunction medications, hypnotic medicines, sedative medicines, painkillers, and dermatology medicines. Approximately, 37,000 counterfeit medical devices were seized, most of which were fake surgical masks and self-administered test kits for HIV or glucose monitoring (INTERPOL, 2020).
2020	Authorities in Los Angeles, California, seized over 1.2 million counterfeit pills that were found to contain fentanyl. Agents noted that counterfeit drugs containing fentanyl were made to look exactly like authentic prescription pills by using the correct shape, size, color, stamping, and imprints (CBS Los Angeles, 2021).
2020	The FDA warned consumers of websites and stores selling treatments and preventative medicines related to the COVID-19 global pandemic (Pitts, 2020).
2021	The CBP seized 44 pounds of Viagra in Cincinnati (CBP, 07/2021).

TABLE 2.3
Recent Laws and Policies Enacted to Reduce the Counterfeit Drug Economy

Year	Case
2010	United States Congress passed the Drug Safety & Accountability Act of 2010 and the FDA Globalization Act (Pitts, 2020).
2013	The U.S. Drug Quality and Security Act was signed into law (Pitts, 2020).
2013	The WHO launched a Global Monitoring and Surveillance System (Pascu et al., 2020).

2.1.6 Drugs Most Prone to Counterfeiting

Several drugs are have been identified as being prone to counterfeiting including, lifestyle drugs, cannabinoids, benzodiazepines, pain medication, anti-cancer agents, antimalarials, and medicines to treat severe acute respiratory syndrome coronavirus-2 (SARS-CoV-2) 2019 (COVID-19).

2.1.6.1 Counterfeit Lifestyle Drugs

One of the primary drug categories targeted by counterfeiters and Internet pharmacies is lifestyle drugs. One such lifestyle drug, Viagra, remains one of the most frequently counterfeited drugs available in online markets (Figure 2.3). In January of 2021, Cincinnati law enforcement seized 33.5 pounds of counterfeit Viagra, consisting of approximately 10,350 pills and another 44 pounds in July 2021 (CBP, 2021). The January seizure had an estimated worth of $238,000, had these counterfeit pills been sold at the manufacturer's suggested retail price (CBP, 2021). The counterfeit Viagra was labeled as having been produced in the United States, a tactic likely to put customers at ease (CBP, 2021). The pills were actually imported from the Middle East, with a possible link to Hong Kong and China (CBP, 2021).

In 2015, the FDA released a statement that a counterfeit version of the lifestyle drug Cialis had entered the United States markets. (FDA, 2015). The FDA had seized fake Cialis tablets, used to treat erectile dysfunction and high blood pressure, prior to them being delivered via mail to an American customer. The drugs were determined to contain a variety of active ingredients that could have caused harm if consumed. The FDA noted slight differences found on the counterfeit version of this drug, including a lack of National Drug Code (NDC) number on the bottle, different patterns and colors on the packaging, and a misspelling of Cialis, among other discrepancies (FDA, 2015).

In a 2020 study regarding the authenticity of medicines sold on the Internet, researchers purchased 45 tadalafil (Cialis) tablets from online pharmacies (Sanada et al., 2020). The components of the Cialis tablets were examined using Raman scattering and near-infrared spectroscopy. Of the 45 tablets analyzed, 23 were

FIGURE 2.3 U.S. Customs and Border Protection official holding a bag of counterfeit Viagra (CBP, 2007).

determined to be fake and nine were unlicensed. Researchers noted discrepancies in APIs, ingredient proportions, and packaging. This study provides a concrete example of the ease with which an individual could unknowingly purchase fake lifestyle drugs from online pharmacies (Sanada et al., 2020).

2.1.6.2 Counterfeit Life-saving Medicines

Another category of drugs that are at risk for counterfeiting is expensive cancer treatment drugs. In February of 2012, the FDA warned consumers that counterfeit Avastin, an injectable cancer medicine, had been found in the United States drug market (Mages and Kubic, 2016). The counterfeit version of Avastin was found to contain cornstarch and acetone but lacked bevacizumab, the active ingredient. Later reports issued in June of 2012 and February of 2013 stated that the risk of encountering counterfeit Avastin was still present (PSM, 2013). During their investigation, the FDA discovered that American physicians had purchased the counterfeit Avastin for lower prices from unregulated suppliers, putting the health of their cancer patients at great risk (Blackstone et al., 2014).

Cancer-treating medications are not the only potentially life-saving drugs that have been targeted by counterfeiters. In December of 2020, Janssen

Pharmaceutical Company (owned by Johnson & Johnson) released a statement that the company had been made aware of counterfeit SYMTUZA being distributed within the United States (FDA, 2020). Three pharmacies had been found to have counterfeit SYMTUZA on their shelves (FDA, 2020). SYMTUZA is used to treat the human immunodeficiency virus type 1 (HIV-1) infection in both adults and children (FDA, 2020). As of 2020, no adverse effects had yet been reported from consuming the counterfeit version of this drug, but infection concern remained high for patients who required the authentic medicine but had instead consumed its counterfeit version (FDA, 2020).

2.1.6.3 Counterfeit Antimalarial Medicines

Antimalarial drugs, along with other anti-infectives, are some of the most counterfeited drugs in the world. In fact, they represent the most commonly used and counterfeited drugs in many developing nations. In Southeast Asia, anywhere from 38% to 53% of artesunate tablets, a type of antimalarial medicine, do not contain any active ingredient. Because approximately 90% of malaria cases are found in Africa, counterfeiters have targeted the antimalarial medicines in African nations. Around 37% of artemisinin-based combination therapies sold in privatized drug shops in Uganda have been proven to be fake (Mhando et al., 2016). The counterfeiting of antimalarials has presented a deadly outcome, as approximately 200,000–450,000 of the one million deaths attributed to malaria have directly resulted from the consumption of bogus antimalarial drugs (Mhando et al., 2016).

2.1.6.4 Counterfeit Pain Medicine

According to the 2019 National Survey on Drug Use and Health, prescription pain relievers represent the second highest illicit drug abused by individuals 12 years and older (SAMHSA, 2020). The high abuse rate of prescription pain relievers has resulted in new guidelines and restrictions aimed at reducing the amount of prescription drug use (DEA, 2020). Unfortunately, this has left many individuals unable to obtain controlled prescription drugs through viable pharmacies and physicians, resulting in an increased number of individuals turning to illegal means to obtain those drugs (DEA, 2020). This has led to an increased number of counterfeit controlled prescription drugs being sold to consumers who often do not suspect that what they are consuming is not an authentic medication (DEA, 2020). These counterfeit drugs may contain anything from fentanyl and carfentanil to heroin and methamphetamine, putting consumers at a heightened risk of overdose (DEA, 2020).

One particular drug that has been widely utilized in the counterfeit drug trade is fentanyl, a synthetic opioid that is 50–100 times more potent than morphine (Elkins, 2019). After consistent warnings of the spread of deadly fentanyl-containing counterfeit pills across the nation, the United States reached an eye-opening milestone in 2020 (PSM, 2019). Until 2019, most counterfeit drugs containing fentanyl remained concentrated east of the Mississippi River, with 28 states accounting for 88% of synthetic opioid-related overdose deaths (Shover et al.,

2020). In October of 2020, however, police in Hawaii seized 400 counterfeit oxycodone pills containing fentanyl, demonstrating the quick spread of these deadly counterfeit drugs to every state (PSM, 2019).

Not only have all 50 states in the U.S. reported the presence of these counterfeit drugs, but 42 states have now reported deaths resulting from the consumption of drugs containing fentanyl (PSM, 2019). The states yet to report fentanyl-related deaths are Wyoming, Maine, Michigan, Hawaii, Nebraska, Missouri, Alabama, and Delaware (PSM, 2019). The number of synthetic opioid-related deaths, typically involving fentanyl, has increased by 10 times since 2013. The year 2018 alone saw over 31,000 deaths from synthetic opioids (Shover et al., 2020). While 2018 saw a significant number of fentanyl-related deaths, it also saw the first drop in drug overdose deaths in the last four decades. Unfortunately, the rapid spread of fentanyl, particularly in fake drugs, reversed that progress; 2019 saw over 50,000 opioid-related deaths, 36,509 of which were related to synthetic opioids (Shover et al., 2020). Currently, fentanyl is the most abundant and deadly in the Northeastern, Midwestern, and Great Lakes regions of the United States (DEA, 2020). In 2017, the most fentanyl-related overdose deaths occurred in Maryland, Massachusetts, Ohio, New Hampshire, and West Virginia (DEA, 2020).

Synthetic opioids have been exploited by counterfeiters due to their potency as well as the simplicity of their production and distribution (Pardo et al., 2019). Fentanyl incited its first crisis in 2006, when fentanyl was mixed into heroin and sold to unknowing addicts as genuine heroin (DEA, 2016). This crisis is ongoing, as the recent studies have found that fentanyl is essentially omnipresent in all heroin samples (as well as some cocaine and methamphetamine samples) seized near the United States borders in Western Canada and Northwestern Mexico (Fleiz et al., 2020). According to Shover et al. (2020), any seized heroin should be assumed to contain fentanyl, if not be solely comprised of fentanyl. There has also been an observed increase in overdoses related to heroin, methamphetamine, and cocaine that involved fentanyl, marking a concerning union of increased stimulant and fentanyl-related overdose rates (Shover et al., 2020).

The current fentanyl crisis got its start with heroin in 2013, but traffickers have since expanded their market to include counterfeit prescription or over-the-counter medicines (DEA, 2016). Reports regarding fake drugs that contained fentanyl first began spreading in 2015, but it is unlikely that anyone could have predicted the rapid spread of counterfeit drug consumption and death that would ensue (PSM, 2019). The pills targeted by counterfeiters have not all been opioids, however. Counterfeit pill molds have been used to produce pills that mimic anti-anxiety medications, diabetes medications, aspirin, and Aleve (PSM, 2019).

The use of fentanyl in counterfeit drugs presents a significant concern, because it is much more dangerous than the drugs it is used to mimic. Fentanyl is a synthetic opioid that has 50–100 times the potency of morphine and as much as 50 times the potency of heroin (DEA, 08/2020, Elkins, 2019). Fentanyl use can easily become a deadly matter, as just two milligrams (approximately the same size as two grains of salt) can produce a fatal dose (DEA, 01/2020). The 2019 National Drug Threat Assessment produced by the DEA listed fentanyl, along with other highly potent synthetic opioids, as the most lethal category of illegal substances that are abused in the United States, with fentanyl being involved in more deaths than any other illegal drug (DEA, 2020). A study performed by the University of California, Davis, determined that counterfeit pills collected from the Sacramento area contained anywhere from 0.6 to 6.9 mg of fentanyl per pill, a steep increase from the just 2 mg necessary to cause death (DEA, 2016). Fentanyl is so potent that some police and first responders must take precautions to avoid exposure to the deadly substance, and many agencies have eliminated on-site testing of suspected drug substances (DEA, 08/2020).

Because consumers cannot predict the potency of fake drugs, those who buy counterfeit oxycodone or Xanax can easily overdose on the drugs, simply because they weren't aware that the pills were laced with fentanyl. For instance, in February of 2020, a 20-year-old individual who experienced anxiety took oxycodone before bed and was found dead in the morning of what turned out to be a fentanyl overdose (DEA, 08/2020). In 2016, nine residents of Pinellas County, Florida, died as a result of consuming counterfeit Xanax pills that contained fentanyl (DEA, 2016). In March and April of 2016, 10 residents of Sacramento, California, died after taking Norco pills later determined to contain fentanyl (DEA, 2016).

In 2018, Mexican law enforcement located a surreptitious carfentanil (a structural analog of fentanyl) pill mill in Mexicali, where a Bulgarian biochemist and Mexican colleague had produced 20,000 counterfeit carfentanil pills using a pill press (DEA, 2020). Later that year, Mexican law enforcement located a fentanyl-related counterfeit pill mill in Azcapotzalco, where they found fentanyl-laced oxycodone M30 pills, fentanyl powder, related chemicals, and a pill press (DEA, 2020). In 2018, officials in Sacramento, California, seized over 15,000 counterfeit oxycodone pills that were stamped with "M30" and suspected of having been produced in Mexico (DEA, 2020). Also in 2018, law enforcement in Atlanta, Georgia, identified a vendor on the dark web using Chinese skincare products to hide shipments of fentanyl to be trafficked into the United States (DEA, 2020). When inspecting a shipment from this vendor, law enforcement found fentanyl concealed inside a gel pack, hidden by surrounding facial mask products (DEA, 2020).

In 2018, San Francisco law enforcement identified a local resident who had pill presses and more than 1,000 grams of methamphetamine used to produce counterfeit Adderall, such as those shown in Figure 2.4 (DEA, 2020). Similarly, in 2019, law enforcement in both Michigan and Florida seized counterfeit Adderall pills; the pills had the same orange/peach color, markings, and morphology of authentic 30 mg Adderall; however, the pills contained methamphetamine, caffeine, and acetaminophen instead (DEA, 2020). In 2019, law enforcement in San Diego, California, arrested 20 individuals for trafficking methamphetamine, cocaine, marijuana, oxycodone, and carfentanil. Several of these individuals were distributors of counterfeit prescription opioids made with carfentanil, which has a potency 100 times that of fentanyl (DEA, 2020). These counterfeit opioids have been directly linked with a fatal overdose (DEA, 2020).

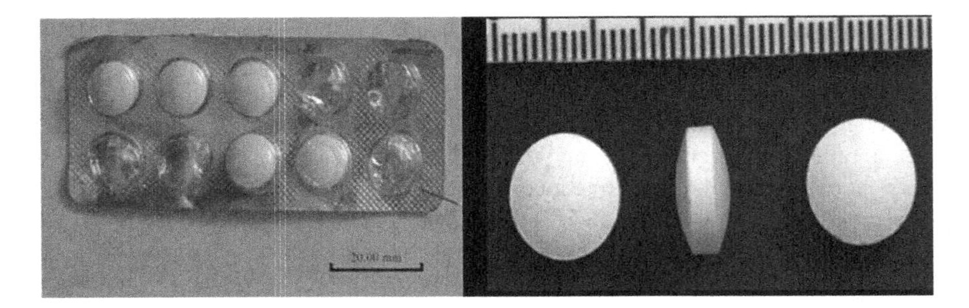

FIGURE 2.4 Counterfeit Teva Adderall tablets (FDA, 2016).

2.1.6.5 Counterfeit Cannabinoids

Synthetic cannabinoids have marked a new trend in counterfeit drugs. These drugs are not only most commonly inhaled, but they can also be pressed into counterfeit pills (DEA, 2020). While synthetic cannabinoids tend to present a lower threat than other abused drugs, they were linked to mass overdoses throughout the United States in 2018 and, as such, should not be dismissed as a threat. For example, in 2018, Arizona officials seized 76 tablets that appeared to be Xanax but actually contained cyclopropylfentanyl, methamphetamine, and FUB-AKB48 (a synthetic cannabinoid) (DEA, 2020).

Another concerning trend in counterfeit drugs is the emergence of counterfeit cannabidiol (CBD) products (CNBC, 2019). The American Association of Poison Control Centers has determined that CBD is an "emerging hazard," as it has a high potential for contamination and mislabeling. A 2017 study on the authenticity of CBD products found that 70% of the studied products were mislabeled (CNBC, 2019). Counterfeiters seem to be taking advantage of the newfound popularity of CBD products by substituting low-cost, illicit synthetic marijuana for authentic CBD. Synthetic marijuana has been found in both CBD vapes and edibles (CNBC, 2019).

The Associated Press commissioned laboratory testing of different CBD vape products and found that 10 of the 30 tested products contained synthetic marijuana, while others contained no CBD whatsoever. The products varied by flavor and purchase location, indicating that counterfeit CBD products may be difficult to trace and identify. Counterfeit CBD products are of particular concern not only because they may not contain CBD oil but also because they present a significant danger to the consumer. For instance, in 2018, a teenager bought CBD vape oil out of boredom, and shortly after using it suffered from acute respiratory failure and ended up in a coma; the CBD vape oil was laced with a potent street drug (CNBC, 2019).

2.1.6.6 Counterfeit Benzodiazepines

Because nonprescribed benzodiazepines have seen an increase in use globally, counterfeiters have recently set their sights on benzodiazepines as a huge source of income. (Maust et al., 2019). Benzodiazepine use has become increasingly prevalent among young people to numb emotions, come down from other drug use, and/or lessen opioid withdrawal symptoms (Mateu-Gelabert et al., 2017). The increased demand for access to benzodiazepines without a prescription has resulted in a heightened presence of these drugs in unregulated markets and an increase of counterfeit benzodiazepines available for purchase (Arens et al., 2016). To further this issue, new psychoactive substance (NPS) benzodiazepine use has also seen a rapid increase (EMCDDA, 2019). NPS benzodiazepines are typically designed to bypass relevant drug laws and imitate the effects of illicit drugs currently on the market (Tracy et al., 2017). The pressing danger of NPS benzodiazepines is that while they may mimic the short-term effects of the target drug, the long-term effects and side effects of their use are completely unknown (Baumann and Volkow, 2016). Notably, 2.4 million NPS benzodiazepine tablets were seized in Europe in 2017 (EMCDDA, 2019).

Xanax (alprazolam), a common benzodiazepine, has been prone to counterfeiting in the recent years (Tobias et al., 2021). Counterfeit Xanax has been determined to contain synthetic opioids such as fentanyl or U-47700, as well as NPS benzodiazepines such as etizolam or flualprazolam in a range of doses (Chapman et al., 2021). This is particularly concerning for benzodiazepine-dependent users, as consumption of counterfeit products may result in benzodiazepine withdrawal and serious adverse health risks, some of which may be life-threatening (Ait-Daoud et al., 2018).

Tobias et al. presented a study aimed at examining counterfeit alprazolam tablets found in unregulated markets in Canada (Tobias et al., 2021). Researchers found that a large portion of the tablets purported to be Xanax was counterfeit and did not contain any of the active ingredient (alprazolam) when examined by drug-checking services (Tobias et al., 2021). The supposed Xanax pills were benzodiazepine-negative over 25% of the time (Tobias et al., 2021). Alternatively, the pills that tested positive for benzodiazepines often contained etizolam, a thienodiazepine, rather than alprazolam (Tobias et al., 2021). The vast majority of pills analyzed contained a variety of active ingredients (Tobias et al., 2021). Some pills contained other NPSs such as synthetic opioids, other benzodiazepines, and synthetic cannabinoids (Tobias et al., 2021).

2.1.6.7 Counterfeit COVID-19 Medicines

The evolution of counterfeit medicines seems to have culminated in such a way that counterfeiters have learned to take advantage of perilous situations and desperate consumers. This has become particularly evident during the COVID-19 global pandemic. The COVID-19 health emergency has created an environment of confusion, vulnerability, and fear that has allowed counterfeiters to expand their predatory practices. It has also exposed vulnerabilities in the American drug supply chain, including pandemic-related issues such as reduced drug manufacturing and exportation due to the closure of some Chinese pharmaceutical suppliers (Tesfaye et al., 2020). Because COVID-19 lockdowns in China and India disrupted the production of many pharmaceutical products (China and India remain two of the biggest pharmaceutical producers), there has been a stronger demand for drugs affected by this disruption (Tesfaye et al., 2020). The shortage of certain medicines has led to an increased presence of their counterfeit versions on the market to appease the demand (Tesfaye et al., 2020).

As misinformation and concern regarding the spread of COVID-19 increased, counterfeiters expanded their methods to take advantage of consumers (Tesfaye et al., 2020). INTERPOL's Operation Pangea XIII has seen the seizure of counterfeit face masks, low-quality hand sanitizer, "coronavirus packages," "corona spray," and unlicensed medicines claiming to treat or prevent COVID-19 (INTERPOL, 2020). Chloroquine, an antimalarial medication, has been particularly targeted by counterfeiters for its rumored ability to help prevent COVID-19 (INTERPOL, 2020). Figure 2.5 shows the results of color tests to detect fake chloroquine. INTERPOL reported an increase of 18% in antiviral drug seizures since the beginning of the pandemic, when compared to the same time frame in 2018. INTERPOL also reported a 100% increase in unauthorized chloroquine, thought to be a result of the COVID-19 pandemic (INTERPOL, 2020).

As COVID-19 vaccines began rolling out, INTERPOL released an Orange Notice, warning the public of the potential for counterfeit vaccines, stolen vaccines, and illegal advertising related to the vaccines. INTERPOL also emphasized the need to ensure supply chain safety and identify unauthorized websites advertising the sale of COVID-19 vaccines (INTERPOL, 12/2020). The FDA released a list of recommendations for consumers to identify the tactics used by those peddling fake COVID-19 products, including the need to search for scientific evidence of product claims rather than customer testimonials and to use caution when presented with "quick fix" or "miracle cure" claims for COVID-19 (FDA, 2021). The FDA has identified several fake "cure-all" products for COVID-19, including essential oils, colloidal silver, tinctures, teas, and other treatments (FDA, 2021).

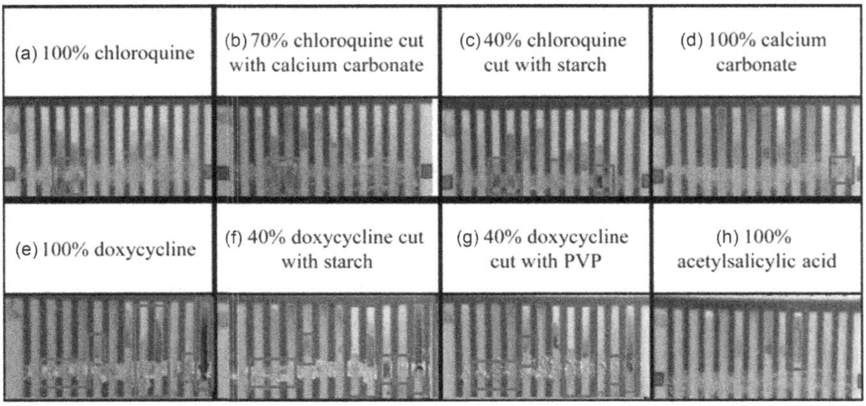

FIGURE 2.5 Color test of the antimalarial drug chloroquine and its counterfeit versions (Weaver et al., 2015).

Pfizer, Inc., one of the primary pharmaceutical corporations responsible for producing COVID-19 vaccines, recently confirmed that counterfeit versions of its COVID-19 vaccine had been discovered (WSJ, 2021). Pfizer, Inc. identified counterfeit versions of the vaccine in both Mexico and Poland, after vials had been seized by authorities. The vials discovered in Mexico had inaccurate labels, and the vials recovered in Poland were thought to actually be skincare treatments. Unfortunately, 80 individuals received the counterfeit vaccine in Mexico prior to its discovery (WSJ, 2021). This report signifies the need for expanded public awareness efforts, particularly during times of great concern and uncertainty such as that of the COVID-19 pandemic.

2.2 CONCLUSION

As sophisticated as many counterfeit drug manufacturers may be, the fight against counterfeit drug trafficking is not an impossible one to win. While the techniques needed to identify fake drugs are complex and sometimes daunting, they are not the only means in the fight against counterfeiting. One study performed in 2016 provided refreshing evidence that the best weapon of all may be consumers themselves (Mhando et al., 2016).

A study aimed to assess general consumer awareness of counterfeit drugs as well as the ability of the average consumer to discern counterfeit drugs from genuine medicines (Mhando et al., 2016). Of the consumers interviewed, over half were able to determine whether a drug was counterfeit or genuine. That number jumped to over 80%, when the interviewed consumers were also healthcare professionals (Mhando et al., 2016). Interestingly, consumers who were aware of the potential health consequences of ingesting counterfeit drugs were almost three times more likely to successfully differentiate fake drugs from genuine medicines (Mhando et al., 2016).

This chapter exemplifies the need for more advanced and widespread consumer education when it comes to fighting the counterfeit drug trade. Counterfeit drugs continue to evade control and be documented in the literature. However, if consumers can be empowered in their ability to distinguish authentic drugs from their counterfeit versions, and new regulations can be enacted to control the spread of counterfeit drugs, the world may well see a drop in counterfeit drug sales and, more importantly, an increase in the number of consumers and patients receiving authentic medicines. Overcoming the sophisticated methods and means found in the counterfeit drug trade, as well as the seemingly constant evolution of drugs of choice for counterfeiters, may appear next to impossible at our current state. Yet, studies like these are educational tools to empower those actively fighting against the counterfeit drug trade to protect consumers worldwide. Several recommendations are diagrammed in Figure 2.6 to overcome the counterfeit drug trade. These include economic development and jobs in developing nations, consumer education and public awareness, increased surveillance of counterfeit drugs, improved chemical detection methods, increased access to quality pharmaceutical medicines, and increased manufacturing of pharmaceuticals in Western countries.

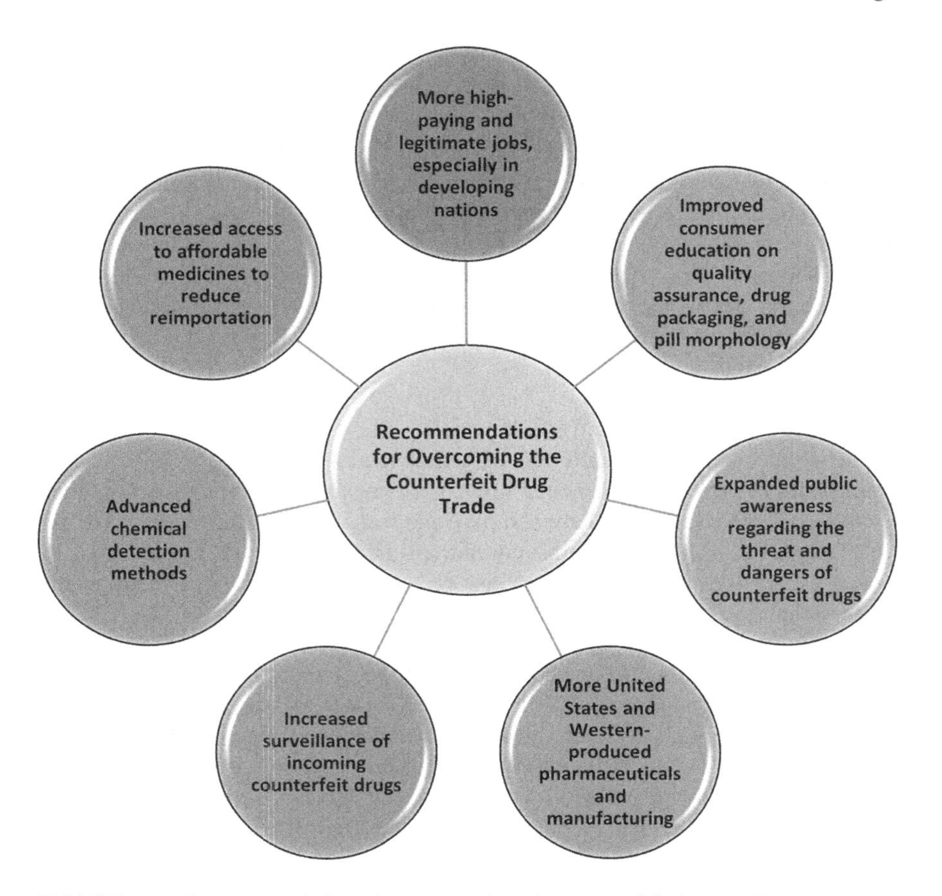

FIGURE 2.6 Recommendations for overcoming the counterfeit drug trade.

REFERENCES

Ait-Daoud, N., Hamby, A.S., Sharma, S., Blevins, D. A review of alprazolam use, misuse, and withdrawal. *J Addict Med.* 2018;12(1):4–10. doi: 10.1097/ADM.0000000000000350.

All of Us Research Hub, National Institutes of Health (NIH). *Drug Exposures*. 2020 https://databrowser.researchallofus.org/ehr/drug-exposures.

Arens, A.M., van Wijk, X.M.R., Vo, K.T., Lynch, K.L., Wu, A.H.B., Smollin, C.G. Adverse effects from counterfeit alprazolam tablets. *JAMA Intern Med.* 2016;176:1554–5. doi: 10.1001/jamainternmed.2016.4306.

Baumann, M.H., Volkow, N.D. Abuse of new psychoactive substances: Threats and solutions. *Neuropsychopharmacol.* 2016;41(3):663–5. doi: 10.1038/npp.2015.260.

Blackstone, E.A., Fuhr, J.P., Pociask, S. The health and economic effects of counterfeit drugs. *Am Health Drug Benefits*. 2014;7(4):216–224.

CBS Los Angeles. DEA agents going online in effort to catch drug dealers, distributors. 2021 Feb 25. https://losangeles.cbslocal.com/2021/02/25/dea-online-drug-dealers-selling-fentanyl-busted/.

Chapman, B.P., Lai, J.T., Krotulski, A.J., Fogarty, M.F., Griswold, M.K., Logan, B.K., et al. A case of unintentional opioid (U-47700) overdose in a young adult after counterfeit Xanax use. *Pediatr Emerg Care.* 2021;37(9):e579–e580. doi: 0.1097/PEC.0000000000001775.

CNBC. Investigation finds illegal synthetic marijuana in vape and edible products sold as CBD. 2019 Sep 16. https://www.cnbc.com/2019/09/16/investigation-finds-illegal-synthetic-marijuana-in-vape-and-edible-products-sold-as-cbd.html.

Cockburn, R., Newton, P.N., Agyarko, E.K., Akunyili, D., White, N.J. The global threat of counterfeit drugs: Why industry and governments must communicate the dangers. *PLoS Med.* 2005;2(4):e100. doi: 10.1371/journal.pmed.0020100. https://journals.plos.org/plosmedicine/article?id=10.1371/journal.pmed.0020100.

Drug Enforcement Administration (DEA). Counterfeit prescription pills containing fentanyls: A global threat. 2016. https://www.dea.gov/sites/default/files/docs/Counterfeit%2520Prescription%2520Pills.pdf

Drug Enforcement Administration (DEA). 2019 National Drug Threat Assessment. 2020 Jan 30. https://www.dea.gov/documents/2020/01/30/2019-national-drug-threat-assessment.

Drug Enforcement Administration (DEA). Alarming spike in fentanyl-related overdose deaths leads officials to issue public warning. 2020 Aug 06. https://www.dea.gov/press-releases/2020/08/06/alarming-spike-fentanyl-related-overdose-deaths-leads-officials-issue.

Elkins, K.M. Introduction to Forensic Chemistry, Taylor & Francis/CRC Press, 2019.

Elkins, K.M., Weghorst, A., Quinn, A.A., Acharya, S. "Color Quantitation for Chemical Spot Tests for a Controlled Substances Presumptive Test Database," *Drug Testing and Analysis*, 2017; 9:306–310.

European Alliance for Access to Safe Medicines (EAASM). Internation Institute for Research Against Counterfeit Medicines (IRACM) launches a new awareness campaign on counterfeit medicines and internet – 2015. http://www.eaasm.eu/index.php?cID=32&cType=news.

European Monitoring Centre for Drugs and Drug Addiction (EMCDDA). European Drug Report: Trends and Developments 2019–2019. Publications Office of the European Union, Luxembourg. https://www.emcdda.europa.eu/system/files/publications/11364/20191724_TDAT19001ENN_PDF.pdf.

Fleiz, C., Arredondo, J., Chavez, A., Pacheco, L., Segovia, L.A., Villatoro, J.A., et al. Fentanyl is used in Mexico's northern border: Current challenges for drug health policies. Addiction 2020;115:778–781. doi: 10.1111/add.14934.

Frosch, T., Wyrwich, E., Yan, D., Domes, C., Domes, R., Popp, J., et al. Counterfeit and substandard test of the antimalarial tablet Riamet by means of Raman hyperspectral multicomponent analysis. *Molecules.* 2019;24(18):3229.

Gautam, C.S., Utreja, A., Singal, G.L. Spurious and counterfeit drugs: A growing industry in the developing world. *Postgrad Med J.* 2009;85(1003):251–256. doi: 10.1136/pgmj.2008.073213.

Glass, B.D. Counterfeit drugs and medical devices in developing countries. *Res Rep Trop Med.* 2014;2014(5):11–22. doi: 10.2147/RRTM.S39354.

INTERPOL. Millions of Medicines Seized in Largest INTERPOL Operation Against Illicit Online Pharmacies – 2017. https://www.interpol.int/en/News-and-Events/News/2017/Millions-of-medicines-seized-in-largest-INTERPOL-operation-against-illicit-online-pharmacies.

INTERPOL. Global operation sees a rise in fake medical products related to COVID-19. 2020 Mar 19. https://www.interpol.int/en/News-and-Events/News/2020/Global-operation-sees-a-rise-in-fake-medical-products-related-to-COVID-19.

INTERPOL. Interpol warns of organized crime threat to COVID-19 vaccines. 2020 Dec 02. https://www.interpol.int/en/News-and-Events/News/2020/INTERPOL-warns-of-organized-crime-threat-to-COVID-19-vaccines.

Mages, R., Kubic, T.T. Counterfeit medicines: Threat to patient health and safety. *Pharm Policy Law*. 2016;18:163–177. doi: 10.3233/PPL-160441.

Markel, H. How the Tylenol murders of 1982 changed the way we consume medication. PBS News Hour 2014 Sep 29. https://www.pbs.org/newshour/health/tylenol-murders-1982.

Martino, R., Malet-Martino, M., Gilard, V., Balayssac, S. Counterfeit drugs: Analytical techniques for their identification. *Anal Bioanal Chem*. 2010;398:77–92. doi: 10.1007/s00216-010-3748-y.

Mateu-Gelabert, P., Jessell, L., Goodbody, E., Kim, D., Gile, K., Teubl, J., et al. High enhancer, downer, withdrawal helper: multifunctional nonmedical benzodiazepine use among young adult opioid users in New York City. *Int J Drug Policy*. 2017;46:17–27. Doi:10.1016/j.drugpo.2017.05.016.

Maust, D.T., Lin, L.A., Blow, F.C. Benzodiazepine use and misuse among adults in the United States. *Psychiatr Serv*. 2019;70:97–106. doi: 10.1176/appi.ps.201800321.

Mhando, L., Jande, M.B., Liwa, A., Mwita, S., Marwa, K.J. Public awareness and identification of counterfeit drugs in Tanzania: A view on antimalarial drugs. *Advances in Public Health*. 2016;2016:6254157. doi: 10.1155/2016/6254157.

Mohamed Ibrahim, M.I. Counterfeit medicines: A quick review on crime against humanity. In: Babar, Z., editor. *Pharmacy practice and clinical pharmacy*. Cambridge, MA:2019:60–76.

Newton, P.N., Green, M.D., Fernández, F.M. Impact of poor-quality medicines in the 'developing' world. *Trends Pharmacol Sci*. 2009;31(3):99–101. doi: 10.1016/j.tips.2009.11.005.

Obi-Eyisi, O., Wertheimer, A.I. The background and history of counterfeit medicines. In: Wertheimer, A.I., Wang, P.G., editors. Counterfeit medicines volume 1: Policy, economics and countermeasures. Glendale, AZ:2012;1–18.

OECD/EUIPO (2020), Trade in Counterfeit Pharmaceutical Products, Illicit Trade, OECD Publishing, Paris, https://doi.org/10.1787/a7c7e054-en.

Ossola, A. The Fake Drug Industry Is Exploding, and We Can't Do Anything About It – 2015. https://www.newsweek.com/2015/09/25/fake-drug-industry-exploding-and-we-cant-do-anything-about-it-373088.html.

Pardo, B., Taylor, J., Caulkins, J.P., Kilmer, B., Reuter, P., Stein, B.D. The future of fentanyl and other synthetic opioids. Santa Monica, CA: RAND Corporation, 2019.

Pascu, G.A., Hancu, G., Rusu, A. Pharmaceutical serialization, a global effort to combat counterfeit medicines. *Seria Medica*. 2020;66(4)132–139. doi: 10.2478/amma-2020-0028.

Pharmaceutical Security Institute (PSI). Measuring Pharma Crime – 2021. https://www.psi-inc.org/pharma-crime (accessed January 26, 2021).

Pitts, P.J. The spreading cancer of counterfeit drugs. *J Commer Biotechnol*. 2020;25(3):20–34. doi: 10.5912/jcb940.

Sanada, T., Yoshida, N., Matsushita, R., Kimura, K., Tsuboi, H. Falsified tadalafil tablets distributed in Japan via the internet. *Forensic Sci Int*. 2020;307:110143. Doi: 10.1016/j.forsciint.2020.110143.

Sanofi. About counterfeiting – 2013. http://fakemedicinesrealdanger.com/web/about-counterfeiting.

Shover, C.L., Falasinnu, T.O., Dwyer, C.L., Benitez Santos, N., Cunningham, N.J., Freedman, R.B., et al. *Drug Alcohol Depend*. 2020;216:108314. doi:10.1016/j.drugalcdep.2020.108314.

Substance Abuse and Mental Health Services Administration (SAMHSA). Key substance use and mental health indicators in the United States: Results from the 2019 national survey on drug use and health. 2020. https://www.samhsa.gov/data/sites/default/files/reports/rpt29393/2019NSDUHFFRPDFWHTML/2019NSDUHFFR1PDFW090120.pdf.

The Partnership for Safe Medicines. Update on fake Avastin – FDA warnings in 28 states, six prosecutions – 2013. https://www.safemedicines.org/2013/02/fda-warnings-in28-states-six-prosecutions-511.html.

The Partnership for Safe Medicines (PSM). Counterfeit drug manufacturing trio from Florida sentenced – 2018. https://www.safemedicines.org/2018/02/counterfeit-drug-manufacturing-trio-from-florida-sentenced.html.

The Partnership for Safe Medicines (PSM). Deadly counterfeit pills found in all 50 U.S. states; Deaths now reported in 42 of them. 2019 Oct 26. https://www.safemedicines.org/2020/10/deadly-counterfeit-pills-found-in-all-50-u-s-states.html.

Tesfaye, W., Abrha, S., Sinnollareddy, M., Arnold, B., Brown, A., Matthew, C., et al. How do we combat bogus medicines in the age of the COVID-19 pandemic? *Am J Trop Med Hyg*. 2020;103(4):1360–3. doi: 10.4269/ajtmh.20-0903.

Tobias, S., Shapiro, A.M., Grant, C.J., Patel, P., Lysyshyn, M., Ti, L. Drug checking identifies counterfeit alprazolam tablets. *Drug Alcohol Depend*. 2021;218:108300. doi: 10.1016/j.drugalcdep.2020.108300.

Tracy, D.K., Wood, D.M., Baumeister, D. Novel psychoactive substances: Types, mechanisms of action, and effects. *BMJ*. 2017;356:i6848. doi: 10.1136/bmj.i6848.

The Wall Street Journal (WSJ). Pfizer identifies fake Covid-19 shots abroad as criminals exploit vaccine demand. 2021 Apr 21. https://www.wsj.com/articles/pfizer-identifies-fake-covid-19-shots-abroad-as-criminals-exploit-vaccine-demand-11619006403.

U.S. Customs and Border Protection (CBP). CBP Cincinnati seizes 44 pounds of sildenafil pills worth $712,756. 2021 Jul 30. https://www.cbp.gov/newsroom/local-media-release/cbp-cincinnati-seizes-44-pounds-sildenafil-pills-worth-712756.

U.S. Customs and Border Protection (CBP). CBP Cincinnati seizes over 33 pounds of Viagra pills. 2021 Jan 13. https://www.cbp.gov/newsroom/local-media-release/cbp-cincinnati-seizes-over-33-pounds-viagra-pills.

U.S. Customs and Border Protection (CBP). 2007 Oct 8. Posted to http://www.cbp.gov/xp/cgov/newsroom/photo_gallery/afc/inspectors_seaports/cs_photo21.xml. Accessed on 2021 May 2 at https://commons.wikimedia.org/wiki/File:CBP_with_bag_of_seized_counterfeit_Viagra.jpg.

U.S. Food & Drug Administration (FDA). 2016 Sep 9. Posted to https://flickr.com/photos/39736050@N02/7296764626. Accessed on 2021 May 2 at https://commons.wikimedia.org/wiki/File:Counterfeit_Teva%E2%80%99s_Adderall_30_mg_Tablets_(7296764626).jpghttps://www.ajtmh.org/view/journals/tpmd/92/6_Suppl/article-p17.xml.

U.S. Food & Drug Administration (FDA). Counterfeit versions of Cialis tablets identified entering the United States. 2015 Jan 21. https://www.fda.gov/drugs/drug-safety-and-availability/counterfeit-versions-cialis-tablets-identified-entering-united-states.

U.S. Food & Drug Administration (FDA). Janssen alerts counterfeit SYMTUZA (darunavir/cobicistat/emtricitabine/tenofovir alafenamide) is being distributed in the United States. 2020 Dec 24. https://www.fda.gov/media/144858/download.

U.S. Food & Drug Administration (FDA). Beware of fraudulent Coronavirus tests, vaccines, and treatments. 2021 Mar 1. https://www.fda.gov/consumers/consumer-updates/beware-fraudulent-coronavirus-tests-vaccines-and-treatments.

Valisure. Valisure Certificate of Analysis (CoA) – 2021. https://www.valisure.com/analysis/.

Weaver, A.A., Lieberman, M. Paper test cards for presumptive testing of very low quality antimalarial medications. *Am J Trop Med Hyg*. 2015;92(6):17–23. doi: 10.4269/ajtmh.14-0384. https://www.ajtmh.org/view/journals/tpmd/92/6_Suppl/article-p17.xml.

Wertheimer, A.I. Identifying and combating counterfeit drugs. *Expert Rev. Clin. Pharmacol.* 2008;1(3):333–6. doi: 10.1586/17512433.1.3.333.

3 Fake Pharmaceuticals

Clandestine Tablets and Counterfeit Medicines

Theresa DeAngelo
Maryland State Police Forensic Sciences Division

CONTENTS

3.1 OVERVIEW OF SEIZED DRUG EVIDENCE AND PARAPHERNALIA

Forensic laboratory drug chemistry units encounter an abundance of controlled dangerous substances (CDS) submissions every year. Submissions they receive include controlled substances contained in baggies, pill form, within paraphilia, and in pharmaceutical tablet shapes. For the submissions that are in a pharmaceutical tablet shape, a drug chemist does not know upon first glance if these submissions are really a prescription drug or if they are instead a counterfeit medicine made to look like the real pharmaceutical drug. An investigator submitting this type of drug evidence may not know the different either with physical observation of the tablets and may be under the assumption that they are submitting a prescriptive drug to the forensic laboratory. These counterfeit medicines, also known as fake pharmaceuticals or clandestine tablets, are produced and sold by drug dealers with the deceptive intent to misrepresent their origin, authenticity, or effectiveness. These fake pharmaceuticals can include drugs that contain no active pharmaceutical ingredient, an inferior quality of the active pharmaceutical ingredient, an incorrect amount of the active pharmaceutical ingredient, contaminants, incorrect active pharmaceutical ingredients, and even repackaged expired pharmaceutical ingredient products. These counterfeit medicines are generally

DOI 10.1201/9781003183327-3

unsafe and can be very harmful to anyone who takes such medicines. So how do these counterfeit medicines make their way to a forensic laboratory as part of CDS evidence submissions?

3.2 COMMONLY COUNTERFEITED DRUGS

First, let us look at common counterfeit medicines that are encountered by consumers. High-demand, expensive medications for consumers, such as various chemotherapeutic drugs, antibiotics, vaccines, erectile dysfunction drugs, weight loss aids, hormones, analgesics, and antianxiety drugs are common counterfeiting medicines that are encountered.[1] These are drugs that are usually prescribed for the treatment of chronic diseases, and many consumers try to be savvy and look for ways to save money on these particular types of prescription drugs, especially if they cannot afford the drugs via health insurance (or lack thereof). Unfortunately, the drugs purchased (usually online) are not the real medication the consumer is looking for but instead are fake medications laced with things such as road paint, antifreeze, or unapproved controlled dangerous substances. Those that are laced with controlled dangerous substances usually end up being confiscated by an illegal drug investigation or by an arrest where a person is in possession of a prescription that does not belong to them. These particular fake pharmaceuticals then end up as submissions to forensic laboratory drug chemistry units. In 2020, the Maryland State Police Forensic Sciences Division received into their CDS Units far more fake pharmaceutical tablet case submissions in comparison to actual real pharmaceutical tablet case submissions.[2] This same laboratory in 2021 has received hardly any real pharmaceutical tablet case submissions, meaning the tablet case submission received are majority counterfeit tablet cases.

Let us take a look at the most common CDS submissions to a forensic laboratory that fall under the category of fake pharmaceuticals or counterfeit medicines. Many forensic laboratories have seen fake drugs such as alprazolam (brand name Xanax), oxycodone (brand names such as Percocet and OxyContin), and clonazepam (brand name Klonopin) in tablet form in their recent drug evidence case submissions[2] (see Figures 3.1–3.3). These counterfeit drugs are usually submitted to forensic laboratories in small quantities for criminal drug possession cases and in larger quantities when the case is related to a drug distribution case, which may include the intent of delivery or sale of the drugs. The difference between a drug possession case and a drug distribution case varies via U.S. state criminal drug laws as well as Federal criminal drugs laws.

For years, the number of counterfeit drug cases submitted to forensic laboratories has risen. For the public, awareness of the crimes of producing and selling counterfeit drugs generally become known only when the perpetrators are caught, and the results of the investigation are publicly announced. However, forensic laboratory drug units have been analyzing such counterfeit medicines since before 2002. The amount of drug case submissions that contain fake pharmaceuticals also depends on the location of the forensic laboratory. For example, in 2009,

20 million pills, bottles, and sachets of counterfeit medicines were seized in an operation coordinated by the International Criminal Police Organization (Interpol) across China and seven of its Southeast Asian neighbors. In this case, 33 people were arrested, and 100 retail outlets were closed.[3] In Maryland, from

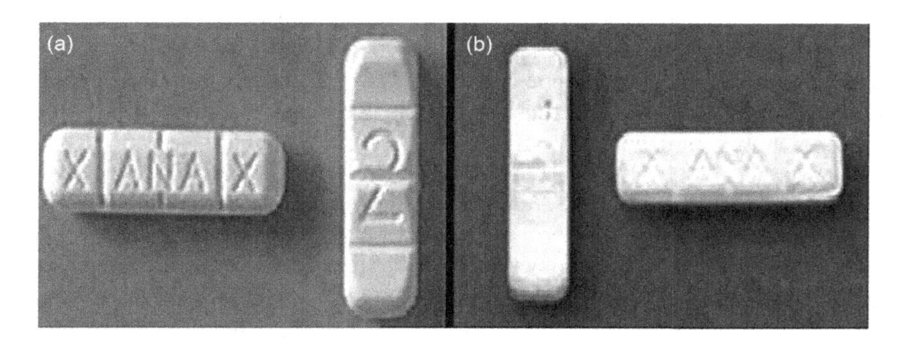

FIGURE 3.1 On the left is real Xanax (Alprazolam), and on the right is fake Xanax.

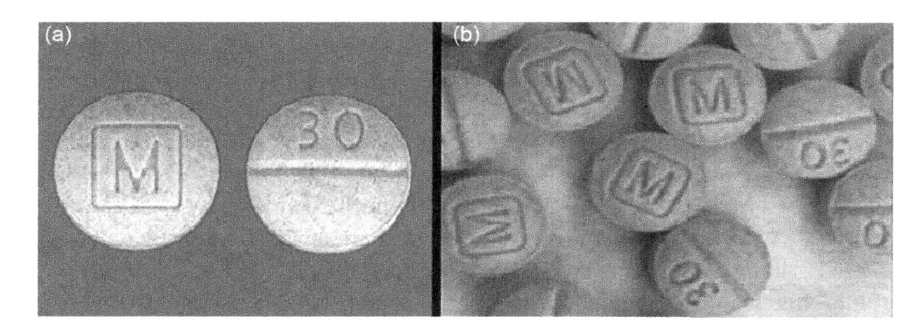

FIGURE 3.2 On the left is real oxycodone tablets, and on the right are the fake oxycodone.

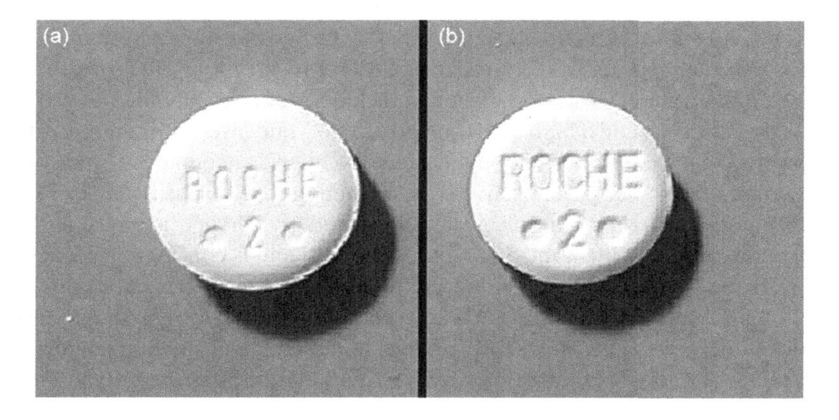

FIGURE 3.3 On the left is real Rivotril (Clonazepam), and on the right is fake Rivotril.

2019 to 2020, only 5%–12% of total drug cases submitted to the Maryland State Police Forensic Sciences Division contained fake pharmaceuticals.[2] However, back in 2018, one of the largest drug distribution cases in Maryland resulted in the arrest of two men who sold more than 920,000 counterfeit Xanax pills on the dark web. Across the country in Colorado, Federal authorities were able to dismantle an international drug trafficking ring in 2021, where 64 people were arrested for transporting large quantities of counterfeit oxycodone pills from Mexico through Mexico to the U.S. For this single case, over 77,000 counterfeit oxycodone pills were recovered from drug trafficking cells in Denver, Colorado Springs, and Adams County, Colorado.[4]

Now let us look at some of the major components that make up these counterfeit drugs and illicit medicines. Alprazolam is classified as a benzodiazepine and is usually prescribed for short-term management of anxiety disorders, specifically panic disorder or generalized anxiety disorder.[5] Alprazolam, sold under the brand name Xanax, is a Schedule IV Controlled Substance. However, in forensic drug chemistry units, what appears to be alprazolam tablets may actually be tablets produced to look like alprazolam and instead are laced with etizolam, flualprazolam, flubromazolam, or clonazolam. Etizolam, a thienodiazepine, which is chemically related to benzodiazepines, is also a Schedule IV Controlled Substance.[5] However, etizolam has a median lethal dose of a toxin that is beyond a recommended dosage for the general public. Etizolam has been found to be a drug of potential misuse and cases of intentional suicide by overdose have been traced back to this particular drug. Flualprazolam is an analog of alprazolam, and is considered a designer benzodiazepine and a novel psychoactive substance. Flualprazolam is not a FDA approved prescription drug and is currently not controlled under the U.S. Controlled Substances Act. Flualprazolam has been linked to many criminal cases regarding impaired driving as well as overdose cases within the U.S.

Flubromazolam is another example of a designer benzodiazepine that is considered to be highly potent. Flubromazolam poses higher risks to users, because of their ability to produce strong sedation and amnesia at very low oral doses. Flubromazolam, just like flualprazolam, is not a FDA approved prescription drug and is currently not controlled under the U.S. Controlled Substances Act; however, it is classified as a Schedule I Controlled Substance in Virginia.[5] The spectrum of adverse effects after taking flubromazolam include confusion, inability to walk or talk, sleep paralysis, and a high addiction potential. Many users of flubromazolam had lowered inhibitions that led to police and psychiatric encounters. Clonazolam is also another designer psychoactive drug belonging to the classification of benzodiazepines. Clonazolam produces anxiolytic, sedative, hypnotic, muscle relaxant, anticonvulsant, depressant, and amnesic effects, and is considered fast acting with users feeling these effects within 20–60 minutes after ingestion. Long-term use of clonazolam may result in increased seizures, aggression, increased anxiety, violent behavior, loss of impulse control, irritability, and suicidal tendencies. Clonazolam is another example of a substance that is not FDA approved and is not controlled under the U.S. Controlled Substances Act; however, this drug is considered a Schedule I Controlled Substance in both Virginia and Minnesota.

The designer benzodiazepines mentioned above (flualprazolam, flubromazolam or clonazolam) are all easily accessible for purchase through online research chemical vendors, where they are sold as designer drugs. Since these are all easily accessible online, it does not require a buyer or user to have a legal prescription in order to purchase these designer drugs. There are many drug dealers that obtain these unapproved drugs in bulk quantities from overseas suppliers for the purpose of pressing the drugs into pills or tablets and distributing them to customers throughout the U.S.[1] The dealers then set up internet-based business websites for customers to purchase products, which are really "misbranded" drugs such as alprazolam. The customers then purchase what they think is a legitimate prescriptive drug (such as alprazolam), and are instead exposed to risks such as toxicity and even fatal overdose.

Another example of a common fake pharmaceutical seen within submissions to forensic drug chemistry units is counterfeit oxycodone tablets. Oxycodone is an opioid analgesic, which is legally prescribed to help relieve moderate to severe pain. Oxycodone is one of the most commonly abused prescription drugs in the U.S., where users start out taking the prescribed amount, but as their body develops a tolerance to the drug, the users need or seek a higher dose to maintain the same relief or high. Since this drug has a high potential for abuse, there have been more regulations placed throughout the U.S. for oxycodone prescriptions. Those that abuse the drug then seek out other ways to obtain the drug through illegal means. This may mean purchasing online what a user thinks is oxycodone, but instead are oxycodone-looking clandestine tablets laced with heroin, fentanyl, xylazine, tramadol, 4-Aminophenyl-1-phenethylpiperidine (4-ANPP), and/or para-fluorobutyryl fentanyl (para-FBF). Heroin and fentanyl are both opioids, where heroin (diacetylmorphine) is a derivative of morphine and fentanyl being a synthetic opioid, is 50–100 times more potent than morphine. Heroin is a Schedule I Controlled Substance, and fentanyl is a Schedule II Controlled Substance. Fake pharmaceuticals sold as oxycodone, but really contain heroin and fentanyl, can be highly dangerous for any user and have led to overdose issues within the U.S. Buyers may think they are purchasing oxycodone, but instead these tablets contain other opioid drugs such as heroin or fentanyl. On the street, these drugs have nicknames like beans, green apples, apples, shady eighties, and greenies.

Xylazine—another substance often found within fake oxycodone tablets—is a veterinarian drug used for sedation, anesthesia, muscle relaxation, and analgesia in animals such as horses.[5] For humans, xylazine causes side effects such as bradycardia, respiratory depression, hypotension, transient hypertension, and other changes in cardiac output. An overdose of xylazine is usually fatal for humans. Tramadol, also found often within fake oxycodone, is a synthetic opioid prescribe to treat moderate to severe pain. Tramadol is a Schedule IV Controlled Substance, but abuse or misuse of tramadol can lead to overdose and death. Those who use or abuse tramadol for long time periods will develop a tolerance for the pain relief provided by the drug, and if dosage is reduced, like other opioids, a person will experience withdrawal symptoms. 4-ANPP is a direct precursor to fentanyl; it is not psychoactive; and this substance is only present in a fake oxycodone tablet because of improper processing of the intended product of the synthesis of fentanyl. Para-FBF is a designer opioid that is an analog of butyrfentanyl (an analog of fentanyl). Side effects of this fentanyl analog include itching, nausea, and potentially serious respiratory depression, which can be lethal. Para-FBF is a Schedule I Controlled Substance, and because dosage of this substance is not known within a fake oxycodone tablet, fatalities due to overdoses is common.

The counterfeit oxycodone tablets are often cheap for users to purchase and look so similar to actual oxycodone tablets (blue in color stamped with M30),[5] but can cause serious harm because they are laced with substances usually 100 times more potent than oxycodone. However, due to an abuser's increased tolerance of opioids, individuals seek out counterfeit pills, because they assume the pills purchased on the street contain fentanyl and other powerful synthetic opioids. These counterfeit pills and fake oxycodone tablets are especially dangerous, because the amount of fentanyl or other synthetic opioids varies from pill to pill, even in the same batch of fake tablets, and will cause the individual an increased risk of death.

A third type of fake pharmaceutical tablets encountered by forensic drug chemistry units are clandestine clonazepam tablets. Clonazepam is another type of benzodiazepine usually prescribed to treat seizures, panic disorders, and akathisia. Just like alprazolam, clonazepam is a Schedule IV Controlled Substance and can be highly addictive to those prescribed this drug, with some individuals becoming addicted in as little as a few weeks. Once a person is addicted to clonazepam, their brain can no longer produce feelings of relaxation and calmness without it, and they struggle to quit and are unable to function normally when they do not have it. This becomes a problem when the drug is no longer legally prescribed, and an individual seeks out illegal means of purchasing clonazepam. Illegally purchased clonazepam tablets, which then end up as submissions to forensic drug chemistry units, can be found laced with clonazolam, etizolam, and even fentanyl.

Clonazolam, a designer drug that is not a Scheduled Controlled Substance, mixed with etizolam, a Schedule IV Controlled Substance, can be very dangerous. Clonazolam is very potent and produces an amnesic effect on the body, where etizolam produces euphoric as well as sedative effects.[5] Together, you have a mixture of a depressant drug substance that slows everything down leading to intense sedation. If you include the highly potent opioid fentanyl in this mixture, it can lead to extreme respiratory depression and then death. Those purchasing the fake clonazepam tablets illegally on the street or online may have no idea that the tablets are counterfeit, laced with such a dangerous substance mixture, and risk overdosing on these tablets and dying.

3.3 SOURCES OF FAKE PHARMACEUTICALS

One may ask what are the original sources of these types of fake pharmaceuticals or how are individuals able to purchase such counterfeit medicines? Overall, counterfeit pills are sold on the black market, either on the street by drug dealers or online via the dark web. Drug dealers obtain the fake pharmaceuticals from drug trafficking organizations that may stem from southern U.S. states and

Mexico. The drug trafficking organizations may obtain their supply from China, Hungary, and other countries from the United Kingdom (U.K.).[1] One must note that there is no quality control in these counterfeit pills, and drug trafficking organizations do not employ scientists or use professional laboratories to create these pills; therefore, they cannot create the safe chemical mixtures that the legitimate pharmaceutical prescription drugs do. If a legitimate doctor did not prescribe the drug, or if the drug is not coming from a pharmacy, it is very likely a counterfeit drug. To a consumer, counterfeit drugs may look slightly strange or be in poor-quality packaging, but they often seem identical to the real thing.

3.4 EXAMPLES OF FAKE PHARMACEUTICALS

For an example, take a look at Figure 3.4, these are three tablets, blue in color that appear to all be the same. However the differences are subtle, and two of the tablets are actually counterfeit tablets. One may notice that the tablet at the top has a slightly longer 'f' in the company logo, while the bottom two tablets are a grayish blue and have rougher edges. The tablet at the top is the genuine medicine, where the two lower tablets are fake. As you can see from this example, it is not always easy to identify counterfeit medicines. These fake tablets could be diluted, contaminated, or even just chalk.

An example of fake pharmaceutical packaging can be seen in Figure 3.5. Variations in the packaging can also be subtle, just like the look of a tablet, where in this example the colored bars are of different lengths. However, other times, fake pharmaceutical packaging is more obvious where there are spelling errors in the company or medicine name.

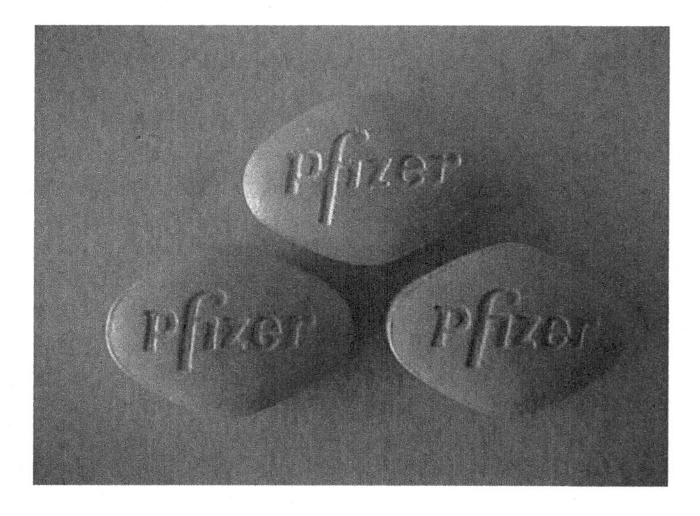

FIGURE 3.4 Three tablets, one is real and two are fake.

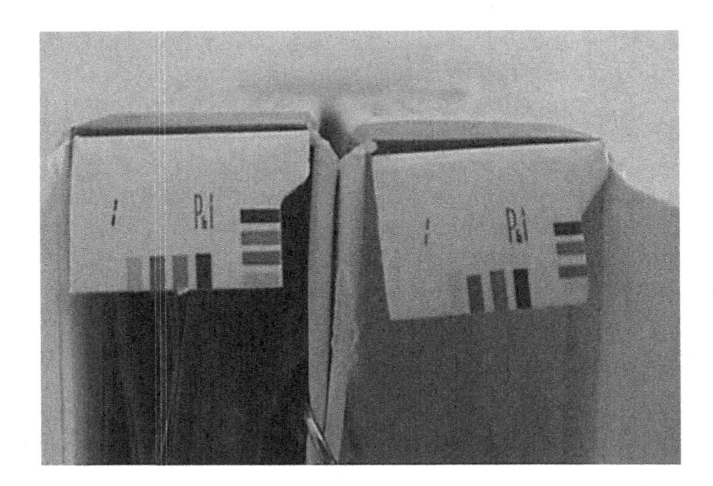

FIGURE 3.5 Fake pharmaceutical packaging examples.

3.5 MONITORING SOURCES OF FAKE PHARMACEUTICALS

Monitoring the sources of fake pharmaceutical and counterfeit drugs is difficult. Those that use the dark web to purchase counterfeit drugs can do so anonymously, and the websites that are set up for purchasing provide limited information on the manufacturer, distributor, and health risks for the products sold on these sites. Online criminal sellers have anonymity, giving them security and an unlimited worldwide market to sell their counterfeit products. For those illegal sellers that are organized, distribution of their fake products can be simple, based on already legitimate supply chains for purchasing and distribution via the internet. Many consumers may be unaware that they are purchasing fake or counterfeit products, because the supply chain is the same for the legitimate pharmaceuticals as well as the illegal produced fake pharmaceuticals. Due to the lack of awareness among consumers, many may be a disadvantage and do not even think to question the authenticity of the pharmaceutical product they are purchasing. However, many consumers are well aware of where to purchase the counterfeit medications and are seeking out places on the dark web to purchase the illegitimate products, knowing they are getting an illegal drug (or drugs) mixed in with the tablets they are purchasing. Overall, the main groups that counterfeit illegal sellers focus on are those individuals looking to avoid doctors and the healthcare systems altogether. These may be consumers looking for lifestyle drugs due to embarrassing personal health conditions such as obesity, hair loss, and impotence, or drug addicts looking for their much-needed doses in a secretive manner.

With the current global economy, plus the expansion and ease of access to the internet, there is unfortunately a secure and very strong foundation of sources for counterfeit medications, and fake pharmaceuticals available for consumers to make purchases. The U.K. tends to be a main source of these counterfeit medications due to their location; they are a convenient transit point from Asian countries to markets in the U.S. and other Western European countries.[1] Criminal gangs, which have already been set up selling illegal drugs such as cocaine or heroin, find that counterfeit medication production and distribution has a lower risk involved with higher profits gained for these gangs. These criminal gangs have already established large networks and vast resources, so they are able to maintain complex structures over long periods of time and avoid additional costs by using already existing trafficking routes, all while maintaining anonymity if operating via the dark web.

Once large supplies of counterfeit medicines reach beyond the realms of the dark web, the supply of these fake pharmaceuticals can be found in street pharmacies or via street drug dealer. Drug dealers will focus on areas where there is limited access to genuine medical products or areas where there is poor structure in the detection of such illegal markets. According to the World Health Organization (WHO), if there is an insufficient medical product supply in the market, within days, the vacuum is filled with falsified versions.[1] Those involved in the production and supply of fake pharmaceuticals are attracted by profit margins, where even low-priced counterfeit medicines can make money for criminals, as long as the sales volume is high enough within their market. Unfortunately, counterfeit medications only come to light when they reach the retailer or the patient/consumer; therefore, it is very hard to trace them back through complex supply chains, or to prove when and where the criminal activity occurred. For most illegal activities involving counterfeit drugs, the investigation of the activity is left in the hands of the police, where some departments may not have extensive expertise in the specialized techniques sometimes needed to investigate fake pharmaceuticals.

Investigations become more complicated because of the international nature of much of the illegal activity for counterfeit medicines and often require cross-border investigations to occur. These investigations rely heavily on forensic laboratory analysis of the fake pharmaceuticals, but there is also a need to follow the paper trail of the products to trace its point of origin for these investigations. Most investigations now lead to paper trails in an electronic format via the internet or dark web. These investigations also face the obstacles of requiring a shared legal understanding when various countries are involved, often involving cooperation treaties, or there may be limited capacity to tackle offshore companies that have complex ownership structures and use foreign bank accounts. Often these investigations show that there are multiple levels involved before the fake pharmaceutical gets to the consumer level. At minimum, those involved with the illegal production of counterfeit medications include producers of raw ingredients, manufacturers of the finished tablet/medical products, transport companies, stock managers, brokers, distributors, retailers, and/or health facilities. So finding accountability in an investigation to then prosecute cases regarding fake pharmaceutical crime can be very difficult.

3.6 CASES AND PROSECUTIONS

Although it may be difficult, there are some specific case examples of fake pharmaceuticals investigations where the arrest and prosecution was successfully carried out for those illegally involved in such cases. Looking at that case again from 2018, the U.S. Department of Justice announced a plea deal for the two men involved from Maryland for manufacturing and distributing over 920,000 fake Xanax pills, and involvement with money laundering between 2013 and 2017. One defendant, Ryan Farace, began manufacturing the fake pills while still living at his parent's house in 2013 and moved his operations to his own home after purchase of the house in 2016.[6] To manufacture the fake pills, Farace used a standalone pill press, a table top pill press, Xanax pills molds, dyes, and other equipment, which was purchased online from eBay. Farace then sold the counterfeit pills on dark website such as Silk Road 2.0, Hansa, and Darknet Heroes League. Farace was traced after the Federal Bureau of Investigation (FBI), and Dutch authorities seized the illegal sites Silk Road 2.0 and Hansa. Shockingly, data from March 2016 through May 2017 showed that in about a year, Farace was able to ship orders totaling over 500,000 fake Xanax pills.[6] Farace would ship the fake pills to customers using the U.S. Postal Service. The other defendant, Robert Swain, was only involved with money laundering under this complex fake pill operation. Numerous other individuals involved with this case were arrested and charged, and this operation resulted in the seizure of weapons, illegal drugs, virtual currency proceeds, U.S. currency, and computer equipment. Farace faced numerous sentences, including a sentence of five years in prison for conspiracy to manufacture, distribute, and possess with the intent to distribute alprazolam; five years in prison for each of three counts of distributing and possessing with intent to distribute alprazolam; and 20 years in prison for maintaining drug-involved premises. Farace and Swain faced a maximum of 20 years in prison for the money laundering conspiracy. Overall, this particular case resulted in extensive work conducted by the FBI, Dutch authorities, the U.S. Drug Enforcement Administration (DEA), Homeland Security Investigations (HSI), the U.S. Postal Inspection Service (USPIS), the IRS-Criminal Investigation, the U.S. Marshals Service, and a local Maryland Police Department in Baltimore County.[6]

Another case example from Washington, D.C. in 2015 involved a local pharmacist who was arrested and charged with operating a rogue internet pharmacy, providing illegal pharmaceutical shipments, and generating about $8.3 million in illegal proceeds.[7] Although not a case involving the sales of counterfeit medications, it involved a fake online pharmacy and providing consumers with illegal prescription drugs. The defendant from D.C., Titilayo Akinyoyenu, worked with a physician from Florida to operate an internet-based pharmacy site, and illegally shipped both controlled and noncontrolled dangerous substances (where both categories required prescriptions) to over 38,000 U.S. consumers. The illegal website was operating between 2005 and 2010, where consumers could order the drugs that were illegally prescribed by the physician from Florida, Alan Saltzman.[7] In the U.S., any physician who writes or authorizes prescriptions without a valid doctor–patient relationship are issuing invalid prescriptions, because the physicians are acting outside the usual course of professional practice. Likewise, any pharmacist who knowingly fills such prescriptions, or who have reasons to know such prescriptions are invalid, are violating the law. Akinyoyenu was arrested based on filling more than 58,000 prescriptions (including refills) for customers based solely on the internet site's medical questionnaire, and Saltzman was charged with approving the entire customers' requests based solely on their answers to that online medical questionnaire. The investigation involved the Department of Justice's Organized Crime Drug Enforcement Task Force, the FBI, the DEA, USPIS, and the FDA's Office of Criminal Investigations. Based on the findings and evidence in this investigation, both defendants were charged with conspiracy to distribute and dispense controlled substances, conspiracy to distribute controlled drugs over the internet, conspiracy to introduce misbranded drugs into interstate commerce, and conspiracy to engage in mail fraud. These charges added up to over 45 years in prison and fines over $2.0 million dollars for both defendants.[7]

A more recent case example comes from Utah in 2020, where Aaron Shamo was sentenced to life in prison after conviction for organizing and directing a drug trafficking organization that distributed more than a half of a million fentanyl-laced oxycodone tablets on the dark web.[8] Shamo's drug trafficking organization imported alprazolam and fentanyl from China, and then created fake oxycodone tablets as well as counterfeit Xanax tablets to distribute across the U.S. In one year alone, Shamo sold over one million fentanyl-laced fake oxycodone pills to unsuspecting consumers, and evidence showed that at least 90 of the known customers died from overdoses. Investigators discovered that Shamo would even receive messages from customers that they were getting sick, and Shamo's response was to just send more fake pills to the complaining customers. Unfortunately, Shamo sold the fentanyl-laced fake oxycodone pills in bulk to redistributors; therefore, investigators could not always locate all of the end users of the pills. This means the total number of victims affected by this illegal operation could be over 10,000 people across the U.S. Shamo took advantage of the opioid crisis in the U.S. and established the dark web storefront, hired employees, took charge of marketing and product placement, and knowingly sold consumers dangerous fake tablets marketed as real oxycodone. This particular investigation highlighted the ongoing struggle that the U.S. faces with opioid addiction and the cooperative efforts needed to dismantle criminal organizations targeting these consumers. Investigators in this case included special agents from the HSI, the DEA, the IRS, the FDA Office of Criminal Investigations, and the USPIS.[4] This is just another case example where numerous law-enforcement officers and special task forces from various Federal agencies cooperated in order to bring justice to the families of the victims and everyone else affected by Shamo's crimes.

3.7 CONCLUSION

Manufacturing and distribution of illegal counterfeit drugs, counterfeit medicines, fake pharmaceuticals, or even clandestine tablets is an ongoing issue for not only the U.S. but globally as well. Many criminals will continue to operate anonymously on the dark web to distribute their products, which are knowingly produced to misrepresent the drug product's origin, authenticity, and effectiveness. These fake products can be harmful to consumer's health and, in some cases, have cause serious long-term health damages or even death for some users. Globally, there are existing international treaties that protect against trafficking women and children, counterfeiting money, and transferring contraband tobacco. However, there is currently not an international treaty targeting counterfeit drug crime and promoting detection, apprehension, extradition, and punishment of the individuals involved in such crimes. Counterfeit drugs or fake pharmaceuticals range from complete fraudulent products that have been made by a counterfeiter to original products with manipulated expiration dates. Unfortunately, consumers, patients, doctors, and/or pharmacists may not be able to distinguish such "medications" from the real ones without a very detailed inspection. Perhaps in the future, with more consumer awareness of the issues that counterfeit medications pose, as well as an increase effort for global prevention of criminal activities involving fake pharmaceuticals, the threat of illegal counterfeit drug activity, and fake pharmaceutical manufacturing and sales will decline.

REFERENCES

1. World Health Organization. WHO Global Surveillance and Monitoring System for substandard and falsified medical products. Geneva: World Health Organization; 2017, ISBN 978–92–4–151342–5.
2. Beck, L., DeAngelo, T., Katz, D., and Kuperus, W. Maryland State Police Forensic Sciences Division 2020 Annual Report, March 2021.
3. World Health Organization. *Bulletin of the World Health Organization*, Volume 88: Number 4, April 2010, 241–320.
4. U.S. Department of Justice, U.S. Attorney's Office. "Colorado Springs Man Pleads Guilty to Providing Deadly Fentanyl to Teen". August 31, 2022. https://www.justice.gov/usao-co/pr/colorado-springs-man-pleads-guilty-providing-deadly-fentanyl-teen.
5. Marnell, Tim. Drug Identification Bible. Amera-Chem Inc. 2020/2021.
6. U.S. Department of Justice, U.S. Attorney's Office. "Maryland Man Sentenced to 57 Months in Federal Prison and Ordered to Forfeit at Least $5.665 Million as Result of His Conviction on Charges Relating to Dark Web Drug Distribution and Money Laundering". November 30, 2018. https://www.justice.gov/usao-md/pr/maryland-man-sentenced-57-months-federal-prison-and-ordered-forfeit-least-5665-million-s.
7. U.S. Department of Justice, U.S. Attorney's Office. "Washington, D.C.-Area Pharmacist Indicted and Arrested On Charges Involving Illegal Pharmaceutical Shipments". March 30, 2015. https://www.justice.gov/usao-dc/pr/washington-dc-area-pharmacist-indicted-and-arrested-charges-involving-illegal.
8. U.S. Department of Justice, U.S. Attorney's Office. "Shamo Sentenced To Life In Prison After Conviction For Organizing, Directing Drug Trafficking Organization". October 15, 2020. https://www.justice.gov/usao-ut/pr/shamo-sentenced-life-prison-after-conviction-organizing-directing-drug-trafficking

4 Results from a Survey of Trends in Seized Counterfeit Drugs

Kelly M. Elkins
Chemistry Department and Forensic Science Program, Towson University

CONTENTS

4.1 INTRODUCTION

Law enforcement routinely seizes suspected controlled substances (Dégardin et al., 2015). Crime labs employ a myriad of physical and chemical investigations to determine the identity of the substances (Elkins, 2019). Most of the time, the materials are identified and classified as controlled substances listed under the U.S. Controlled Substances Act (CSA). Controlled substances include illegal drugs that have no accepted medical use and illegal possession of pharmaceutical medicines with accepted medical uses. A recent review of counterfeiting in Ukraine found that the most commonly identified counterfeit drugs that reused expired drugs' repackaging were produced in nonlicensed labs and contain nonpharmacological substances instead of the active ingredient (Lohvynenko et al., 2019).

In addition to controlled substances, law enforcement personnel seize evidence of counterfeit medicine (fake pharmaceuticals) and evidence of production including tablets, pill presses, and packaging. Labs also discern if pharmaceutical drug products have been adulterated. Fake and adulterated medicines pose a global threat to public health (Wertheimer, 2008). Forensic labs can identify the authenticity of the pills and packaging using microscopy and chemical instrumentation methods. For example, Dégardin et al. (2015) analyzed commonly counterfeited capsules and packaging from 33 drug seizures using analytical

DOI 10.1201/9781003183327-4

methods. The results of the packaging and chemical analyses are gathered by forensic and national labs. In the current study, forensic labs in the United States were surveyed in 2021 regarding the counterfeit drugs that they encounter, and what testing they perform to confirm or refute suspected counterfeit drugs.

Surveys are a widely used tool for gathering information from groups of individuals. Several groups around the world have conducted surveys to capture the prevalence, and better understand the drug problem in their area or network (Elkins et al., 2017; Khan et al., 2011; O'Neal et al., 2000). Surveys have been used in forensic studies for a variety of applications, including determining wholesaler encounters with counterfeit medicines (Khan et al., 2011), counterfeit medical devices and injectables (Wang et al., 2020), and use of color tests for drug testing by labs (Elkins et al., 2017; O'Neal et al., 2000), to name a few.

Preliminary methods that law enforcement professionals and consumers can use to detect counterfeit pills include websites and apps such as drugs.com, WebMD (https://www.webmd.com/pill-identification/default.htm), Pill+ app, Pill Identifier and Drug List app, Pill Identifier Mobile app, Pill Identifier app by drugs.com, Pill Identification, Smart Pill ID, Drug Facts by PillSync.com, PolicePal, and iNarc Pill Finder and Identifier to aid in visual identification (Elkins et al., 2015). In addition to online verification, other digital detection methods include mobile detection, radio frequency identification, and blockchain technology (Mackey and Nyyar, 2017).

Counterfeit drugs and packaging materials can be detected and identified using a myriad of chemical instrumentation methods. In a review of methods for analyzing counterfeit drugs, chromatography and spectroscopy were both reported to be used (Dégardin et al., 2014). Gas chromatography–mass spectrometry (GC–MS) and liquid chromatography–mass spectrometry/mass spectrometry (LC–MS/MS), attenuated total reflectance Fourier transform infrared (ATR FT–IR) spectroscopy, Raman spectroscopy, nuclear magnetic resonance spectroscopy (NMR), neutron activation analysis (NAA), infrared microscopy, gas chromatography–infrared spectroscopy (GC–IR), thin-layer chromatography (TLC), and visual inspection and chemical color tests can be used to detect counterfeit drugs (Elkins, 2019; Bolla et al., 2020).

Color test paper cards were used to detect counterfeit antimalarial drugs, including chloroquine, doxycycline, quinine, sulfadoxine, pyrimethamine, and primaquine (Weaver and Lieberman, 2015). The color tests could detect 60%–100% filler but not 30% filler (Weaver and Lieberman, 2015). Interestingly, the color tests could detect low quantities of substitute pharmaceuticals from the other colors produced by the reactions on the cards (Weaver and Lieberman, 2015). Visual inspection followed by TLC was used to detect counterfeit sulfadoxine/pyrimethamine tablets in Malawi; confirmatory tests were performed using HPLC (Khuluza et al., 2016). The tablets were collected by a mystery shopper at health facilities and found to contain a wrong active ingredient (Khuluza et al., 2016).

GC–MS and LC–MS/MS can provide specific and definitive identification of drug substances and components, including "designer drugs" as well as inks used in packaging and imprinted on pills and capsules (Neves and Caldas, 2017; Abdel-Megied and El-Din, 2019). GC–MS was used to quantitatively detect anabolic steroids from tablets, as well as aqueous suspensions and oils (Neves and Caldas, 2017). In a 2021 case of a patient intending to take alprazolam, GC–MS detected only caffeine but LC quadrupole time-of-flight MS and LC–MS detected and identified U-47700—a synthetic designed drug, and its metabolites in urine and blood samples (Chapman et al., 2021). Abdel-Megied and El-Din (2019) developed a novel LC–MS/MS method for the detection and quantification of tramadol hydrochloride in mislabeled counterfeit drugs, including benzodiazepines alprazolam and diazepam, as well as other commonly used drugs chlorpheniramine maleate, diphenylhydramine, and paracetamol (acetaminophen).

^{1}H NMR was used to detect adulteration of anabolic steroids in seized drugs; adulteration was found in 80% of the samples (de Moura Ribeiro et al., 2018). Previously, Krummel et al. (2015) used ^{1}H NMR spectroscopy to detect adulteration of synthetic cathinones with sugars.

Portable near infrared spectroscopy (NIR) devices have been used in the field to screen heroin, cocaine, and cannabis street samples and can produce results in as little as five seconds (Coppey et al., 2020). Portable Raman spectrometry was used to detect fake chloroquine phosphate in Cameroon, Democratic Republic of Congo, and Niger (Waffo Tchounga et al., 2021). The results were confirmed by hyperspectral Raman imaging and HPLC (Waffo Tchounga et al., 2021). In the pills that contained chloroquine, the dose was too low, and others did not contain chloroquine but substituted drugs including low-dose metronidazole and paracetamol (Waffo Tchounga et al., 2021). Surface-enhanced Raman spectroscopy (SERS) with handheld instruments was used to detect trace-level opioids (e.g., fentanyl, hydrocodone, oxycodone, and tramadol) in pharmaceutical tablets suspected of being counterfeit (Kimani et al., 2021). Morphologically directed Raman spectroscopy, which combines automated particle imaging with Raman microspectroscopy, was used to perform physical and chemical analysis of illicit and counterfeit drug samples (Koutrakos et al., 2018). ATR FT–IR spectroscopy and micro-ATR FT–IR spectroscopic imaging were used to evaluate counterfeit pharmaceutical tablet cores; micro-ATR imaging proved more effective for sourcing the formulation, while macro-ATR spectroscopy was more amenable to screening (Lanzarotta et al., 2011).

Very recently, ion beam analysis (IBA) and NAA were used to characterize authentic and illegal Viagra® and sildenafil-based products (Romolo et al., 2021). Particle-induced X-ray emission (PIXE) and MeV range primary ion secondary ion mass spectrometry (MeV-SIMS) characterized and differentiated drug products (Romolo et al., 2021).

To update our knowledge of counterfeit drugs confirmed in forensic evidence, a survey was sent to crime lab directors and personnel to gauge the prevalence of counterfeit drugs encountered in forensic labs, including the types of submitted evidence samples, substances encountered, and testing performed by the lab.

4.2 METHODS

A seven-question survey was developed and sent to 79 laboratory directors and personnel from forensic laboratories across the United States in February 2021. Respondents were queried using SurveyMonkey. Email addresses were obtained through an Internet search and by using a forensic database of the National Clearinghouse for Science, Technology, and the Law at Stetson University College of Law (http://www.ncstl.org/resources/laboratories), which is funded through a grant from the Bureau of Justice Assistance, Office of Justice Programs, U.S. Department of Justice and regional contacts.

The survey included questions regarding the location of the lab, counterfeit drug evidence received and tested, types of counterfeit drug substances encountered, and which lab tests are performed on suspected counterfeit drugs (Table 4.1). The survey questions and project design were approved by the Institutional Review Board for the Protection of Human Participants (IRB) at Towson University (IRB#1379). Results were analyzed using the report generated by SurveyMonkey, with additional analysis performed in Microsoft Excel. Figures were created using Excel, Mapchart.net, and freewordcloudgenerator.com.

TABLE 4.1
Survey Questions

Number	Question
1	In which state is your lab located?
2	Does your lab receive suspected counterfeit drugs (fake or adulterated pharmaceuticals) for testing? (Yes/No)
3	Approximately, how many cases of counterfeit drugs does your lab process annually?
4	Are the number of cases of suspected counterfeit drugs increasing, remaining the same, or decreasing? (Increasing/Decreasing/No change)
5	Which of the following counterfeit drug substances does your lab encounter? Select all that apply. (Fake pharmaceuticals/Adulterated pharmaceutical drug substances/Other—please specify)
6	What are the three most commonly identified counterfeit drug products at your lab?
7	Which of the following lab tests does your lab perform on suspected counterfeit drugs? Select all that apply. (Visual inspection/Microscopy/Color tests/GC–MS/LC–MS/Raman spectroscopy/FT–IR spectroscopy/No testing/Other—please specify)

National Institutes of Health (NIH) All of Us survey data on drug exposures was retrieved on March 5, 2021 (https://databrowser.researchallofus.org/ehr/drug-exposures). Data was compared to common counterfeit pharmaceutical products reported by forensic labs.

4.3 RESULTS

The survey was sent to 79 laboratory directors and forensic practitioners from forensic laboratories across the United States; 15 responses were received for a 19% response rate. Respondents represented labs in the Northeast, Midatlantic, Midwest, West, the South, and Alaska. Survey responses from 14 states were received including Alaska, Arkansas, Arizona, Illinois, Kansas, Louisiana, Maryland, New Jersey, Maine, New Mexico, Ohio, South Carolina, Utah, and Wyoming (Figure 4.1).

Of the respondents, all indicated that their labs receive suspected counterfeit drugs (fake or adulterated pharmaceuticals) for testing. All of the labs (100%) indicated that the number of cases of suspected counterfeit drugs are increasing. On average, the respondents indicated that their labs receive over 300 estimated counterfeit drugs cases annually, on average, for testing with a median of 300 and a range of unknown to 500.

All of the reporting labs encounter fake pharmaceuticals, and 47% of the labs encounter adulterated pharmaceutical drug substances (Figure 4.2).

The respondents indicated that their labs perform a variety of lab tests on suspected counterfeit drugs, including visual inspection, color tests, GC–MS, FT–IR spectroscopy, Raman spectroscopy, TLC, and GC–IR spectroscopy (Figure 4.3).

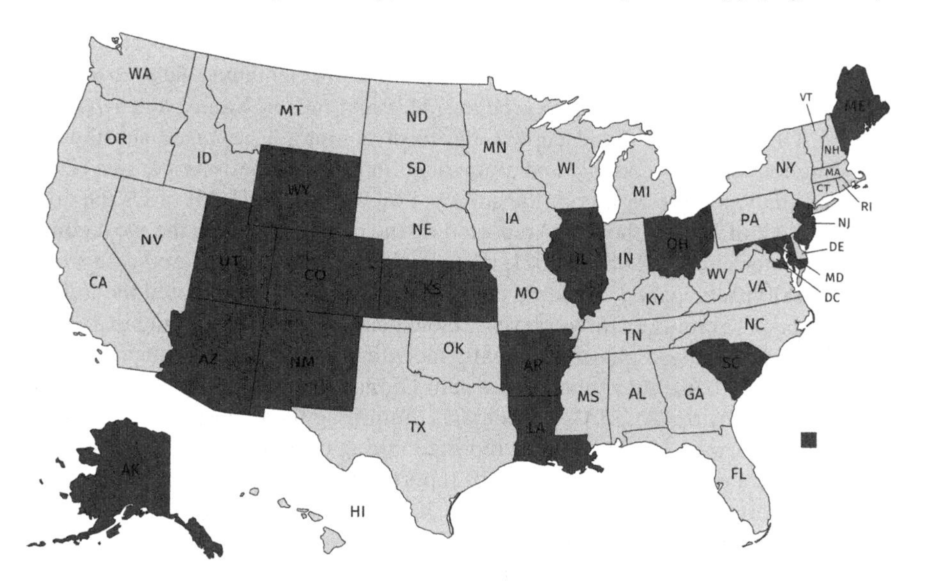

FIGURE 4.1 Map of states where respondent labs are located https://mapchart.net/usa.html.

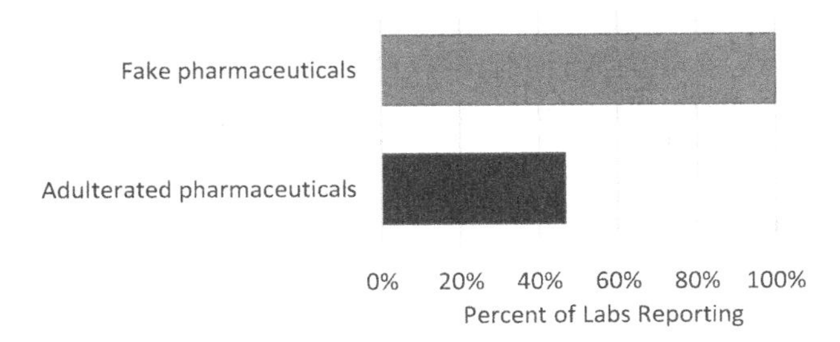

FIGURE 4.2 Percentage of labs reporting that they encounter fake or adulterated pharmaceuticals.

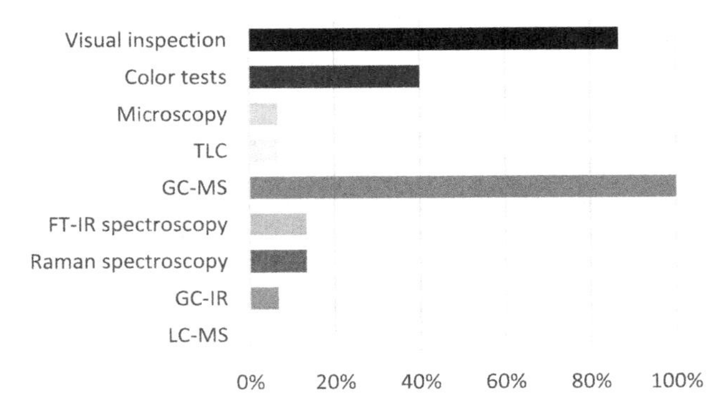

FIGURE 4.3 Methods used for identifying counterfeit drugs by percentage reported.

The most commonly encountered drug substances that were reported were alprazolam, benzodiazepines, clonazolam, etizolam, fentanyl, flualprazolam, hydrocodone, MDMA, opioids, oxycodone (M30 and A 215), Xanax, and 4-ANPP (Figure 4.4). The word cloud depicts the most commonly reported substances including fentanyl, oxycodone, and alprazolam in proportionally larger text size.

After analyzing the results of the survey, I viewed the NIH All of Us data on 3/5/2021 (All of Us) for the drugs reported by the crime labs to be the most commonly counterfeited. As of 3/4/2021, the All of Us database contained data from over 371,000 participants, including over 273,000 who have completed the initial steps of the program (i.e., three surveys, biological donation). The drug exposure data showed that those pharmaceuticals, including oxycodone, fentanyl, hydrocodone, and the benzodiazepine alprazolam (Xanax), were among the most used drugs reported by the All of Us participants. Summing the public data (groups of 20 individuals), over 134,360 people reported taking oxycodone tablets or oxycodone with acetaminophen, over 115,780 reported taking fentanyl products, over 104,800 reported taking hydrocodone with 325 mg or 500 mg acetaminophen, and over 17,300 reported taking alprazolam (Xanax) of various tablets (e.g., 0.25 mg, 0.5 mg, 1 mg, 2 mg, 5 mg). It is clear from this report that counterfeit drugs pose a significant risk to the U.S. population.

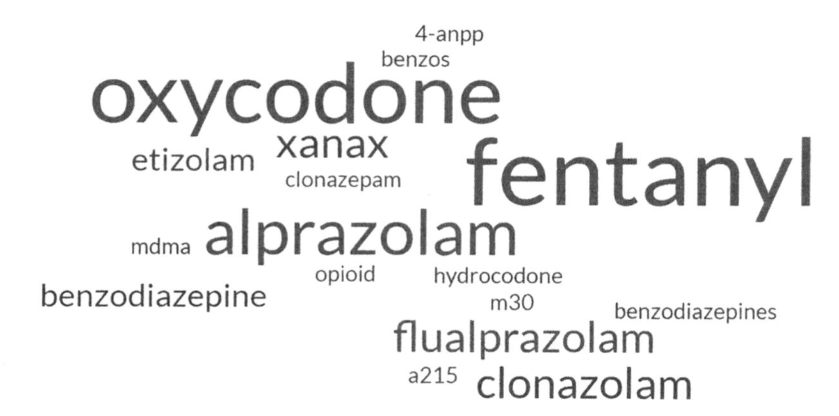

FIGURE 4.4 Word cloud of top three reported counterfeit substances by each respondent lab. Created using https://www.freewordcloudgenerator.com/generatewordcloud.

4.4 DISCUSSION

The survey response rate was low but typical of survey response rates. For comparison, the response rate for a survey of the use of color spot tests forensic laboratories was 37% (121 of 325) (O'Neal et al., 2000). In another study, the author of this study surveyed at least two labs in each state and the United States Drug Enforcement Administration (DEA) labs using rosters from the American Society of Crime Laboratory Directors (ASCLD), a Regional Association of Forensic Scientists and the Criminalistics Section of the American Academy of Forensic Sciences (AAFS), and observed an 8% response rate when CSIs and lab directors were surveyed about the testing time and skill level of interpreting colorimetric drug tests (Elkins et al., 2017). A survey of forensic science practitioners netted 73 responses (percentage not indicated) when asked about certification (Melbourn et al., 2019).

Survey data can be very informative. A survey of American Society for Dermatologic Surgery and the American Society for Laser Medicine and Surgery members regarding counterfeit medical devices and injectables received 765 responses (total members the survey was sent to was not indicated) (Wang et al., 2020). Of the respondents, 37.4% reported encountering counterfeit medical devices, including 71.9% in the United States, 46.4% outside the U.S., and 93.2% on the Internet (Wang et al., 2020). Respondents (41.1%) reported encountering counterfeit injectables (Wang et al., 2020). A 2009 survey of 62 managing executives (83.8%) of registered "modern medicine" wholesalers in Cambodia reported that 12.9% of the wholesalers had encountered counterfeit medicines, which they largely classified as unregistered medicines, while 8.1% did not know the definition of counterfeit medicines (Khan et al., 2011). The wholesalers reported performing numerous checks, including lot numbers, expiration dates, intactness, manufacture dates, and analytical certificates (Khan et al., 2011).

The results of the survey in this study showed that counterfeit drugs are encountered in forensic casework all over the United States, and the number of cases is increasing. The top three methods employed to identify fake drugs were GC–MS, visual identification, and color tests. Labs also reported using ATR FT-IR spectroscopy, Raman spectroscopy, TLC, microscopy, and GC–IR in their identification schemes. The most commonly encountered counterfeit substances at forensic labs were benzodiazepines, opiates, and pills, tablets and capsules containing fentanyl, mirroring the pharmaceutical drugs Americans most frequently use as reported in the All of Us database.

Internet pharmacies, home labs, and unauthorized manufacture of drugs in factories all contribute to counterfeit drugs being seized by law enforcement and analyzed by local, state, and national labs. Counterfeit drugs pose a significant health threat for Americans as well as people worldwide. Counterfeit drugs offer significant revenue and profits for counterfeiters and street dealers (Dégardin et al., 2014). The DEA reported seizing 50,000 fake oxycodone pills with fentanyl in Las Vegas in 2019 (Constantino, 2021). Home, garage, and apartment manufacturers have been reported to be flooding the underground market with counterfeit pills in Minnesota (Raiche, 2020). The pills, which appeared to be oxycodone but were actually fentanyl, were reported to fetch prices of $100 each on the street (Raiche, 2020). Online markets including the darkweb and sales through social media are booming, and counterfeit pharmaceuticals such as OxyContin "M30" pills have been determined by the DEA to contain fentanyl and filler substances (CBSLA, 2021; Broséus et al., 2017). Snapchat communicated that they are working to improve its algorithm to detect illicit drug sales (CBSLA). Fentanyl is an especially troubling ingredient due to its potency – 2 mg of the synthetic opiate is deadly, as it is 50–100 times more potent than morphine (Elkins, 2019; CBSLA, 2021). The "M 30" pills described in the news article were highlighted by the respondents of the survey in this study as commonly encountered counterfeits containing fentanyl. Similarly, forensic labs reported encountering large quantities of fake oxycodone tablets "A 215" containing fentanyl and "XANAX" tablets containing designer benzodiazepines.

In another study, the legitimacy of Viagra pills purchased via the Internet was evaluated (Campbell et al., 2012). Of 22 sample tablets purchased, 17 (77%) were determined to be counterfeit, and only four (18%) were authentic Viagra medicines (Campbell et al., 2012). One pill (5%) was an illegal generic (Campbell et al., 2012). The active ingredient, sildenafil citrate, was detected in only 30%–50% of the labeled ingredient (Campbell et al., 2012). In addition, the counterfeit products lacked product information and safety warnings (Campbell et al., 2012). The shipment origins of the pills were tracked: 11 shipped from Hong Kong, 6 from the United States, 2 from the United Kingdom, and 1 each shipped from Canada, China, and India (Campbell et al., 2012). Four Internet pharmacies that claimed to be situated in Canada did not ship from a Canadian address (Campbell et al., 2012).

In the survey of dermatology medical professionals, a majority of the respondents (73.7%) reported no familiarity or only some familiarity with the FDA's rules and regulations concerning counterfeits, and responded that stricter and better enforced rules and regulations are needed (Wang et al., 2020).

4.5 CONCLUSION

Fake pharmaceuticals and adulterated pharmaceuticals are a widespread and significant problem for consumers as well as law enforcement. Law enforcement encounters counterfeit drugs in property searches and seizures and in collecting evidence for cases. Forensic labs reported that they receive and process increasing numbers of counterfeit drug cases and use a variety of chemical detection methods, but primarily GC–MS and color tests, to determine the providence of pills, tablets, powders, packaging, and drug information sheets and detect counterfeits. Strong regulatory systems and routine detection of counterfeit pills and production is an important step for law enforcement to disrupt counterfeit networks. Use of standard pharmaceutical distribution channels was noted as one of the most efficient ways to counteract counterfeiting. Consumers should exercise extreme caution when purchasing pharmaceuticals via the Internet.

ACKNOWLEDGMENTS

Asia Quick is acknowledged for researching and compiling the emails of the survey recipients. The author thanks the lab directors or their representatives that responded to the survey for their responses, Katie McDougall for discussions about the All of Us data, and Haley Fallang and Rachel Elkins for feedback on the manuscript.

REFERENCES

Abdel-Megied, A.M., and K.M.B. El-Din. "Development of a novel LC-MS/MS method for detection and quantification of tramadol hydrochloride in presence of some mislabeled drugs: Application to counterfeit study." *Biomedical Chromatography* 33, no. 6 (June 2019): e4486. doi: 10.1002/bmc.4486.

Bolla, A.S., Patel, A.R., and R. Priefer. "The silent development of counterfeit medications in developing countries – A systematic review of detection technologies." *International Journal of Pharmaceutics* 587 (September 25, 2020): 119702. doi: 10.1016/j.ijpharm.2020.119702.

Broséus, J., Rhumorbarbe, D., Morelato, M., Staehli, L., and Q. Rossy. "A geographical analysis of trafficking on a popular darknet market." *Forensic Science International* 277 (August 2017): 88–102. doi:10.1016/j.forsciint.2017.05.021.

Campbell, N., Clark, J.P., Stecher, V.J., and I. Goldstein. "Internet-ordered viagra (sildenafil citrate) is rarely genuine." *The Journal of Sexual Medicine* 9, no. 11 (November 2012): 2943–251. doi:10.1111/j.1743-6109.2012.02877.x.

CBSLA. "DEA Agents Going Online In Effort To Catch Drug Dealers, Distributors." (February 25, 2021). Accessed May 27, 2021. https://losangeles.cbslocal.com/2021/02/25/dea-online-drug-dealers-selling-fentanyl-busted/.

Chapman, B.P., Lai, J.T., Krotulski, A.J., Fogarty, M.F., Griswold, M.K., Logan, B.K., and K.M. Babu. "A Case of Unintentional Opioid (U-47700) Overdose in a Young Adult After Counterfeit Xanax Use." *Pediatric Emergency Care* 37, no. 9 (2021): e579–e580. doi: 10.1097/PEC.0000000000001775.

Constantino, M. "50K fake oxycodone pills with fentanyl seized in Las Vegas." Las Vegas Review-Journal (February 19, 2021). Accessed May 27, 2021. https://www.reviewjournal.com/crime/50k-fake-oxycodone-pills-with-fentanyl-seized-in-las-vegas-2284684/.

Coppey, F., Bécue, A., Sacré, P.-Y., Ziemons, E.M., Hubert, P., and P. Esseiva. "Providing illicit drugs results in five seconds using ultra-portable NIR technology: An opportunity for forensic laboratories to cope with the trend toward the decentralization of forensic capabilities." *Forensic Science International* 317 (December 2020): 1–13. doi: 10.1016/j.forsciint.2020.110498.

Dégardin, K., Roggo, Y., and P. Margot. "Understanding and fighting the medicine counterfeit market." *Journal of Pharmaceutical and Biomedical Analysis* 87 (January 2014): 167–175. doi: 10.1016/j.jpba.2013.01.009.

Dégardin, K., Roggo, Y., and P. Margot. "Forensic intelligence for medicine anti-counterfeiting." *Forensic Science International* 248 (March 2015): 15–32. doi: 10.1016/j.forsciint.2014.11.015.

Drug Exposures. NIH All of Us. Accessed March 5, 2021. https://databrowser.researchallofus.org/ehr/drug-exposures.

Elkins, K.M. *Introduction to Forensic Chemistry*. Boca Raton, FL: CRC Press/Taylor & Francis, 2019.

Elkins, K.M., Gray, S.E., and Z.M. Krohn. "Evaluation of Technology in Crime Scene Investigation." *CS Eye* (April 2015): 24–29. https://www.csofs.org/write/MediaUploads/Publications/CSEye/CSEye_April_2015.pdf

Elkins, K.M., Weghorst, A., Quinn, A.A., and S. Acharya. "Color Quantitation for Chemical Spot Tests for a Controlled Substances Presumptive Test Database." *Drug Testing and Analysis* 9, no. 2 (February 2017): 306–310. doi: 10.1002/dta.1949.

Koutrakos, A.C., Leary, P.E., and B.W. Kammrath. "Illicit and Counterfeit Drug Analysis by Morphologically Directed Raman Spectroscopy." *Methods in Molecular Biology* 1810 (2018): 13–27. doi:10.1007/978-1-4939-8579-1_2.

Khan, M.H., Akazawa, M., Dararath, E., Kiet, H.B., Sovannarith, T., Nivanna, N., Yoshida, N., and K. Kimura. "Perceptions and practices of pharmaceutical wholesalers surrounding counterfeit medicines in a developing country: a baseline survey." *BMC Health Services Research* 11 (November 11, 2011): 306. doi: 10.1186/1472-6963-11-306.

Khuluza, F., Kigera, S., Jähnke, R.W.O., and L. Heide. "Use of thin-layer chromatography to detect counterfeit sulfadoxine/pyrimethamine tablets with the wrong active ingredient in Malawi." *Malaria Journal* 15 (April 14, 2016): 215. doi:10.1186/s12936-016-1259-9

Kimani, M.M., Lanzarotta, A., and J.S. Batson. "Trace level detection of select opioids (fentanyl, hydrocodone, oxycodone, and tramadol) in suspect pharmaceutical tablets using surface-enhanced Raman scattering (SERS) with handheld devices." *Journal of Forensic Sciences* 66 no. 2 (March 2021): 491–504. doi: 10.1111/1556-4029.14600.

Krummel, J.N., Russell, L.N., Haase, D.N., Schelble, S.M., Tsai, E., and K.M. Elkins. "Application of High-field NMR Spectroscopy for Differentiating Cathinones for Forensic Identification." *Colonial Academic Alliance Undergraduate Research Journal* 5, no. 1 (2015) http://publish.wm.edu/caaurj/vol5/iss1/1/.

Lanzarotta, A., Lakes, K., Marcott, C.A., Witkowski, M.R., and A.J. Sommer. "Analysis of counterfeit pharmaceutical tablet cores utilizing macroscopic infrared spectroscopy and infrared spectroscopic imaging." *Analytical Chemistry* 83, no. 15 (August 1, 2011): 5972–5978. doi: 10.1021/ac200957d.

Lohvynenko, B.O., Sezonov, V.S., and T.A. Frantsuz-Yakovets. "Tendencies for the falsification of medicinal products in ukraine: general analysis and areas of counteraction." *Wiadomosci Lekarskie* (Warsaw, Poland) 72, no. 12 (2019): 2478–2483.

Mackey, T.K., and G. Nayyar. "A review of existing and emerging digital technologies to combat the global trade in fake medicines." *Expert Opinion on Drug Safety*, 16, no. 5 (April 7, 2017): 587–602. doi: 10.1080/14740338.2017.1313227

Melbourn, H., Smith, G., McFarland, J., Rogers, M., Wieland, K., DeWilde, D., Lighthart, S., Quinn, M., Baxter, A., and L. Quarino. "Mandatory certification of forensic science practitioners in the United States: A supportive perspective." *Forensic Science International. Synergy* 1 (2019): 161–169. doi: 10.1016/j.fsisyn.2019.08.001.

de Moura Ribeiro, M.V., Boralle, N., Felippe, L.G., Pezza, H.R., and L. Pezza. "1H NMR determination of adulteration of anabolic steroids in seized drugs." *Steroids* 138 (2018): 47–56. doi: 10.1016/j.steroids.2018.07.002.

Neves, D.B.J., and E.D. Caldas. "GC–MS quantitative analysis of black market pharmaceutical products containing anabolic androgenic steroids seized by the Brazilian Federal Police." *Forensic Science International* 275 (June 2017): 272–281. doi:10.1016/j.forsciint.2017.03.016.

O'Neal, C.L., Crouch, D.J., and A.A. Fatah. "Validation of twelve chemical spot tests for the detection of drugs of abuse." *Forensic Science International* 109(2000): 189.

Raiche, R. "DEA: 'Garage manufacturers' flooding underground drug market with fake pills." 5abcNews (November 24, 2020). Accessed May 27, 2021. https://kstp.com/news/dea-lsquogarage-manufacturersrsquo-flooding-underground-drug-market-with-fake-pills-november-23-2020/5932790/.

Romolo, F.S., Sarilar, M., Antoine, J., Mestria, S., Rossi, S.S., Gallidabino, M.D., de Souza, G.M.S., Chytry, P., and J.F. Dias. "Ion beam analysis (IBA) and instrumental neutron activation analysis (INAA) for forensic characterisation of authentic Viagra® and of sildenafil-based illegal products." *Talanta* 224 (March 1, 2021): 121829. doi:10.1016/j.talanta.2020.121829.

Waffo Tchounga, C.A., Sacre, P.Y., Ciza, P., Ngono, R., Ziemons, E., Hubert, P., and R.D. Marini. "Composition analysis of falsified chloroquine phosphate samples seized during the COVID-19 pandemic." *Journal of Pharmaceutical and Biomedical Analysis* 194 (February 5, 2021): 113761. doi: 10.1016/j.jpba.2020.113761.

Wang, J.V., Hattier, G., Rohrer, T., Zachary, C.B., and N. Saedi. "Experiences With Counterfeit Aesthetic Medical Devices and Injectables: A National Survey." *Dermatologic Surgery* 46, no. 10 (October 2020): 1323–1326. doi: 10.1097/DSS.0000000000002307.

Weaver, A.A., and M. Lieberman. "Paper test cards for presumptive testing of very low quality antimalarial medications." *The American Journal of Tropical Medicine and Hygiene* 92, no. 6 Suppl (June 2015): 17–23. doi: 10.4269/ajtmh.14-0384.

WebMD. Pill Identifier. Accessed May 27, 2021. https://www.webmd.com/pill-identification/default.htm.

Wertheimer, A.I. "Identifying and combatting counterfeit drugs." *Expert Review of Clinical Pharmacology* 1, no. 3 (2008): 333–336. doi: 10.1586/17512433.1.3.333

5 Fentanyl Adulteration of Drugs and Seizure Trends

Kelly M. Elkins and Haley Fallang
Chemistry Department and Forensic Science Program, Towson University

CONTENTS

5.1 INTRODUCTION

Combatting counterfeit drugs requires the collaboration of many entities, including health professionals, laboratory analysts, and law enforcement. Law enforcement routinely seizes suspected controlled substances and submits the items to labs for analysis (Dégardin et al., 2015). Crime labs identify and quantify the substances submitted for analysis. The High-Intensity Drug Trafficking Area (HIDTA) program funded by the Office of National Drug Control Policy collects data from law enforcement and forensic labs, and analyzes data trends. HIDTA is a drug reduction program run by the United States Office of National Drug Control Policy that was established in 1990 following the passage of the Anti-Drug Abuse Act of 1988.

One definition of counterfeit drugs is substances containing adulterants that corrupt or make the original substance impure. Adulterated drugs can cause adverse drug reactions or be deadly to users. Fentanyl has been identified in seized heroin as an adulterant since 2006 (DEA, 2016). Fentanyl is an "ultrapotent" synthetic opioid reportedly 50–100 times more potent than heroin (Elkins, 2019). The percentage of drugs reported to contain fentanyl or carfentanil increased from 3.4% in 2014 to 48.6% in 2017 (Zibbell, 2021). Whereas heroin was the original target of fentanyl adulteration, now adulterated cocaine and methamphetamine have been identified, and reports of increased trafficking and use of methamphetamine continue to grow.

DOI 10.1201/9781003183327-5

The United States Centers for Disease Control and Prevention (CDC) reports that there were over 70,000 drug overdose deaths in 2019 (CDC). Whereas the CDC reports that heroin overdose led to over 14,000 deaths in the United States in 2019, fentanyl adulteration of heroin is making it even more deadly and is reported to cause the most drug overdose deaths. Specifically, heroin overdoses have increased sevenfold from 1999 to 2019. Cocaine and methamphetamine adulterated with fentanyl also pose significant threats to users. According to the CDC, illicitly manufactured fentanyls, which include fentanyl and fentanyl analogs, heroin, cocaine, and/or methamphetamine, led to nearly 85% of overdose deaths in 24 states and the District of Columbia from January to June 2019 (CDC). More recently, 100,000 Americans were reported to have died of drug overdoses over 12 months during the pandemic (Keating and Bernstein, 2021). Using data provided to us by HIDTA, we report herein upon trends in seizures of heroin, cocaine, methamphetamine, new psychoactive substances (NPS), and fentanyl-adulterated substances in the United States from 2018 through the first half of 2021.

5.2 METHODS

HIDTA provided drug trafficking data for drugs including fentanyl, heroin, methamphetamines, and new psychoactive substances (NPS) from 2018 through the first half of 2021 with approval of the HIDTA Program Performance Management Process (PMP) Research Guidance & Procedures, April 2021. Data was provided in a spreadsheet file which contained deidentified data consisting of seized substance with date, location, and quantity. Pure or adulterated substances containing fentanyl were classified as fentanyl in the data set.

5.3 RESULTS

HIDTA provided three and a half years of data for the analysis in this chapter the analysis in this chapter. Drug trafficking and seizure trends were analyzed for fentanyl, heroin, methamphetamines, and NPS. Table 5.1 and Figures 5.1–5.3 show the trend of increasing fentanyl in various types of seized drug evidence, including kilograms, volume, and dosage units, respectively. The quantity of seized fentanyl in kilograms increased from 1,795 kg in 2018 to 4,310 kg in 2020, reflecting a 240% increase. In the same time period, the volume of seized fentanyl increased from 1 milliliter to 49 milliliters. A total of 289,304 dosage units was seized in 2018, and 4,148,042 were seized in 2020 or over 14 times as many doses. Already for the first half of 2021, the seized dosage units (3,564,401) are almost as much as in all of 2020. The dosage units for heroin also increased by a large number from 310 in 2018 to 37,583 in 2020 (an over 12,000% increase) (Figure 5.4). The volume of heroin increased dramatically from 50 mL in 2018 to 821.1 mL through June 2021 after decreases in the intervening years (Figure 5.5). The kilogram mass of seized heroin decreased in 2020 to 5,965 kg from 6,607 kg in 2019 or by just under 10% (Figure 5.6).

TABLE 5.1

Drug Trends for Fentanyl, Heroin, and Methamphetamines from 2018 through the First Half of 2021

Year	Kilograms			Dosage Units			Liters		
	Fentanyl	Heroin	Methamphetamines	Fentanyl	Heroin	Methamphetamines	Fentanyl	Heroin	Methamphetamines
2018	1,795.7585	6,687.6341	41,876.4082	289,304	310	2,488	0	0.05	4,869.044
2019	3,251.6677	6,607.4919	128,562.5741	1,572,675	21,815	4,712	0.001	0	864.7709
2020	4,309.8475	5,965.0069	69,045.7503	4,148,042	37,583	16,225	0.049	0.017	762.7654
2021	2,030.0308	1,561.1386	24,775.8442	3,564,401	2,280	1,964	0.034	0.8211	29.1341

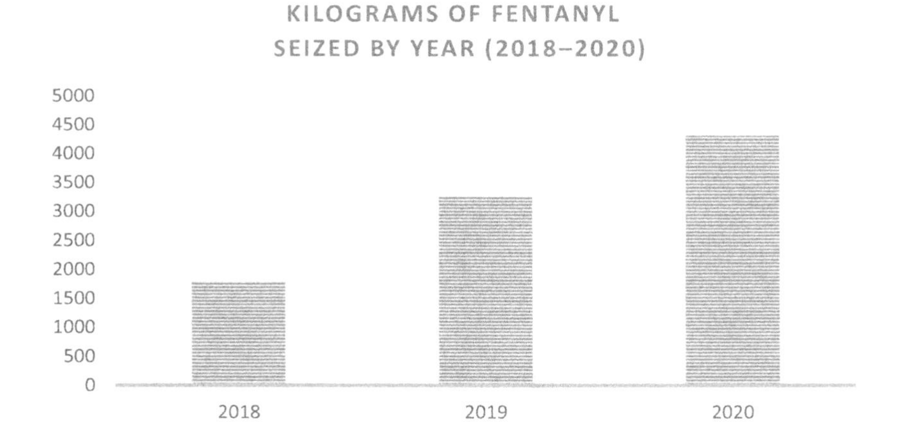

FIGURE 5.1 Trend of increasing fentanyl in kilograms of seized drug evidence (2018–2020).

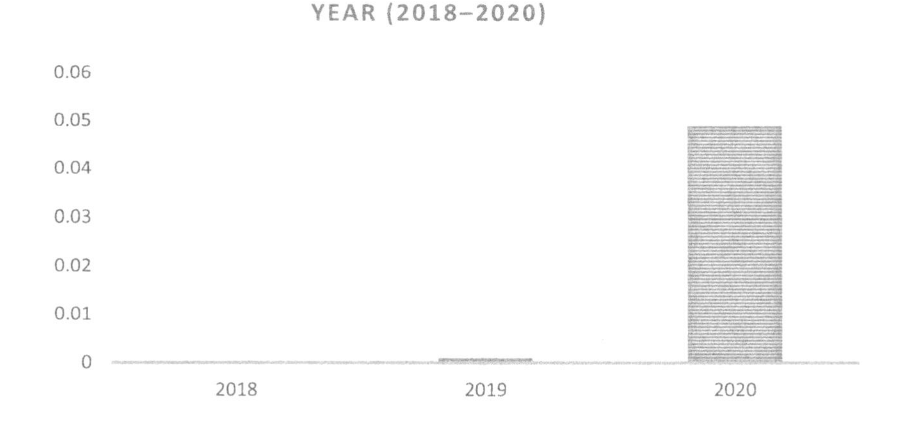

FIGURE 5.2 Trend of increasing fentanyl in liters of seized drug evidence (2018–2020).

The mass of seized methamphetamines also has not been constant over the time period analyzed. In 2018, 41,876 kg of amphetamines were seized, which increased to 128,562 kg in 2019. The quantity decreased to 69,045 kg in 2020 and 24,775 kg for the first half of 2021 (Table 5.2). In contrast to the volume of fentanyl and heroin increasing, the volume of methamphetamines seized decreased each year observed in the dataset from 4,869 L in 2018 to 762 L in 2020 (Figure 5.7) and 29 L in the first half of 2021. The dosage unit of seized methamphetamines increased from 2,488 to 16,225 in 2020 (Figure 5.8). By the first half of 2021, a decreased number of methamphetamines' 1,964 dosage

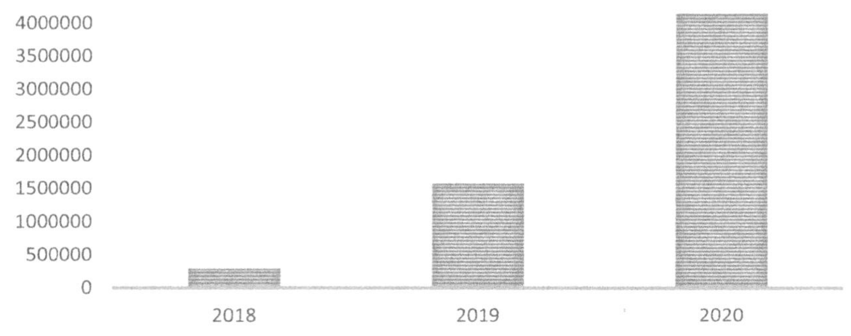

FIGURE 5.3 Trend of increasing fentanyl in dosage units of seized drug evidence (2018–2020).

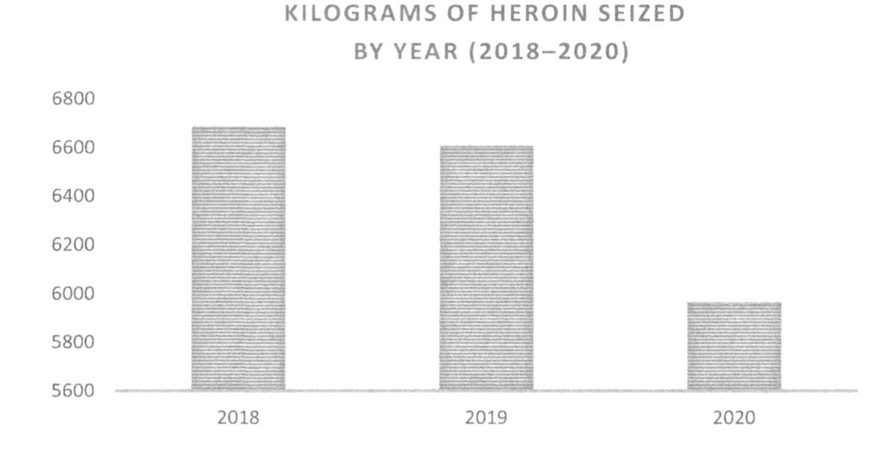

FIGURE 5.4 Trend of decreasing heroin in kilograms of seized drug evidence (2018–2020).

units were seized. The quantity of cocaine seized each year from 2018 to 2020 declined slightly from 181,200.39 kg to 147,079.10 kg (Figure 5.9).

In the three-year period, fentanyl was seized in every state (Figure 5.10). The drugs and distribution have permeated the United States. New, more rural seizure locations outside of major cities and coastal ports were observed in 2021 (Figure 5.11–5.13).

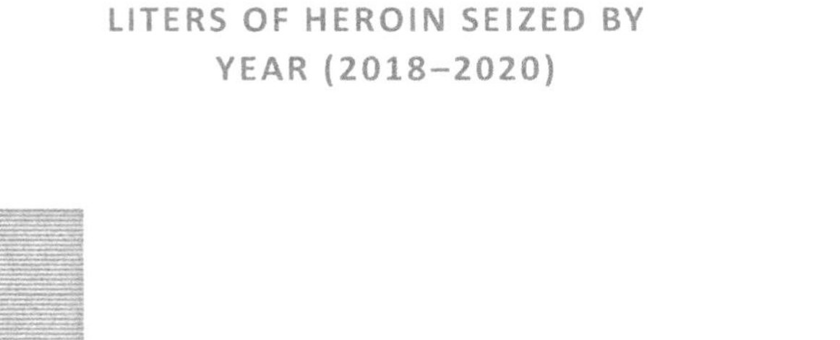

FIGURE 5.5 Trend of decreasing heroin in volume of seized drug evidence (2018–2021).

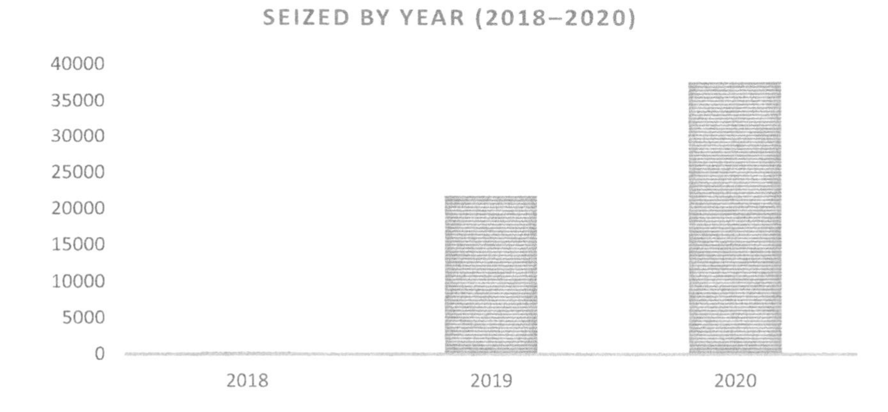

FIGURE 5.6 Trend of increasing heroin in dosage units of seized drug evidence (2018–2020).

TABLE 5.2
Drug Trends in Kilograms for Fentanyl, Heroin, Methamphetamines, and Cocaine from 2018 through the First Half of 2021

Year	Kilograms			
	Fentanyl	Heroin	Methamphetamines	Cocaine
2018	1,795.76	6,687.63	41,876.41	181,200.39
2019	3,251.67	6,607.49	128,562.57	167,663.93
2020	4,309.85	5,965.01	69,045.75	147,079.10
2021	2,030.03	1,561.14	24,775.84	90,223.65

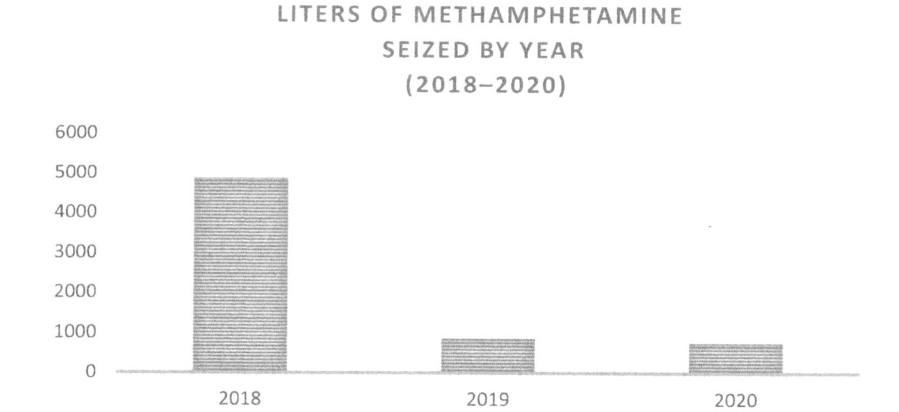

FIGURE 5.7 Trend of decreasing methamphetamines in liters of seized drug evidence (2018–2020).

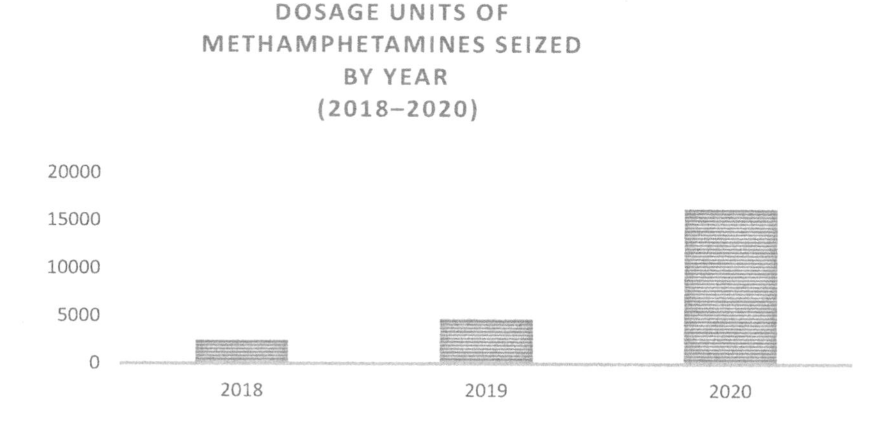

FIGURE 5.8 Trend of increasing methamphetamines in dosage units of seized drug evidence (2018–2020).

New psychoactive substance seizures varied from 2018 to 2021. Seized NPS included 1-(2-phenylethyl)-4-phenyl-4-acetoxypiperidine, 1-dimethylamino-1, 2-diphenylethane, 4-chloro-2,5-dimethoxyamphetamine, 5-DMA, alpha-PHP, alpha-PVP, 1-(2-phenylethyl)-4-phenyl-4-acetoxypiperidine, 4-fluoroamphet-amine, 5-methoxy-3,4-methylenedioxyamphetamine, and bath salts. The dosage units of NPS decreased from 633,692 to 46 from 2018 through the first half of 2021. The mass of NPS increased from 1,136kg in 2018 to 24,210kg in 2020. In the first half of 2021, 340kg had been seized.

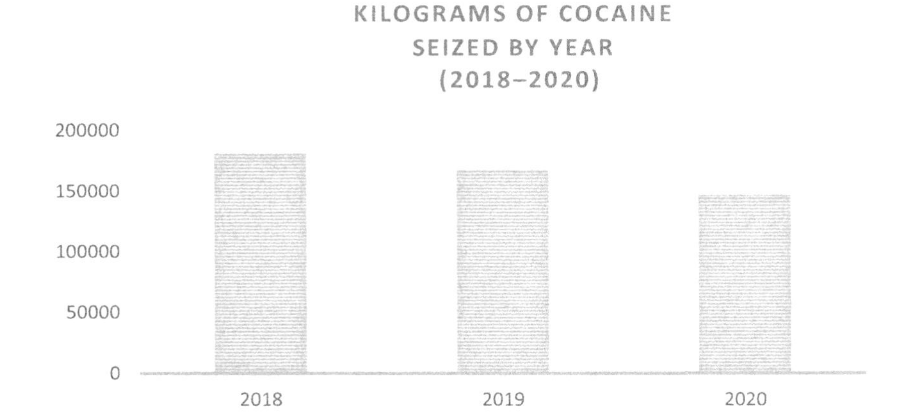

FIGURE 5.9 Trend of decreasing cocaine in kilograms of seized drug evidence (2018–2020).

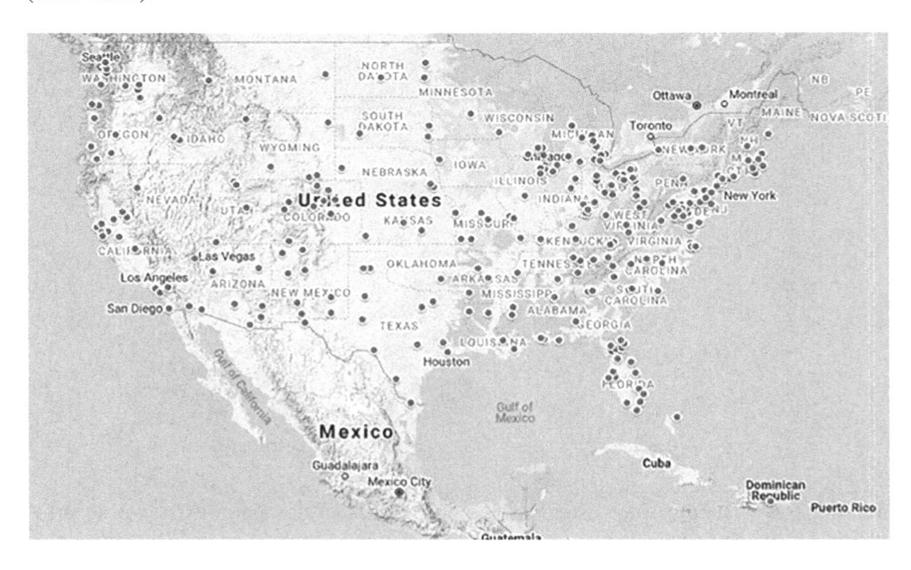

FIGURE 5.10 Map of fentanyl seizures in the continental United States in 2018.

5.4 DISCUSSION

The mission of HIDTA is as follows (https://www.hidta.org/about-hidta/):

> To reduce drug trafficking and misuse by improving interagency collaboration, promoting accurate and timely information and intelligence sharing, and providing specialized training and other resources to its law enforcement, intelligence, treatment, and prevention initiatives. To accomplish its mission, the Washington/Baltimore HIDTA will strategically apply its resources to initiatives designed to save lives, prevent initiation of drug use, and apprehend drug traffickers and money launderers.

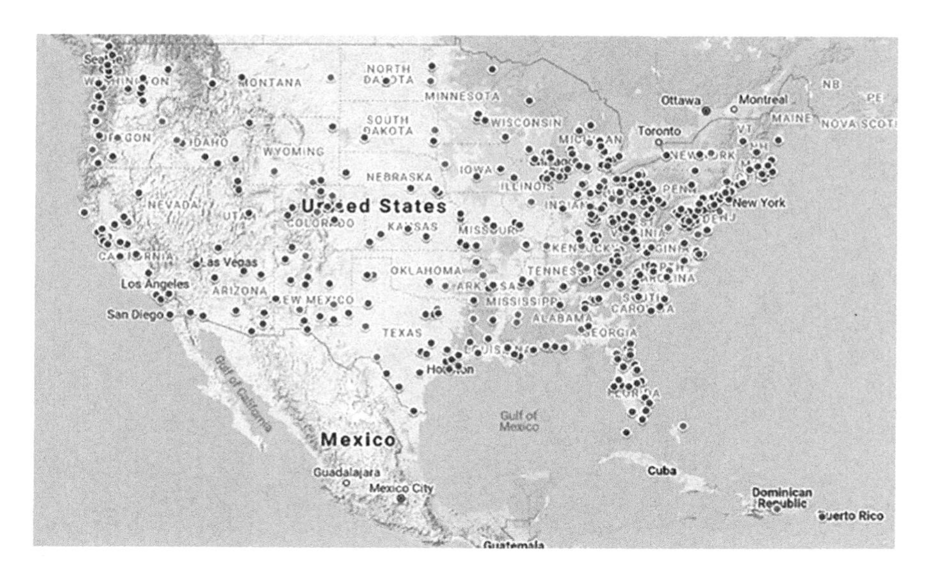

FIGURE 5.11 Map of fentanyl seizures in the continental United States in 2018 and 2019.

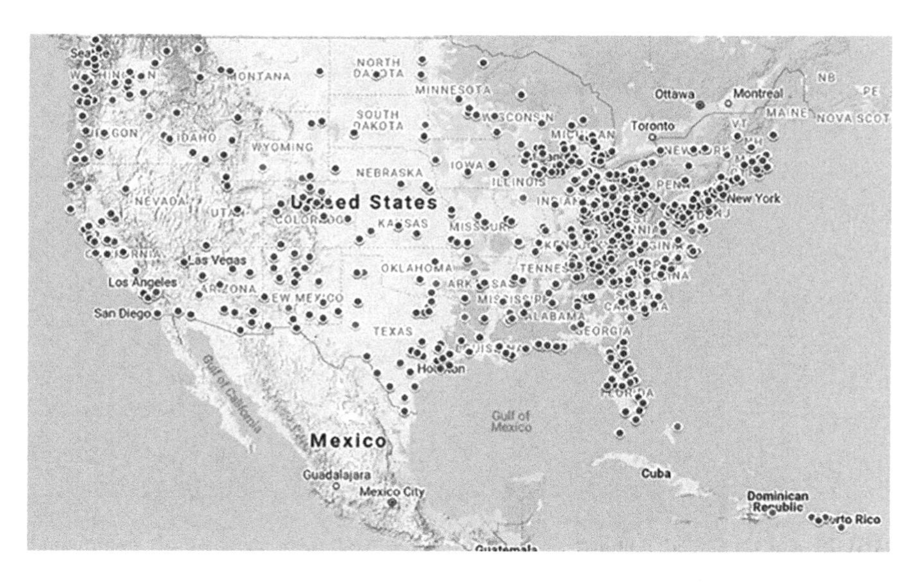

FIGURE 5.12 Map of fentanyl seizures in the continental United States during 2018–2020.

HIDTA created the Case Explorer web-based tool that enables law enforcement to share data. The tool is funded by a grant from the Office of National Drug Control Policy. Law enforcement and HIDTA can use the data to detect and monitor drug trafficking, seizure, and analysis trends. The drug seizure quantity and location data collected by HIDTA shows the importance of law enforcement sharing their

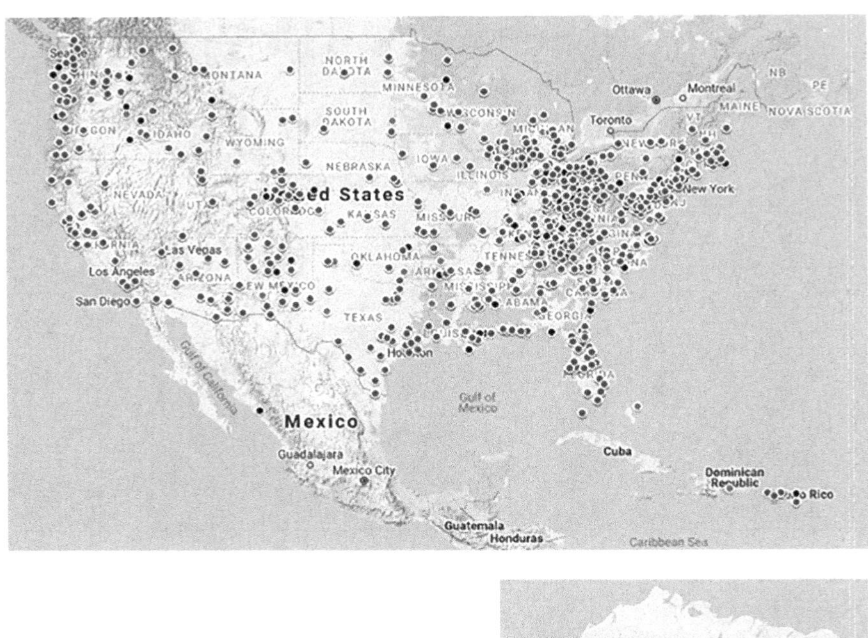

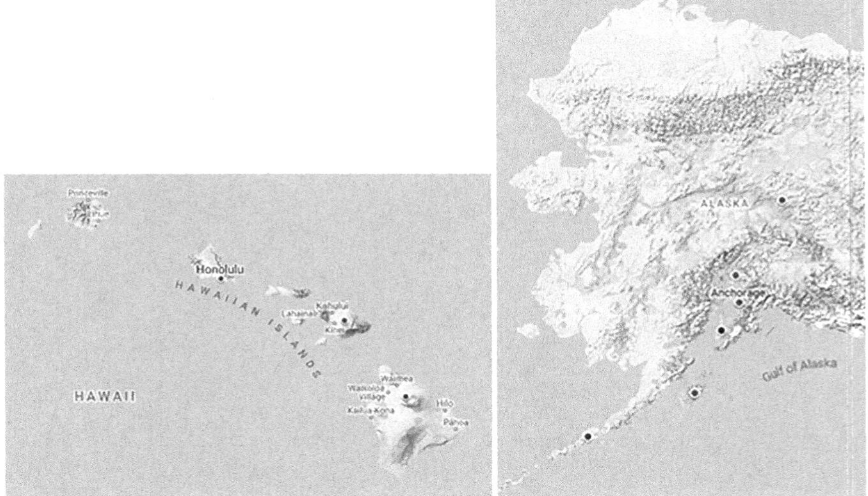

FIGURE 5.13 Seized fentanyl or fentanyl-adulterated substances by all continental United States locations and Alaska and Hawaii during 2018–2020 (Google Maps).

data and working as a team to reduce counterfeiting. The data shows that drugs are being trafficked and sold all over the United States from large cities and urban areas to small towns and rural areas. Law enforcement continued to seize large quantities of heroin and methamphetamines.

The data collected by HIDTA shows that the prevalence of fentanyl increased from 2018 to 2021. Fentanyl has been seized across the United States, and new seizure locations emerged in 2020 and 2021. Increases were most prominent in Midwest and mountain West. There is a variation in quantities of seized mass of drug and dosage unit that varies somewhat by year, depending the product law enforcement has seized. It is expected that many counterfeit adulterated products continue to evade seizure.

Collective data trends may lead to the arrest of suspected counterfeiters, their distributors, and networks and aid law enforcement in developing new control strategies, especially for emerging fentanyl analogs (Weedn et al., 2021).

5.5 CONCLUSION

Due its potency and the adulteration of drugs with the substance, fentanyl is associated with driving up the death rate of users who overdose on drugs. Data collected by the CDC shows that fentanyl-adulterated drug deaths include heroin and extend to cocaine and methamphetamines. Due to their potency, illicitly manufactured and trafficked fentanyls are responsible for most of the deaths.

ACKNOWLEDGMENTS

We acknowledge Tom Carr and HIDTA for providing us with data for analysis and reported in this study and Tom Carr and Jerry Daley for helpful discussions.

REFERENCES

A record 100,000 overdose deaths in 12 months, driven by opioids, fentanyl. The Washington Post Accessed December 14, 2021. https://www.wsj.com/articles/drug-overdose-deaths-fueled-by-fentanyl-hit-record-high-in-u-s-11637161200.

ACSH 'New' Opioid Crisis. Dr. Vanila Singh Warns Of Adulterated Street Drugs Accessed July 9, 2021. https://www.acsh.org/news/2019/11/29/new-opioid-crisis-dr-vanila-singh-warns-adulterated-street-drugs-14412.

Baltimore Man Pleads Guilty To Federal Drug Conspiracy Charge. Accessed July 9, 2021. https://www.justice.gov/usao-md/pr/baltimore-man-pleads-guilty-federal-drug-conspiracy-charge

CDC Heroin Overdose. Data Accessed July 9, 2021. https://www.cdc.gov/drugoverdose/deaths/heroin/index.html.

CDC Overdose Deaths and the Involvement of Illicit. Drugs Accessed July 20, 2021. https://www.cdc.gov/drugoverdose/featured-topics/VS-overdose-deaths-illicit-drugs.html.

Drug Enforcement Administration (DEA). Counterfeit prescription pills containing fentanyls: A global threat. 2016. https://www.dea.gov/sites/default/files/docs/Counterfeit%2520Prescription%2520Pills.pdf.

Dégardin, K., Roggo, Y., and P. Margot. "Forensic intelligence for medicine anti-counterfeiting." *Forensic Science International* 248 (March 2015): 15–32. doi: 10.1016/j.forsciint.2014.11.015.

Elkins, K.M. *Introduction to Forensic Chemistry*. Boca Raton, FL: CRC Press/Taylor & Francis, 2019.

HIDTA Accessed July 9, 2021. https://www.hidta.org/.

How Fentanyl Affects the Brain: More than 100,000 people died from drug overdoses in a 12-month period for the first time, a majority of which were fueled by the potent drug fentanyl. The Washington Post Accessed December 14, 2021. https://www.washingtonpost.com/health/2021/11/17/overdose-deaths-pandemic-fentanyl/.

Keating D., and L. Bernstein. 100,000 Americans died of drug overdoses in 12 months during the pandemic. Accessed November 17, 2021. https://www.washingtonpost.com/health/2021/11/17/overdose-deaths-pandemic-fentanyl/.

Weedn, V.W., Elizabeth Zaney, M., McCord, B., Lurie, I., and Baker, A. "Fentanyl-related substance scheduling as an effective drug control strategy." *Journal of Forensic Sciences* 66 (July 2021): 1186–1200. https://doi.org/10.1111/1556-4029.14712.

Zibbell, J.E. "The Latest Evolution of the Opioid Crisis: Changing Patterns in Fentanyl Adulteration of Heroin, Cocaine, and Methamphetamine and Associated Overdose Risk." RTI International Accessed July 9, 2021. https://www.rti.org/insights/latest-evolution-opioid-crisis-changing-patterns-fentanyl-adulteration-heroin-cocaine-and.

6 Detection and Analysis of Counterfeit Drugs

Brooke W. Kammrath
University of New Haven

Pauline E. Leary
Noble

Richard A. Crocombe
Crocombe Spectroscopic Consulting, LLC

CONTENTS

DOI: 10.1201/9781003183327-6

6.1 INTRODUCTION

The celebrated 1949 film, "*The Third Man*" (Greene, 1949), is set in post-war Vienna, when it, like Berlin, was still under four-power control. An author of cheap western novels investigates the suspicious death of his friend, Harry Lime, the titular Third Man. Lime has been running a diluted-penicillin racket, charging £70 a tube, and causing death in gangrene patients, women in childbirth, and infant meningitis cases. His organization is apparently being protected by the Soviet authorities, so this is really an example of organized crime. A British Army major has been building a case against him. The author of the screenplay, Graham Greene, had close ties to British Intelligence and based that element of the plot on actual cases (García Sánchez, García Sánchez, & Merino Marcos, 2006; Newton & Timmermann, Fake Penicillin, The Third Man, and Operation Claptrap, 2016; Riegler, 2020). So, then as now, organized crime is involved in falsified pharmaceuticals. The public health, regulatory, forensic, and law-enforcement problems caused by falsified pharmaceuticals are centered on the ability to detect falsified pharmaceutical products, remove them from the supply chain, trace them to the perpetrators, and achieve justice. In this chapter, we will see how analytical testing in the field and in the laboratory are used in this endeavor. The size and scope of the falsified-drug problem is briefly described below and has been described in more detail in other chapters of this book.

There is no doubt that counterfeit drugs are a global problem, but it is important to recognize that the challenges faced as a result of counterfeit drugs are dependent upon geography. For example, the U.S. Food and Drug Administration (FDA) reports that the United States drug supply is among the safest in the world due to federal and state laws that create a closed drug-distribution system (U.S. Food & Drug Administration, 2021). As a result, much of the threat from counterfeit drugs in the United States is limited to specific populations and types of drugs, such as drug abusers exposed to counterfeit fentanyl-laced oxycodone tablets (U.S. Attorney's Office, District of Idaho, 2021) or adult men exposed to counterfeit lifestyle drugs like sildenafil citrate (United States Customs and Border Protection, 2020; Pfizer, 2013). This is a much different situation than in other parts of the world where the risk posed by counterfeit drugs is to the population at large and for treatment of deadly diseases. The World Health Organization (WHO) reported in 2017 that an estimated one in ten medical products circulating in Low- and Middle-Income Countries (LMICs) is either substandard or falsified. Based on this estimate, a modeling exercise developed by the University of Edinburgh predicted that between 72,430 and 169,271 children may be dying each year from pneumonia due to substandard and falsified antibiotics (World Health Organization, 2017). Globally, pneumonia is one of the main causes of child mortality (Liu et al., 2015) and a primary reason for the prescription of antibiotics. A second model developed by the London School of Hygiene and Tropical Medicine, and based upon this one-in-ten estimate, calculated that approximately 116,000 additional deaths from malaria could be caused every year by substandard and falsified anti-malarials in sub-Saharan Africa (World Health Organization, 2017).

This variability in the types of counterfeit and other drugs encountered means that analytical testing requirements may also be variable. When considering the testing of a counterfeit version of a lifestyle drug like sildenafil citrate, used to treat erectile dysfunction and sold in the United States, the analytical workflow would likely be different than if tests were being performed on a substandard anti-malarial drug formulation in a third-world country. In both cases, comparative testing may be performed in the field to establish that the drug is not authentic. Then, chemical components might be identified to verify that no dangerous substances are present, and the amount of active pharmaceutical ingredient (API) may be determined. In some instances, this testing would be performed in the laboratory; in others, such as in cases where controlled substances are involved, this testing might be performed in the field. From here testing pathways might diverge to support further case-specific investigative and/or adjudicative actions—testing which is almost always performed in a laboratory. In the sildenafil citrate example, adjudication would likely be in civil court, and testing would be focused on supporting infringement or licensing violations more common in developed countries with stringent Intellectual Property (IP) rights enforcement. In the anti-malarial example, adjudication may be in criminal court, and testing would support such legal action. Interestingly, in 2006, Newton et al. (2006) reported that although the death of patients with untreated falciparum malaria in Southeast Asia, resulting from unwittingly taking fake artesunate (a drug used to treat malaria), is hidden in the inadequately documented mortality statistics of the relatively voiceless rural poor, there is no doubt that such deaths occur, and they are probably common. The authors identify counterfeiters of these fake drugs as criminals worthy of manslaughter or even murder charges and make a compelling argument to support these charges. For many reasons, most important being the need to use analytical testing to help delineate some of the legal, public health, and other implications of these drugs, we start this chapter by defining the term *counterfeit drug* and the other terms used to identify drugs discussed throughout this chapter.

According to the Collins English Dictionary, the definition of the adjective *counterfeit* is "Made in imitation of something genuine with the intent to deceive or defraud" (Counterfeit, 2014). As such, a *counterfeit drug* may be defined as a drug made in imitation of another drug, that other drug being either a brand-name or a generic drug, intended to deceive or defraud. This derived definition is valid, but a more nuanced understanding of the term, especially how it is sometimes used, is required. As defined here, the term is limiting and does not consider all the different types of drugs that are frequently, and sometimes incorrectly, classified or referred to as counterfeit. The global pharmaceutical industry is quite complex. In addition, counterfeit versions of controlled substances such as those Scheduled by the Drug Enforcement Administration (DEA) in the United States are routinely encountered. As a result, it is important that we expand the types of drugs considered in this chapter to not only include counterfeit drugs but also some of the larger classes of drugs to which these counterfeit drugs may be a part, such as falsified, substandard, and illegal imitation drugs. It is outside the scope of this chapter to discuss the details of these classifications of drugs at length, but it is important to understand some general distinctions among them. For this reason, Table 6.1 summarizes the broader classes of drugs, of which counterfeit drugs may be a part that are described in this chapter, and also includes definitions of relevant drug terms used throughout this chapter. In this table, we define these classes for how they are considered here and provide some examples of each type. It is important to note that these classes are not mutually exclusive, as a drug received for analysis may be classified in more than one group.

TABLE 6.1
Classes of Drugs and How Are They Defined for use in This Chapter

DRUG CLASSIFICATION	Definition	Examples
AUTHENTIC DRUG	Drug that accurately represents its identify, composition, and source, and is licensed for sale in the country of distribution	• Brand-name drug licensed for sale in the country of distribution • Generic drug licensed for sale in the country of distribution
BRAND-NAME DRUG	New drug developed and licensed for sale within a regulated drug environment.	• Drugs approved for sale in the US via a New Drug Application (NDA), which is the vehicle through which drug sponsors formally propose that the FDA approve a new pharmaceutical for sale and marketing in the US (U.S. Food & Drug Admininstration, 2019)
COUNTERFEIT DRUG	Drug made in imitation of another drug that other drug being either a brand-name drug or a generic drug, intended to deceive or defraud	• Physically indistinguishable imitations of brand-name or generic drugs • Physically and chemically indistinguishable imitation of brand-name or generic drugs
FALSIFIED DRUG	Drug that deliberately or fraudulently misrepresents its identity, composition, or source	• Counterfeit drugs • Legally manufactured (whether brand-name or generic) drugs diverted from the legitimate supply chain • Drugs of abuse manufactured clandestinely
GENERIC DRUG	Drug licensed for sale as a non-brand-name version (generic) of a brand-name drug	• Drugs approved for sale in the US via an Abbreviated New Drug Application (ANDA), which contains data submitted to FDA for the review and potential approval of a generic drug product (U.S. Food & Drug Admininstration, 2019)
ILLEGAL IMITATION DRUG	Drug created to be the same as an existing approved brand-name drug in dosage form, safety, strength, route of administration, quality, and performance characteristics, but not authorized for manufacture in the country of distribution (U.S. Food & Drug Administration, 2018)	• Counterfeit drugs • Generic versions of brand-name drugs currently under IP-rights protection within a country or jurisdiction • Generic version of drugs not authorized for sale within a country or jurisdiction

(Continued)

TABLE 6.1 (*Continued*)
Classes of Drugs and How Are They Defined for use in This Chapter

DRUG CLASSIFICATION	Definition	Examples
SUBSTANDARD DRUG	Drug that fails to meet either quality standards or specifications, or both (World Health Organization, 2018)	• Poor-quality counterfeit drugs • High-quality counterfeit drugs that have expired • Authentic drugs (whether brand-name or generic) that have expired • Drugs stored incorrectly causing degradation • Drugs that have incorrect amounts of API • Drugs that contain no API • Drugs that contain the wrong active API

Examples of each class are included.

It is clear from the breadth of the drug classifications, definitions, and examples in Table 6.1 that exposure to many different types of counterfeit or undesirable drugs is possible within a global pharmaceutical industry. There are many reasons that analytical testing of these goods may be performed. As examples, they may be used to provide a quick go/no-go result at the sample site in one setting; in another setting, they may be used to provide the most detailed level of physical and chemical characterization performed in a laboratory setting to adjudicate deadly criminal counterfeiting networks. Regardless of the specific reason testing is performed, detection and characterization of these drugs must be performed using analytical methods that are informative, reliable, and designed to answer critical case-specific questions. This is necessary to adequately address the challenges these drugs pose to public health, pharmaceutical innovation, criminal justice, forensic science, law enforcement, and IP-rights protection. In this chapter, we review analytical testing methods used to detect and identify counterfeit, substandard, falsified, and illegal imitation drugs.

6.2 FIELD TESTING VERSUS LABORATORY TESTING

Analysis of counterfeit and other drug substances of interest, as well as of the packaging and other identifying features of these samples, is an important capability. Testing is used to detect these drugs in the field as well as in the laboratory so that counterfeits are identified, and criminals are investigated and adjudicated. Methods used can be broadly classified as (1) field methods performed using easy-to-use devices or portable spectrometers at the sample site, and (2) laboratory methods using benchtop analytical instruments performed after sample collection, storage, and transport from the field to a fixed laboratory.

A primary purpose of field testing is to screen drugs at the sample site to remove suspect material from the supply chain as quickly as possible (Table 6.2).

TABLE 6.2
Potential Purposes of Field and Laboratory Testing Based upon Suspect Drug Type

SUSPECT DRUG TYPE	Field Testing	Laboratory Testing (Public Health)	Laboratory Testing (Criminal Legal)	Laboratory Testing (Civilian Legal)
COUNTERFEIT DRUGS (NON-CONTROLLED SUBSTANCE)	• Screening	• Lack of efficacy • Dangerous substances • Development of drug resistance	• Prosecution of counterfeiters and criminal networks • Death and/or injury for exposure to dangerous substances	• IP infringement • Wrongful death/injury
COUNTERFEIT DRUGS (CONTROLLED SUBSTANCE)	• Screening • Illicit drug identification	• Lack of efficacy • Dangerous substances	• Prosecution of drug traffickers • Prosecution of counterfeiters and criminal networks • Overdose death and/or other injuries	• IP infringement • Wrongful death/injury
SUBSTANDARD DRUGS	• Screening	• Lack of efficacy • Hazardous degradation products • Hazardous impurities • Development of drug resistance	• Prosecution of counterfeiters and criminal networks • Death and/or injury for exposure to dangerous substances	• IP infringement • Wrongful death/injury
FALSIFIED DRUGS (NON-CONTROLLED SUBSTANCE)	• Screening	• Lack of efficacy • Dangerous substances • Development of drug resistance	• Prosecution for theft of property • Prosecution of counterfeiters and criminal networks • Death and/or injury for exposure to dangerous substances	• IP infringement • Wrongful death/injury
FALSIFIED DRUGS (CONTROLLED SUBSTANCE)	• Screening • Illicit drug identification	• Lack of efficacy • Dangerous substances	• Prosecution of drug traffickers • Prosecution of counterfeiters and criminal networks • Overdose death and/or other injuries • Theft of property	• IP infringement • Wrongful death/injury

(*Continued*)

TABLE 6.2 (*Continued*)
Potential Purposes of Field and Laboratory Testing Based upon Suspect Drug Type

SUSPECT DRUG TYPE	Field Testing	Laboratory Testing (Public Health)	Laboratory Testing (Criminal Legal)	Laboratory Testing (Civilian Legal)
ILLEGAL IMITATION DRUGS (NON-CONTROLLED SUBSTANCE)	• Screening	• Lack of efficacy • Dangerous substances • Development of drug resistance	• Prosecution of counterfeiters and criminal networks • Death and/or injury for exposure to dangerous substances	• IP infringement • Wrongful death/injury
ILLEGAL IMITATIONS (CONTROLLED SUBSTANCE)	• Screening • Illicit drug identification	• Lack of efficacy • Dangerous substances	• Prosecution of drug traffickers • Prosecution of counterfeiters and criminal networks • Overdose death and/or other injuries	• IP infringement • Wrongful death/injury

As such, the task is instant identification of fakes or confirmation of identity. This is different from work performed in the laboratory intended to provide confirmation of field results, or additional testing to support investigative and/or adjudicative actions. As such, the task of laboratory testing is usually done to perform a thorough physical and chemical characterization of a suspect drug/drug product and/or its packaging. It may also be done to perform experiments and testing that work to establish attribution of the sample.

6.2.1 Field Testing

Field testing is frequently performed by non-scientific operators at the sample site. Ideally, a field device would be handheld or easily portable, not require any sample handling or preparation, be non-invasive, able to interrogate the sample through packaging, have minimal accessories, and complete an analysis in under a minute. Cost is also important, including both initial purchase and sustainment costs. Power requirements should be minimal, and this is especially important in LMICs, where the availability of power and/or batteries may be very limited. The device should also be non-destructive to the sample. As we shall describe, some spectroscopic approaches fulfill these criteria (e.g., Raman and near-infrared (NIR) spectroscopies), albeit with limitations. A longer list of features and benefits of portable counterfeit screening tools is shown in Table 6.3.

As previously mentioned, the primary goal of field testing is to screen drugs and remove counterfeits or other suspect drugs from the supply chain as quickly

TABLE 6.3
Features and Benefits of an Ideal Counterfeit-Drug Screening Tool

Feature	Benefit
Handheld or easily portable	Enables analysis at the point of need
Ideally waterproof, dustproof, and ingress protection-rated to define the level of sealing effectiveness of electrical enclosures against intrusion from foreign bodies (tools, dirt, etc.) and moisture, or tested against military standards like MIL-STD 810G (which considers the influences that environmental stresses have on materiel throughout all phases of its service life).	Operates reliably at points of need which are outside climate-controlled buildings and frequently in aggressive environments
Rugged	Protection from drops and rough handling during transport, and when used in aggressive environments
Long battery life during operation	Can be used at the point of need for extended periods
Short measurement time	Enables mission completion in aggressive or kinetic environments; improves ergonomics, since it can be difficult to hold a portable device in a fixed position for a long time
Built-in data system with result generation	Single unit generates actionable data (presumptive evidence) for non-scientist operators. Note: smartphone-operated instruments are becoming more common
Conformance with IT standards such as US FDA 21 CFR Part 11	Data integrity; meets industry standards and requirements
Metadata such as time, data, cameras, GPS location can be included/added	Enables better quality record keeping
Interrogate a sample through packaging	Non-invasive; protect the operator and preserve evidence
Non-destructive	Preserves the sample for further analysis, and as evidence
No consumables	Improves deployment capability; minimizes operating expense, especially for consumables that expire
Uses protected analytical methods	Data integrity; ease of operation
Simple operation	Minimizes expensive training requirements; especially important for users like the military where turnover may be high
Low cost	Must meet funding availability; always important and especially in LMICs
Future cost reduction	Will be required for widespread deployment, especially in LMICs

as possible. To do this, the general approach is to compare testing results of authentic drugs with those of counterfeit or other suspect drugs. If significant differences between the two samples are detected, a suspect drug can be removed

TABLE 6.4
Summary of Methods of Analysis Used to Detect and Identify Counterfeits and Other Drugs of Interest and Their Packaging in the field and in the Laboratory

Technology Type	Method	Field Method	Laboratory Method	Analysis of Drugs and Drug Product	Analysis of Packaging or Other Identifiers
Non-chemical	Track and trace	•			•
Non-chemical	Visual inspection	•	•	•	•
Non-chemical	Physical inspection	•	•	•	•
Chemical	Color tests	•		•	
Chemical	TLC and paper chromatography	•		•	
Chemical	Mid-IR spectroscopy	•	•	•	•
Chemical	Raman spectroscopy	•	•	•	•
Chemical	NIR	•	•	•	
Chemical	HPLC-UV		•	•	
Chemical	LC–MS, LC–MS/MS		•	•	
Chemical	GC–MS	•	•	•	
Chemical	HPMS	•		•	
Chemical	IMS	•		•	
Chemical	XRF	•	•	•	•
Chemical	NMR	•	•	•	

from the supply chain. Methods used to achieve this goal may detect chemical and/or non-chemical differences between the authentic and suspect drugs and/or packaging and are summarized in Table 6.4. These methods are described in detail in Section 6.4, "Methods of analysis."

6.2.2 Laboratory Testing

The desirable features of laboratory instrumentation, which are typically operated by trained scientists, are different from those of field systems. Minimization of size, weight, and power (SWaP) requirements, critical for field devices, are not important considerations for laboratory instruments. What is important for laboratory equipment, aside from cost (both for the initial purchase and for lifetime sustainment), is the ability of the instrument to achieve a sufficient level of analytical performance to enable a thorough evaluation of the physical and/or chemical properties of the sample. For example, criteria such as resolution and signal-to-noise ratio (SNR) are important for laboratory spectrometers.

For microscopes, resolving power and magnification are important criteria. Instrumentation must be able to generate results that can be used to directly compare the properties of the suspect sample with the range of acceptable limits established by the drug manufacturer.

Chemical and physical analysis of suspect counterfeits is usually forensic in nature, whether the analysis is being performed by the laboratory analysts of the authentic drug manufacturer or by forensic scientists working in a crime laboratory. Unless the quality is exceptional, it is usually possible for the authentic drug manufacturer to differentiate their product from suspect drugs. In some cases, a visual observation of the sample will suffice to achieve this goal; in others, further testing may be required. Because authentic drugs for sale in regulated environments have stringent manufacturing and distribution requirements, established ranges of acceptable limits for relevant physical and chemical properties measured using approved and validated analytical methods are known by the authentic drug manufacturer. These methods of analysis can be performed on suspect drugs and drug products to distinguish them from authentic versions. Since counterfeits are not usually manufactured identically to those of the authentic manufacturers, even small deviations in the manufacturing process, formulation, or raw materials can result in detectable differences in physical and chemical profiles of the drug product. In addition, physical and chemical differences in packaging (which is specific for authentic drugs), as well as lot-number information, etc., may provide useful information about the history of a sample.

Analysis of suspect drugs in crime laboratories may be similar to that performed by laboratory analysts of authentic drug manufacturers but may also include more traditional forensic methods with analyses that attempt to classify, identify, and individualize samples. As mentioned, laboratory analysis may also be used to establish attribution of a sample. This can help law enforcement during their investigation as well as during prosecution of defendants. As is the case with all forensics testing, case-specific factors will inform analytical testing performed. Table 6.4 summarizes methods and technologies used in the laboratory for the analysis of counterfeit and other drugs of interest. These methods of analysis are described in detail in Section 6.4, "Methods of analysis."

6.3 DEPLOYMENT APPROACHES FOR TESTING SOLUTION PLATFORMS

6.3.1 Portable Platforms

Field methods can be categorized as non-chemical methods, such as tracking and tracing, visual inspection, and physical inspections, as well as chemical methods (Roth, Biggs, & Bempong, Substandard and Falsified Medicine Screening Technologies, 2019), including color tests, paper chromatography, thin-liquid chromatography (TLC), mid-infrared spectroscopy (mid-IR), Raman spectroscopy, NIR spectroscopy, gas chromatography–mass spectrometry (GC–MS), high-pressure mass spectrometry (HPMS), and ion mobility spectrometry (IMS). Tracking refers to secure-packaging approaches where the drug packaging itself is tracked throughout the supply chain. Visual approaches rely upon visual inspection to detect non-authentic drugs and packaging. Physical approaches rely upon physical properties of the drug such as disintegration and dissolution, and may also include evaluations of the physical properties of packaging, tools and dies, as well as how these tools and dies impress detail into dosage forms. Chemical methods can be chromatographic and/or spectroscopic, and it is only these latter two which confirm the identity of the pharmaceutical product and/or API. In all of these areas, it should be realized that pharmaceutical companies do not divulge all of the means they use to secure their distribution chain—to do so would just provide counterfeiters with more resources.

Regardless of whether tracking, visual, physical, and/or chemical methods are used, there are a number of different ways methods and technology can be deployed to the field. We review these deployment strategies.

The cost and capabilities of deployed solutions are frequently dependent upon geographic location and national income level. In some cases, a mobile-laboratory approach such as the GPHF-Minilab™ is used. The Global Pharma Health Fund (GPHF) is a charitable organization initiated and funded exclusively by donations from Merck Germany (Global Pharma Health Fund, 2021), The GPHF-Minilab (Cummings, 2018) focuses its work on a range of 100 drug compounds that are instantly life-threatening when therapeutic doses are diluted. Drugs included have been selected on the basis of prevailing prescription practices, public health interest, and existing counterfeit case reports. The current short list consists of common antimicrobials, anthelminthic medicines, anti (retro) viral medicines, anti-malarial medicines, antimycobacterials, and some other medicines including their appropriate fixed-dose combination products. The list is regularly extended (Global Public Health Fund, n.d.).

The GPHF set out to develop and supply inexpensive field test kits with simple test methods for rapid drug quality verification and counterfeit medicines detection. It is intended to bridge the capacity gap in regular drug quality monitoring on the national level in low-income countries to overcome the limited access to regular drug-quality testing of public, private, and faith-based drug supplies in these countries. Priority disease programs frequently use GPHF-Minilabs to monitor the quality of essential medicines in malaria-, tuberculosis- (TB-) and acquired immune deficiency syndrome (AIDS-) endemic countries. Other GPHF-Minilab users are, for example, public drug procurement agencies and faith-based drug supply organizations in Nicaragua and Bolivia, medicines regulation authorities in Indonesia and Papua New Guinea, consumer protection institutions in India and Pakistan, diocesan hospitals in Ghana and Cameroon, scientific societies and business institutions in Laos and Thailand, as well as the Bill and Melinda Gates Foundation in cooperation with the London School of Hygiene and Tropical Medicine for rolling back malaria projects in East Africa. Overall, nearly 900

GPHF-Minilabs have been supplied to almost 100 countries. Sufficient quantities of supplies and consumables to perform about 1,000 assays are included with the GPHF-Minilab, while ensuring that the total material costs for one test run do not exceed €3 (Global Pharma Health Fund, n.d.).

The GPHF-Minilab is comprised of a heavy-duty flight case containing a field test kit with simple disintegration, color-reaction tests, and easy-to-use TLC tests for rapid drug detection and drug potency verification. It includes a full range of glassware for sample extraction, preparation, pipetting and spotting, high-performance chromatographic plates, developing and detection chambers, an electronic pocket balance, ultraviolet (UV) lamps with different wavelengths, a hot plate, and caliper rules. Even pens and pencils are included. This provides a physical inspection scheme of dosage forms and associated packaging material for an early rejection of the more crudely presented counterfeits and comprises four steps: (Global Pharma Health Fund, n.d.)

1. A physical inspection scheme of dosage forms and associated packaging material for an early rejection of the more crudely presented counterfeits
2. A quick check of the fill and total weight serves as an early indicator for the detection of false information related to the drug content
3. A simple tablet and capsule disintegration test in order to verify label claims on enteric-coating and other modified-release systems
4. Easy-to-use TLC as chemical test for a rapid verification of label claims regarding drug identity and content (Global Pharma Health Fund, n.d.).

The methods used in the GPHF-Minilab are valid methods for the screening of counterfeit and other drugs of interest for which the kit provides control-reference standards but are not intended to provide more than a comparison result between the analytical result of the test and control-reference samples. While the information provided is limited, the GPHF-Minilab can be applied very effectively to screen for specific drugs in regions where inexpensive screening solutions for medications used to treat diseases that when not controlled have specific public health consequences are required. According to their website (Global Public Health Fund, n.d.), 102 compounds are now covered, taking into account medicines prevailing in developing countries for priority infectious diseases and mother and child care.

Other deployment approaches to screening counterfeits provide more specific analytical results. Interestingly, just as pharmaceutical companies are loathe to disclose the security measures they employ to protect their supply chain, government agencies (e.g., customs and border protection agencies) and regulatory bodies (food and drug regulators) also don't always publicize the methods they employ. However, a reasonable amount of information has been published regarding anti-counterfeiting efforts in Nigeria. In 2009, following a survey (Bate, Ayodele, Tren, Hess, & Sotola, 2009), the National Agency for Food and Drug Administration and Control (NAFDAC), the Nigerian equivalent of the FDA in the United States, deployed portable Raman spectrometers for the detection of counterfeit pharmaceuticals. Raman spectroscopy provides chemical information about a sample, and while it can be used to provide a screening result, it will also provide molecular-specific information about the drug or drug product.

NAFDAC used these systems at the sample site such as at border entries and retail pharmacies to detect non-authentic pharmaceuticals quickly and reliably, and worked with authentic drug manufacturers, to create libraries of authentic medications, so the system generated automatic pass/fail results (green screen/red screen) when suspicious drug products were tested. As a result, the quality of drugs in the retail chain was improved. This effort was covered in a *Dan Rather Reports* televised program in 2010, which illustrated the nature of retail pharmacies is Nigeria and the scale of the problem at ports (Rather, 2010).

In a case study (Bate & Mathur, The Impact of Improved Detection Technology on Drug Quality: A Case Study of Lagos, Nigeria, 2011), it was reported that for each of the three drugs they tested, a higher percentage passed sample testing after the deployment of the technology than before. For instance, in 2007, only 57% of the artemisinin monotherapies passed the spectrometry test, whereas 88% passed in 2010. Within the artemisinin combination therapies, approximately 96% of the samples passed the spectrometry test in 2010, as opposed to 86% in 2007. Within the sulfadoxine–pyrimethamine pharmaceuticals, the success rate climbed from 50% to 85% between 2007 and 2010. The use of the portable spectrometers was a primary reason for the improvement.

As of 2021, Thermo Fisher Scientific estimated that there are more than 120 of their portable Raman instruments in the field for this application in Africa and the Middle East (Michael Gallagher, personal communication, June 2021), with, for instance, 14 instruments at ports of entry in Saudi Arabia. But portable Raman instruments are still comparatively expensive for less-developed countries and most instrument purchases have been funded by organizations like the World Bank, United Nations Development Fund, and GPHF.

Interestingly, when looking at the criminal consequences of drug counterfeiting, John Clark, then Global Head of Security for Pfizer, when interviewed by Dan Rather in 2010, stated that his team at Pfizer will typically perform 80% of investigations required to bring charges against counterfeiters and will then turn their case over to authorities for ultimate prosecution. As such, brand owners may work with local communities and authorities to help them secure their pharmaceutical supply chain, which can be very challenging and even dangerous in regions of the world where counterfeit drugs are commonly encountered.

Nigeria has also added other programs that work with the use of portable spectrometers to tackle the counterfeit-drug trade (Holt, Millroy, Mmopi, & Matthews, 2017; Rotinwa, 2018; Adeyeye, 2019), including mobile authentication devices using unique 12-digit codes to protect the country's most heavily counterfeited drugs, including vaccines, antibiotics, anti-malarials, and diabetes medications. In addition, NAFDAC recently announced a new partnership with the Pharmacists Council of Nigeria (PCN) to eradicate production and sale of counterfeit drugs in the country through enforcement.

It is also possible to deploy other types of portable spectrometers (in addition to Raman) to enable both the molecular and elemental characterization of a sample at the sample site, and these systems are described below in Section 6.4, "Methods of analysis." These systems are useful in developed countries like those in North America and Europe, because screening solutions like the GPHF-Minilab are not relevant for many of the drugs counterfeited in these regions; a spectroscopic analysis using a portable system may provide the fastest and most reliable detection in the field in these locations. They are more expensive than other testing options.

An advantage of this approach is the ability to generate more specific information about the sample, even identifications of components, prior to submission to the laboratory. This can be especially useful in situations where controlled substances such as oxycodone, heroin, methamphetamine, and the synthetic drug fentanyl are present. When controlled substances are encountered, they complicate field response due to the legal restrictions on how these substances must be handled and transported. Identifications at the scene using portable spectrometers can streamline these processes. In addition, exposure to fentanyl and its analogues can be dangerous and presents serious emergency-responder and public health risks due to the potent nature of these drugs. The lethal ingested dose of fentanyl is reported by the United States DEA to be just over 2 mg (United States Drug Enforcement Administration Strategic Intelligence Section, 2019), and Figure 6.1 shows this minute quantity of fentanyl alongside a United States penny. To minimize exposure, especially when these drugs are in powder form, which may be inadvertently inhaled, fast identification of fentanyl at the scene is desirable. Some spectrometers can identify fentanyl in seconds at relatively low concentrations (Leary et al., 2020). Fentanyl has become an epidemic-level problem in the United States (Centers for Disease Control and Prevention, 2021; Felter, 2021), where in 2020, over 92,000 people died from drug overdoses. This is the highest number ever recorded, with an almost 30% increase in deaths from the year 2019 (Ahmad, Rossen, & Sutton, 2021). Illicit fentanyl, primarily manufactured in

FIGURE 6.1 Fatal dose of fentanyl (~2 mg) as compared with a US penny (United States Drug Enforcement Administration, n.d.). Source: https://www.dea.gov/galleries/drug-images/fentanyl

foreign clandestine labs and smuggled into the United States through Mexico, is being distributed across the country and sold on the illegal drug market. It is being mixed in with other illicit drugs to increase the potency of the drug, sold as powders and nasal sprays, and increasingly pressed into pills made to look like legitimate prescription opioids (United States Drug Enforcement Administration, n.d.).

In addition, portable spectrometers can also be used to support other on-site testing required such as the identification of chemical hazards at clandestine laboratories where counterfeit drugs are being manufactured. Portable Raman instruments are the primary tool here, with HPMS becoming more important for fentanyl and its derivatives (see Section 6.4, "Methods of analysis").

A primary disadvantage of a spectroscopic analysis (aside from cost) is the lack of electronic, onboard databases of authentic drug products preloaded or available from the spectrometer manufacturer: The end user may have to develop their own database of tablet, capsule, and drug-product formulations to improve screening capabilities, which can be a daunting challenge. Therefore, in many counterfeit drug situations, portable spectrometers are used as identifying and characterization devices rather than screening devices.

Figure 6.2 shows a timeline of "firsts" within the field of portable spectrometry, demonstrating that over the last 25 years, expansion in the number and types of portable spectrometers available to scientists and non-scientists performing analysis at the sample site is significant. Although many of these technology platforms have been deployed to assess counterfeit and other drugs of interest in the field as part of research projects, much of the real-world screening using portable spectrometers performed in the field is carried out using portable vibrational spectroscopic methods, including Raman and NIR. This is likely because these two technologies meet most of the requirements of a desirable portable spectrometer detailed in Table 6.3. The details of these technologies and how they are applied are discussed later in Section 6.4, "Methods of analysis."

Portable spectroscopic instruments have dramatically increased their capabilities over the past 25 years, while becoming smaller and lighter. This covers every aspect of the instrument: the size, weight, and form factor; the continued shrinking size yet improved performance; the environmental and ruggedness

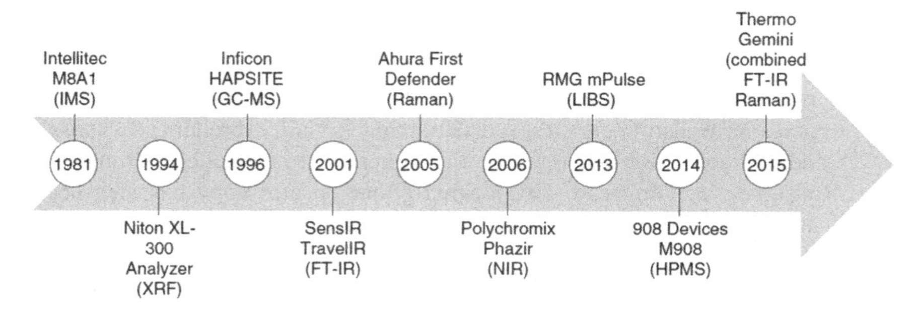

FIGURE 6.2 Timeline of portable spectrometer "firsts," based upon technology type or spectrometer performance (Leary, Crocombe, & Kammrath, 2021).

requirements; the user interface and workflow; communications (i.e., wireless, WiFi, Bluetooth); GPS; data integrity and encryption; algorithms; result reporting, etc. On one hand, this is due to all the developments in consumer electronics and computing power; on the other, to ongoing research and development, manufacturing experience and innovation in the companies producing the instruments.

There may not be much scope for further shrinking Fourier transform mid-IR (FT-IR) spectrometers, because of the need to maintain the current level of optical throughput to generate reasonable SNR spectra in an acceptable time. However, "palm-sized" or "mouse-sized" NIR spectrometers exist today, using smartphones as their data systems (InnoSpectra Corporation, n.d.; Beć, Grabska, & Huck, 2021; Adegoke, Kochan, Heraud, & Wood, 2021), and a smartphone has sufficient computing power and data storage to perform the required tasks, with the added benefit of being able to communicate to "the cloud" if additional work is required. These palm-sized spectrometers have a useful spectral range (e.g., 1,000–1,700 nm) and generate reasonable SNR spectra in a few seconds. These instruments may challenge the current domination of the field counterfeit detection arena by Raman instruments, because their cost is considerably less, in the region of $2000 (USD). Instruments that are only slightly larger and more expensive can cover the visible range as well as the NIR (BaySpec, Inc., n.d.). The advent of spectrometers marketed directly to the general public, operating in the visible-NIR (VNIR) region (~400–1,000 nm), and costing under $1000 (USD) raises both possibilities and questions. Bands in that region arising from molecular vibrations are exceedingly weak, but nonetheless some success has been observed in constructing calibrations (McVey, Gordon, Haughey, & Elliott, n.d.; Dégardin, Guillemain, Guerreiro, & Roggo, 2016).

Over the years portable Raman instruments have diminished in size while maintaining or even improving their performance (i.e., the SNR for the same collection time and resolution). This was initially facilitated by the change from designs using reflective gratings and fiber-coupled components, to free-space coupling, and then to the use of transmission gratings (Rathmell, Bingemann, Zieg, & Creasey, 2021). Now, there are even smaller instruments, a little larger than a pack of playing cards, with yet improved SNR, possibly as much as a factor of 10, made possible via transmission-grating designs (Creasey, Sullivan, Paul, & Rathmell, 2018; Owen, 2006), and even smaller instruments are possible today: a Raman instrument about 1"×1"×1" in size. It is worth noting that in order to perform a quick screening of a sample in the field, low resolution (e.g., 15 cm^{-1}) may be perfectly adequate. Low-cost approaches to the construction of a Raman spectrometer have also been surveyed (Emmanuel, Nair, Abraham, & Yoosaf, 2021), and another possibility is to use the camera in a smartphone (Bonvallet, Auz, Rodriguez, & Olmstead, 2018). Smartphone spectroscopy has also been investigated for visible-region spectroscopy. This technique has been described (Scheeline, 2021), and the issues regarding "hidden" data processing within the phone have been shown in detail (Burggraaff et al., 2019). One promising application here, outside the realm of pharmaceutical counterfeit detection, is their use as "readers" for field clinical analyzers (Peveler & Algar, 2021).

Portable GC–MS systems remain larger and more expensive than their vibrational-spectroscopy brethren, mainly due to pumping needed to achieve vacuum requirements. Nevertheless, they have improved significantly in recent years. Some manufacturers employ a toroidal ion-trap design, which increases the operating pressure (reducing pumping requirements and improving SWaP characteristics); in other systems, quadrupole systems are operated on battery using a high-performance turbomolecular pump (Leary, Kammrath, & Reffner, Field-portable Gas Chromatography–Mass Spectrometry, 2018; Leary, Kammrath, & Reffner, Portable Gas Chromatography–Mass Spectrometry: Instrumentation and Applications, 2021). Regardless, the analytical performance of these portable GC–MS systems is impressive. They are frequently capable of achieving unit mass resolution or better, which makes them directly searchable against commercially available libraries such as the NIST (National Institute of Standards and Technology) MS database. It is important to recognize that the NIST-MS database contains quadrupole spectra, so when ion-trap data is compared against this library, mass analyzer type should be considered when interpreting GC–MS results. In some instances, library-search results can be improved by comparing GC–MS data with proprietary libraries, either developed by the manufacturer or developed by the end user using the GC–MS system deployed to the end user.

Interestingly, as is discussed in Section 6.4, "Methods of analysis," both HPMS and IMS systems have been deployed to address analytical challenges in the field associated with counterfeit drugs. Both of these technologies are deployed with proprietary libraries developed and controlled by the instrument vendor. This is limiting when these systems are deployed to scientific users in the field, because reviewing the data manually is not possible; the operator must rely on the instrument result. However, in situations where these systems are deployed to non-scientific operators using these systems as screening tools, they provide fast results with no data interpretation required at the scene. In these instances, evaluation of system performance may be done by third-party vendors to help establish reasonable expectations for library matching and instrument performance criteria.

X-ray fluorescence (XRF) is a mature technique for elemental analysis (Jenkins, 1999), with portable XRF instruments available for more than 20 years. These systems have been widely deployed (Potts & West, 2008). XRF is a quantitative technique, employing both "empirical" and "fundamental parameter" calibrations (Piorek, 2021) performed at the factory. A limitation in the pharmaceutical industry is that its elemental range is from magnesium to uranium; it is insensitive to light elements, especially in a portable format. However, it can be used to analyze for the presence of inorganic elements in both solid dosage forms and metallic packaging (see Section 6.5, "X-ray fluorescence (XRF) spectroscopy for elemental analysis").

6.3.2 Laboratory Platform

Laboratory testing of counterfeit and other drugs of interest is typically performed to confirm preliminary identifications of screening results in the field, or to provide investigative or adjudicative support to law-enforcement or other members of the criminal justice system. To confirm preliminary identifications, laboratory testing can be quite simple. For instance, stereomicroscope images of an authentic versus a counterfeit drug are sometimes sufficient to demonstrate that it is not possible for a suspect sample to be authentic. Either tooling or other visual indicators such as color, dimensions, or shape, in general, can be different and not within the specification of the authentic product. Sometimes it is difficult to detect subtle or even significant differences in the field; but when observed within a laboratory under reasonable magnification with appropriate light and other conditions, this is usually the most straightforward way to demonstrate that a drug is not authentic. In other cases, chemical analysis might be required either to confirm the identity and concentration of the API, or of the excipients and non-active ingredients. Figure 6.3 shows an authentic Viagra® tablet alongside a counterfeit version. These two tablets are easily differentiable from each other when visually compared.

As the quality of counterfeits improve, additional and more subtle methods of differentiation will likely be required. For instance, dissolution assays, content-uniformity analysis, and particle size and distribution of components may be evaluated and compared.

To confirm screening results, the general analytical goal is to differentiate suspect samples from authentic versions. The easiest way to achieve this goal is to test samples using the authentic manufacturer's validated methods for the drug/drug product and determine whether the suspect drug/drug product is within the acceptable range for this property established by the manufacturer for their authentic product. If the analytical testing results of the suspect sample do not fall within the acceptable range for a specific property, any reasonable explanation for the failure should be factored into determination about whether the sample could be authentic. As an example of a reasonable failure, one property of a suspect sample that may be measured is its residual-solvent profile using GC–MS. Residual solvents are solvents that are present in an authentic drug product at or below a specified quantitative range as a result of either impurities in the raw material or due to the manufacturing process. If the residual-solvent profile of a

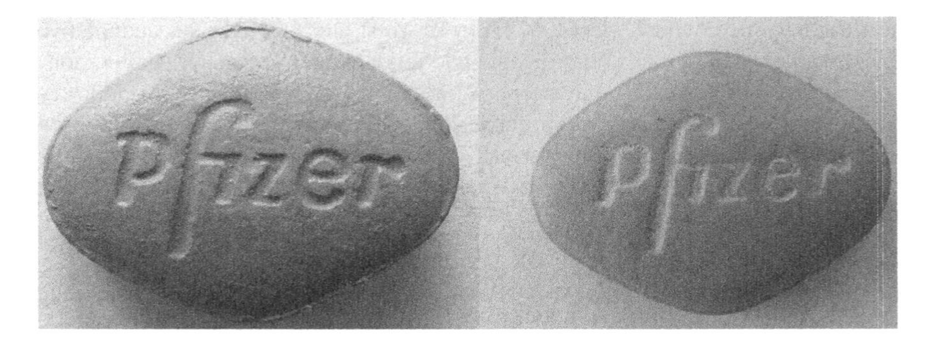

FIGURE 6.3 Stereomicroscope images on the same scale of authentic Viagra (left) and counterfeit Viagra (right) (Leary, 2014).

suspect sample is similar to the acceptable profiles of the authentic drug, then the suspect sample cannot be distinguished from the authentic sample based upon residual solvents. In addition, if the residual solvents are the same solvents as in the authentic product, but at a lower concentration, then if it is possible that sample history (such as exposure to hot environments) caused the solvents to evolve from the suspect sample, then the sample cannot be distinguished from the authentic based upon this method. However, if a solvent is present in the suspect drug that is not present in the residual-solvent profile of the authentic drug, the suspect sample can be differentiated based upon the presence of that solvent.

Some typical validated methods that may be available from the manufacturer include dissolution methods based upon high-performance liquid chromatography (HPLC), residual solvents based upon GC–MS, impurity profiles using liquid chromatography–mass spectrometry/mass spectrometry (LC–MS/MS), and identification based upon mid-IR, Raman, and/or NIR spectroscopy, etc. If a manufacturer's validated method is not available for use, differentiation is still possible based upon routine methods applied in forensic drug labs by comparing suspect samples with authentic samples. It is important though, in that circumstance, that since an acceptable range of anticipated analytical results is not available from the manufacturer as a quantifiable measurement of similarity between the test and suspect sample, a representative range of accepted intra- and inter-lot variability of authentic drug product is defined.

6.4 METHODS OF ANALYSIS

In this section, we outline the methods used to defect falsified pharmaceutical products: both non-chemical and chemical approaches. Table 6.4 shows a summary of many of the field and laboratory analytical technologies, as well as whether they are used to analyze drugs/drug products, and/or packaging and other identifying features. The ability to detect and identify counterfeit drugs is dependent upon the ability to differentiate the test sample from an authentic version.

6.4.1 Non-Chemical Approaches

6.4.1.1 Track and Trace

Track and Trace refers to a variety of measures to track pharmaceutical packages electronically (Bate, PHAKE. The Deadly World of Falsified and Substandard Medicines, 2014). These include the record of batch and lot numbers; unique identification codes on each package, which could be a bar code, radio frequency identification device (RFID), or a one-time code revealed by scratching off a layer, similar to a lottery ticket. In some cases, the unique code can be transmitted to a database via a text message and a reply provided confirming the genuine nature of that package. In addition, blockchain technology has been discussed as a secure means of identifying a package. "One-time" use of serial numbers guards against re-use of authentic packaging. However, unless the contents (e.g., blister packs) are similarly labeled, substitution of genuine product with falsified is still possible.

6.4.1.2 Visual

There are packaging approaches derived from the secure document industry (e.g., passports, banknotes). These include items that are difficult to reproduce, like color-shifting inks, holograms, and microprinting. Inks may be designed to have unique properties when viewed under UV light. Technologies also exist to fabricate "chemical bar codes" that are invisible to the naked eye (Zhang et al., 2020). This could be a unique mixture of rare earth compounds, which have sharp and distinctive lines in the NIR region, or other chemical identifiers. It is possible, therefore, to distinguish fake packaging from genuine articles by visual inspection, often with the aid of a variety of light sources (Lanzarotta et al., 2015; Dégardin et al., 2018; Platek, Ranieri, & Batson, 2016; Andria, Fulcher, Witkowski, & Platek, 2012).

The US Food and Drug Administration (FDA) has developed two, low-cost, counterfeit detection devices. The CD-3 is a compact handheld device with a series of light-emitting diodes (LEDs) of various wavelengths to illuminate a tablet (U.S. Food & Drug Administration, n.d.; Ranieri et al., 2014). A sample is illuminated by a selected wavelength, which may cause fluorescence; the image is visualized and compared with an authentic sample. Samples may be visualized without compromising the blister pack containing the tablets. The CD-3 is relatively inexpensive (< $1000 USD), robust, accurate, and easy to operate. The CD-3 can also be used to distinguish counterfeit packaging, because inks of the same apparent color may have quite different fluorescence behavior. Figure 6.4 shows results from the CD-3 of tablets, blister packaging, and external packaging differentiating authentic and counterfeit anti-malarials.

The CoDI (Green et al., 2015) is a battery-operated handheld device incorporating a 405 nm laser and detector to measure the amount of light passing through the tablet. The detected light is affected by tablet thickness, tablet density, and fluorescence. Colored filters are used to additionally characterize fluorescence. This is a low-cost (< $100 USD), simple-to-use, device and provides rapid results.

6.4.1.3 Physical

Disintegration and dissolution behavior are two physical properties that are important for the performance of pharmaceutical solid dosage tablets and capsules. Disintegration is the process of breaking a substance into smaller fragments, particles, or molecules to improve its solubility. Dissolution is the process through which a solid, liquid, or gaseous substance dissolves into a solvent to produce a solution. These physical properties are carefully controlled by pharmaceutical manufacturers so that the formulation either releases its API immediately following oral administration (immediate-release tablets) or releases the API after a coordinated time frame (modified-release tablets) with the aim of achieving improved therapeutic efficacy, reduced toxicity, and/or improved patient compliance and convenience (Markl & Zeitler, 2017). Specialized apparatuses are used for disintegration and dissolution testing in the laboratory, with both being used to precisely measure the time required for complete disintegration or

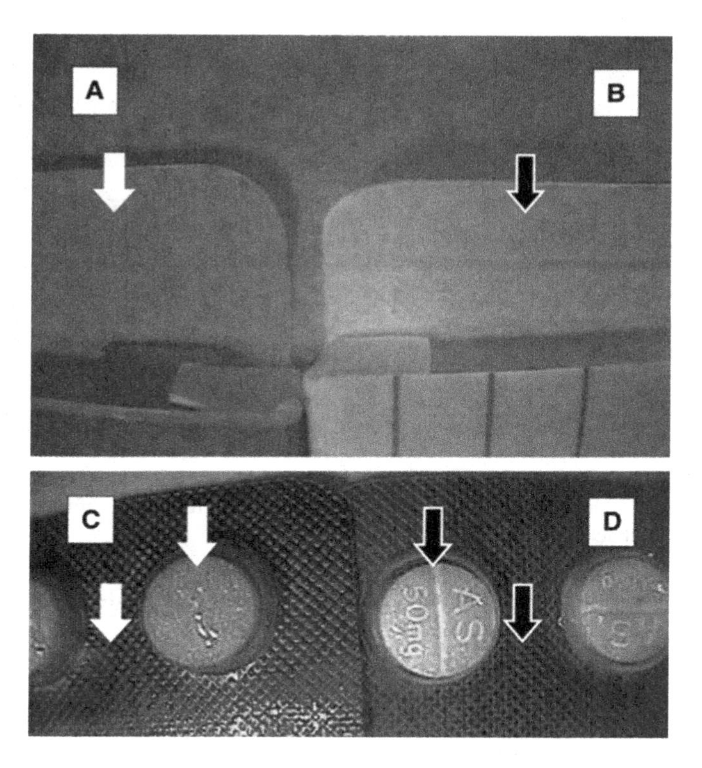

FIGURE 6.4 Examples of differences observed between counterfeit and authentic artesunate (anti-malarial) tablets and packaging. In general, clear differences between the suspect and the authentic tablets and packaging, and many of the comparison features could not be detected with the naked eye. A and B show the interior of the cardboard box dark versus bright (regions are highlighted by white and black arrows, respectively); differences were observed between the (a) counterfeit and (b) authentic carton end flap using a 375-nm wavelength setting on the CD-3. C and D show the (c) absence of tablet debossing details for tablets in AS 50 mg and (d) enhanced tablet debossing details on AS 50 mg. Differences in tablet color and blister pack surface patterns (regions highlighted by white and black arrows, respectively) were observed between the (c) counterfeit and (d) authentic tablets in blisters using a 470 nm wavelength setting on the CD-3 (Ranieri, et al., 2014).

dissolution in a certain volume of solvent. Although these properties are carefully controlled in the formulations of authentic pharmaceuticals and are used to evaluate equivalence when comparing generic to brand-name drug products, neither dissolution nor disintegration appear to be controlled by counterfeit drug manufacturers. Consequently, comparative studies or dissolution and/or disintegration are capable of differentiating authentic from counterfeit drugs (Amin, Snow, & Kokwaro, 2005; Deconinck, Bothy, Courselle, & De Beer, 2014; Galia, Horton, & Dressman, 1999; Gaudiano et al., 2007; Gwaziwa, Dzomba, & Mupa, 2017; Nair, Strauch, Lauwo, Jähnke, & Dressman, 2011). Of particular note is that a field deployable disintegration test included in the GPHF-Minilab is used

FIGURE 6.5 (left) A collection of dies recovered from the scene of a clandestine laboratory, as well as (right) an up-close image of the impressing area on one of the dies. (Photo courtesy of Michael Cashman.)

"to verify label claims on enteric-coating and other modified-release systems" (Global Pharma Health Fund, n.d.).

Tooling used to manufacture illegal tablets and other dosage forms are also important pieces of evidence frequently recovered from clandestine drug laboratories. These tools and dies leave impressions in the dosage forms they are used to manufacture. These impressions and other tool marks created during manufacture can be used for the classification and identification of counterfeit drugs. This pattern-matching type of analysis is commonly used in forensic science laboratories to evaluate tools used in the commission of crimes. Figure 6.5 shows dies recovered from a clandestine laboratory; Figure 6.6 shows a tooling kit used to create illegal Xanax®, as well as one of the Xanax tablets recovered with the equipment. In these instances, the ability to characterize this equipment, and how these tools and dies impress tool marks and other features into dosage forms, enables the identification of counterfeit and other illegal drugs and can also be used to establish provenance of samples. These tools and dies have become more commonly encountered in recent years due to the proliferation of edible gummies into the recreational drug trade (Michael Cashman, personal communication, May 2021).

6.4.2 Chemical Approaches

6.4.2.1 Class Analysis

Classification methods are those that categorize a material based on shared qualities or characteristics, such as the presence of a specific functional group. Although these methods do not provide confirmatory identifications, they are often rapid, inexpensive, and provide valuable information about the chemistry of a sample that can be used to direct the investigation, including determinations regarding additional analyses, if deemed necessary.

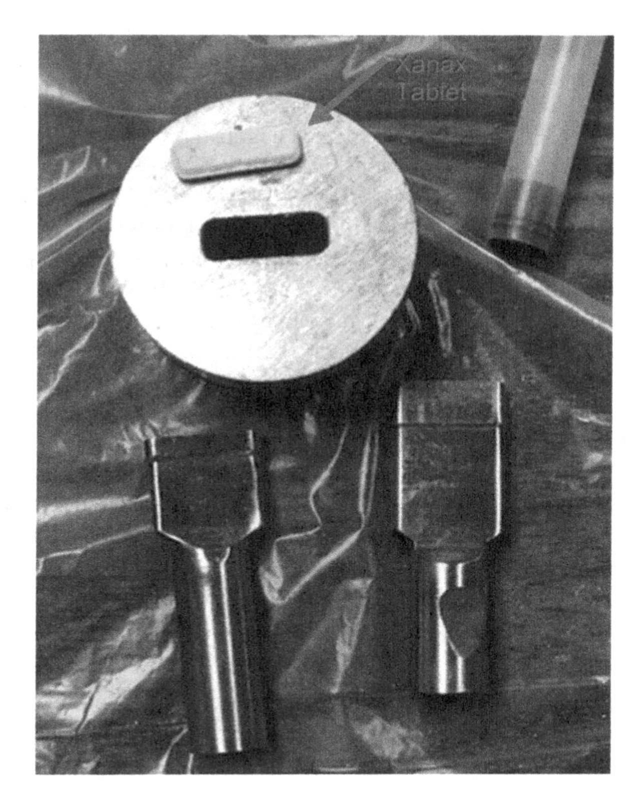

FIGURE 6.6 Tooling kit used to create illegal Xanax tablets, along with an illegal Xanax tablet manufactured using this equipment. (Photo courtesy of Michael Cashman.)

6.4.2.1.1 Color Tests

Colorimetric-based analyses are presumptive or preliminary tests that are used to indicate that a certain class of drug is present in a sample but not the presence of a specific drug. A color test relies on the reaction of the reagent(s) with a particular moiety of the molecule being screened, resulting in a characteristic change of color. This color change may be dramatic or slight, and is observable in mixtures. They are easy to use, inexpensive, rapid, and simple to perform in the field; thus color tests are valuable for the screening of counterfeit drugs, especially in low-income countries (Singh, Parwate, & Shukla, 2009; Green, Mount, Wirtz, & White, 2000; Patil, Pandit, Pore, & Chavan, 2012; Koesdjojo, Wu, Boonloed, Dunfield, & Remcho, 2014). Although color tests are not unique to a specific drug, multiple color tests can be performed on separate aliquots of an unknown sample to reduce the potential for false positives and thus increase the selectivity of testing (Singh, Parwate, & Shukla, 2009). Color tests are included as a test in the GPHF-Minilab kit, originally as stand-alone tests, while more recently they have been used for the visualization of mixture components separated by TLC (Jähnke & Dwornik, 2020).

Recent advances offer a different approach to color tests. Analyses of complex samples such as tablets is performed when tested using paper millifluidic devices that incorporate biochemical selectivity (enzymatic reactions, antibody/antigen binding) (Martinez, Phillips, & Whitesides, Diagnostics for the Developing World: Microfluidic Paper-Based Analytical Devices, 2010; Risha et al., 2008; Martinez, Phillips, & Whitesides, Three-Dimensional Microfluidic Devices Fabricated in Layered Paper and Tape, 2008; Martinez et al., 2010; Osborn et al., 2010; Lutz, Trinh, Ball, Fu, & Yagera, 2011; Abe, Suzuki, & Citterio, 2008; Apilux, Ukita, Chikae, Chailapakul, & Takamura, 2013; Fu et al., 2012). These paper analytical devices (PADs) are intended to detect counterfeit drugs. A solid drug product substance such as a tablet is "swiped" across the PAD about 1–1.5 cm below the top of the lanes, depositing at least 0.5 mg of the solid in each lane. Once the sample is applied, the base of the PAD is set in small level of water up and held upright for about 4 minutes. Capillary action carries water up the hydrophilic test lanes and brings together test reagents and analyte. The outcome is a "color bar code" at the top-third of the PAD, providing the basis of sample characterization (Weaver et al., 2013). They are fast, easy to use, and inexpensive. Figure 6.7 shows a schematic detailing how the test is performed.

These PAD devices were developed for users in low-resource settings such as in third-world countries, meeting the following specific requirements:

1. The reaction must work at room temperature. Controlled heating is not common in many field settings, and a chemical heater would increase the cost of the PAD.
2. All reagents and the overall reaction must tolerate water, as nonaqueous solvents (even ethanol) are not reliably available in the developing world. Including solvents in a kit form would increase costs and introduce hazard and waste disposal issues.
3. All reagents must be stable during storage on paper, which eliminates use of concentrated acids (which eat away the cellulose); volatile materials, such as iodine; or highly reactive species, such as nitrous acid.
4. The quantity of reagents on each PAD must be small enough that the PAD would not be treated as hazardous waste according to US EPA standards, limiting use of most heavy metals and highly toxic reagents.
5. Finally, the array of reagents used in the available lane tests has to provide good sensitivity and specificity for identification not only of the targeted APIs but also for excipients and substitute APIs that might be present in low-quality pharmaceuticals (Weaver et al., 2013).

6.4.2.1.2 Thin-Layer Chromatography (TLC)

TLC is one of the simplest and most versatile methods of mixture separation due to its simplicity, low cost, rapid development time, reproducibility, and sensitivity. The TLC method involves spotting a small amount of dissolved sample onto a silica gel (stationary phase)-coated glass plate and inserting the plate into a sealed chamber containing the developing solvent (mobile phase). Through capillary

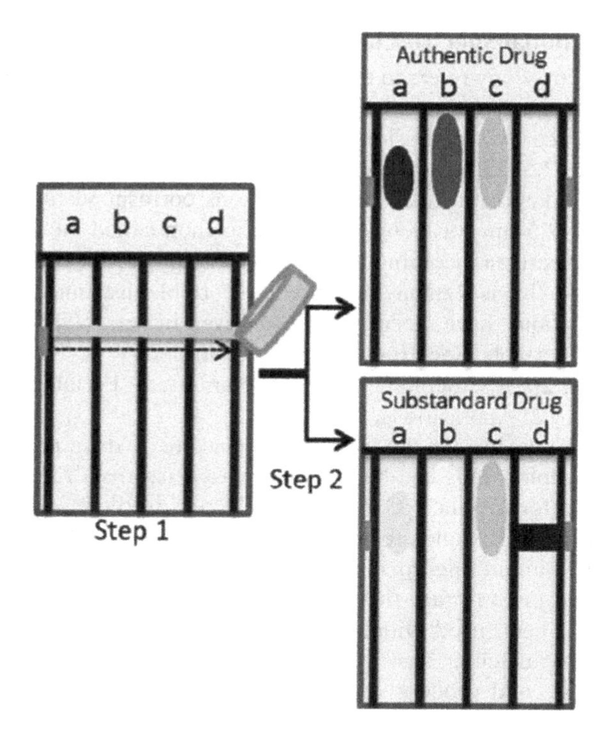

FIGURE 6.7 Schematic of PAD testing procedure in which sample is "swiped" to deposit material for analysis in several lanes (step 1). The bottom edge of the device is dipped in water to bring together reagent and analyte by capillary flow (step 2). Colors produced by an authentic versus a substandard drug will vary in one or more lanes, reflecting differences in chemical makeup of samples (from Weaver, et al., 2013).

action, the solvent travels up the plate carrying the analytes with it at different rates (due to the partitioning of the analyte between the mobile and stationary phases). When the solvent has reached near to the top of the plate, it is removed from the chamber, dried, and visualized using ambient and UV light. Different applications have created modified protocols for TLC, which include pretreating the sample, changing the stationary and mobile phases, the development techniques, and method of detection and visualization, or by coupling TLC to other techniques (such as Raman spectroscopy and MS). Qualitative, semiquantitative, and quantitative TLC methods have been researched and used in both laboratory and field analysis of counterfeit pharmaceutical products (Sherma, 2007; Martino, Malet-Martino, Gilard, & Balayssac, 2010; Khuluza, Kigera, Jähnke, & Heide, 2016; Sherma & Rabel, 2019). A field deployable TLC system is also included in the GPHF-Minilab with the goal of rapidly identifying drug identity and content and comparing that to label information if available (Jähnke & Dwornik, 2020). For additional information on current practices and developments in TLC for the analysis of counterfeit drugs, there are two excellent review articles on this topic by Sherma, 2007 and Sherma and Rabel, 2019.

6.4.2.2 Vibrational Spectroscopy

Vibrational spectroscopy refers to the detection of electromagnetic radiation as a molecule undergoes a transition from one vibrational state to another. A single-level transition occurs with emission or absorption of a photon in the mid-IR region of the spectrum (2.5–25 μm wavelength, or 4,000–400 cm^{-1} in frequency units). Overtones and combinations of these transitions correspond to higher energy NIR photons (1–2.5 μm wavelength). A phenomenon called inelastic scattering can also occur where an incoming photon loses energy corresponding to a vibrational transition—this is Raman spectroscopy. Portable instruments of this type, and their applications, have recently been surveyed in detail (Crocombe, Leary, & Kammrath, Portable Spectroscopy and Spectrometry 1: Technologies and Instrumentation, 2021; Crocombe, Leary, & Kammrath, Portable Spectroscopy and Spectrometry 2: Applications, 2021).

For organic molecules in the condensed phase (e.g., drug products), vibrational spectroscopic bands can be considered to arise from functional groups in the molecule, for instance, C=O, C-N, C-H, and in the mid-IR the characteristic pattern of these bands form an effective "fingerprint" of the molecule. However, like a human fingerprint, to identify the molecule from its spectrum requires matching to a library that contains a reference spectrum. All of these techniques have been used widely in the pharmaceutical industry, and NIR spectroscopy in particular has been employed in Quality Assurance/Quality Control (QA/QC) and process applications (Ciurczak & Igne, 2015; Igne & Ciurczak, 2021; Simpson, 2010). Raman spectra are similarly specific, but NIR spectra tend to be highly overlapped, leading to broader, non-specific bands. For those reasons, the methods for extracting chemical information from these spectra is slightly different.

Given the choice of vibrational spectroscopic methods, it's appropriate to compare their attributes, as they apply to counterfeit detection. See Tables 6.4 and 6.5 for details. There are a number of papers in the literature that address this as well (Fernandez, Green, & Newton, 2007; Bate, Tren, Hess, Mooney, & Porter, 2009; Martino, Malet-Martino, Gilard, & Balayssac, 2010; Kalyanaraman, Dobler, & Ribick, Portable Spectrometers for Pharmaceutical Counterfeit Detection, 2010; Kovacs et al., 2014; Luczak & Kalyanaraman, Analysing the Analytical Techniques for Counterfeit Drug Detection, 2015; Assi, Evaluating Handheld Spectroscopic Techniques for Identifying Counterfeit Branded and Generic Medicines Worldwide, 2016; Kaale, Hope, Jenkins, & Layloff, 2016; Vickers et al., 2018; Roth, Nalim, Turesson, & Krech, 2018; Bottoni & Caroli, 2019; Roth, Biggs, & Bempong, Substandard and Falsified Medicine Screening Technologies, 2019; Deidda et al., 2019; Ciza et al., 2019; Voelker et al., 2021; Lanzarotta et al., 2021). Due to the nature of the chemical phenomena that NIR and Raman characterize, one conclusion is that NIR will be superior in detecting physical differences (particle size, tablet compression, etc.), while Raman spectroscopy will be better in detecting the chemical signatures of APIs and excipients.

TABLE 6.5
Attributes of Portable Vibrational Spectrometers for Counterfeit Detection

Property	Raman	NIR	Mid-IR
Handheld or easily portable	Yes	Yes	Yes
Operates reliably under a wide range of environmental conditions (temperature, humidity)	Yes	Yes	Yes
Rugged	Yes	Yes	Yes
Battery-powered for ~8 hours	Yes	Yes	Yes
Short measurement time (few seconds)	>1 minute	Yes	Yes
Built-in data system with result generation	Yes[a]	Yes[a]	Yes
Traceable data	Yes	Yes	Yes
Can add metadata	Yes	Yes	Yes
Non-invasive	Yes	Yes	No
Non-destructive	Yes	Yes	No
No consumables	Yes[b]	Yes	Yes
Uses protected analytical methods	Yes	Yes	Yes
Safe to operate	Yes[c]	Yes	Yes
Low cost (<$10,000 USD)	No	Yes	No
Future cost reduction	Yes	Yes	Unlikely

[a] Smartphone-based data systems becoming available.
[b] Surface-Enhanced Raman Spectroscopy (SERS) has consumables.
[c] Will incorporate a laser, operating at 785 nm or 1,064 nm, with power in the range 50–300 mW. Laser safety precautions required.

6.4.2.2.1 *Mid-Infrared (mid-IR) Spectroscopy*

Laboratory mid-IR instruments have been used to examine pharmaceuticals from the time that commercial instruments were introduced in the late 1940s. The first portable FT-IR (with attenuated total reflection (ATR sampling)) was introduced in 2001 (Leary, Crocombe, & Kammrath, Introduction to Portable Spectroscopy, 2021), and the current generation of smaller and more rugged instruments around 2007. These do not employ "point-and shoot" sampling, but use a diamond ATR interface (Schiering & Stein, 2021), and therefore require sample handling—removal from the packaging or container, placement of the sample onto the ATR crystal interface, and application of pressure to attain good contact. This would obviously be an issue for high-potency pharmaceuticals. Thus, although mid-IR spectra are specific, provide good "fingerprints," with large spectral libraries available for reference, this technique is not as attractive for field use as NIR and Raman spectroscopies. In addition, both NIR and Raman can be used directly on samples contained in glass vials, with Raman spectroscopy being especially suitable for aqueous solutions.

6.4.2.2.2 Raman Spectroscopy

As with NIR, Raman spectroscopy has been used for the analysis of pharmaceuticals for many years (Frank, 1999). Indeed, an early example of an application of Fourier transform (FT) Raman spectroscopy was the examination of pharmaceutical tablets in blister packs (Amin, Snow, & Kokwaro, 2005; Compton & Compton, 1991). The first portable Raman instruments, really designed for industrial use and applications with hazardous materials, were introduced in 2005. Following that, papers on laboratory and field work on counterfeit identification appeared in the literature (Dégardin, Roggo, Been, & Margot, 2011; Olds et al., 2011; Diehl et al., 2012; Hajjou, Qin, Bradby, Bempong, & Lukulay, 2013; Luczak & Kalyanaraman, Portable and Benchtop Raman Technologies for Product Authentication and Counterfeit Detection, 2014; Peters, Luczak, Ganesh, Park, & Kalyanaraman, 2016; Assi & Fortunato, Optimizing the Method for Authenticating Pharmaceutical Products of Different Dosage Forms Using Dual Laser Handheld Raman Spectroscopy, 2017; Dégardin, Desponds, & Roggo, Protein-Based Medicines Analyiss by Raman Spectroscopy for the Detection of Counterfeits, 2017; Lanzarotta, Lorenz, Batson, & Flurer, 2017; Dégardin, Guillemain, & Roggo, Comprehensive Study of a Handheld Raman Spectrometer for the Analysis of Counterfeits of Solid-Dosage Form Medicines, 2017; Kakio, et al., 2017; Frosch, et al., 2019; Handzo, Luczak, Huffman, Peters, & Kalyanaraman, 2020; Lanzarotta et al., 2020; Sanada, Yoshida, Kimura, & Tsuboi, 2021; Kalyanaraman, Dobler, & Ribick, Portable Spectrometers for Pharmaceutical Counterfeit Detection, 2010). The use of these instruments in the field for counterfeit detection has been described in the literature (see Section 6.3.1, "Portable Platforms," for their use in Nigeria).

Despite the improvements in portable Raman instruments over the past 15 years, some challenges remain, notably the potential for samples to fluoresce, with fluorescence potentially obscuring the Raman signal. Excitation at 785 nm has been a good compromise between long wavelength excitation, while still using low-cost, low-noise, silicon-based detectors. There has been a recent trend to move to 1064 nm excitation, although a penalty is paid in terms of measurement time for comparable SNR spectra (Christesen, Guicheteau, Curtiss, & Fountain, 2016). If the fluorescence signal is modest, and the Raman signal still detectable on top of it, then that fluorescence can be modeled and "subtracted". Spatially offset Raman spectroscopy (SORS) (Matousek, et al., 2005) enables the measurement of samples through translucent packaging (Hargreaves, 2021).

6.4.2.2.3 Near-Infrared (NIR) Spectroscopy

As noted above, NIR spectroscopy has been widely used in the pharmaceutical industry. As such, the potential for NIR to detect counterfeits was realized and demonstrated in laboratory studies more than 20 years ago (Dempster, MacDonald, Gemperline, & Boyer, 1995; Ulmschneider & Pénigault, 2000; Scafi & Pasquini, 2001; Olsen, Borer, Perry, & Forbes, 2002; Yoon, 2005; Shin, Jee, Chae, & Moffat, 2005; Rodionova, et al., 2005; Feng & Hu, 2006) with many more examples in the recent years (O'Neill, Lee, Lee, Charvill, & Moffat, 2008; Kalyanaraman, Dobler,

& Ribick, Near-Infrared (NIR) Spectral Signature Development and Validation for Counterfeit Drug Detection Using Portable Spectrometer, 2011; Alcalà, et al., 2013; Moffatt, Assi, & Watt, 2010; Guillemain, Dégardin, & Roggo, 2017; Wilson, Kaur, Allan, Lozama, & Bell, 2017; Lawson, Ogwu, & Tanna, 2018; Rodionova, Titova, Balyklova, & Pomerantsev, 2019; Rodionova, Titova, Demkin, Balyklova, & Pomerantsev, 2019; Assi, Robertson, Coombs, McEachran, & Evans, 2019; Sroka, Ishizaki, & Barańczuk, 2020; Chen, van Berkel, Luo, Sarsenbayeva, & Kostakos, 2020; Assi, Arafat, Lawson-Wood, & Robertson, 2021; Cui et al., 2021). NIR spectra tend to display non-specific broad bands, which aren't easily relatable to functional groups, and therefore chemometric methods are widely employed. However, NIR spectroscopy enables "point-and-shoot" sampling, which is a huge advantage. As the spectroscopic bands are relatively broad, high resolution (e.g., better than 2–5 nm) is not required, in sharp contrast to elemental visible and NIR spectroscopy where very sharp lines are displayed and resolutions of 0.1 nm are required. Indeed, in condensed-phase molecular NIR spectroscopy, a wider spectral range tends to be more important than high resolution. In contrast to Raman and mid-IR spectroscopies, many different spectrometer technologies are employed in the NIR (Crocombe, Portable Spectroscopy, 2018) with no one technology being dominant, especially in portable instruments.

An early example of mobile deployment of NIR instruments for this purpose was performed by the National Institute for the Control of Pharmaceutical and Biological Products of China (NICPBP), who equipped 360 vans with NIR spectrometers, TLC, and color tests to test for counterfeits in rural China (Mukhopadhyay, 2007; Chu, 2015; Hu, Feng, & Yin, 2015; Moffatt, Assi, & Watt, 2010).

In practice, in the field, apart from the NICPBP example, Raman spectroscopy has been the major technique employed, most likely due to the more specific details in Raman spectra, as compared to NIR spectra, and despite potential difficulties due to sample fluorescence. However, this situation may change as a new generation of portable NIR spectrometers, from a variety of vendors, become available (Crocombe, The Future of Portable Spectroscopy, 2021; Crocombe, Leary, & Kammrath, Packing Light: Spectroscopy Goes Mobile, 2021), at prices of $2000 (USD) or less, and have been evaluated for this application (Eady, Payne, Sortijas, Bethea, & Jenkins, 2021; Wang, Keller, Baughman, & Wilson, 2020). Indeed, a recent press release announced a collaboration between a NIR manufacturer and a major pharmaceutical company in this area (Nynomic, 2021).

6.4.2.2.4 Method Development Vibrational Spectroscopy: Algorithms and Libraries

As noted above, the goal in screening pharmaceuticals in the field is to rapidly remove clearly falsified and suspicious items from the supply chain, and these items can later be subjected to a more thorough laboratory analysis. As such, it is therefore best to err on the side of no false negatives (i.e., not allowing falsified items to remain in the field), rather than no false positives (i.e., flagging some authentic material as fake). The instrument operator is not called upon to examine the spectrum itself and come to a judgment—that decision is made by an algorithm. Consequently, we need to understand the decision-making process in the algorithms used.

The most reliable qualitative or matching algorithms account for every feature in the spectrum, model noise, account for mixtures (Yaghoobi, Wu, Clewes, & Davies, 2016), and in the case of Raman spectra, also model fluorescence; these are statistical-based or probabilistic approaches (Brown, & Rhodes, United States of America Patent No. US 7,254,501 B1, 2007; Green, Hargreaves, & Gardner, 2013). In the past 10 years, a variety of approaches have been developed, and these are described in a review article and book chapter (Gardner & Green, 2014; Zhang, Lee, & Schreyer, 2021).

Without delving into the details of the algorithms that are used, we can give an overview of the process. The instrument generates a spectrum, which is then compared to one or more reference spectra in a database, and a numerical figure of merit is generated. This is commonly between 0 and 1, with 1 indicating a perfect match. The designer of the analysis, which could be the instrument manufacturer or the laboratory scientist in charge, has to set a threshold for flagging an item as falsified. Some spectra will arise from clearly fake products and some from clearly genuine. A "PASS" or green screen on the analyzer indicates that there's no evidence in the spectrum to believe that the sample is anything but genuine. A "FAIL" or red screen indicates that there is spectroscopic evidence of a discrepancy. Those in the middle can be regarded as "suspicious." It's important to realize that there will always be errors (false positives or false negatives). The efficacy of these algorithms, and the effect of setting different threshold values, can be evaluated by examining the false positive and false negative rates, and using "Receiver Operator Curves" (ROC) to model and visualize that data (Swets, 1988; Brown & Davis, Receiver Operating Characteristics Curves and Related Decision Measures: A Tutorial, 2006; Brown & Green, Performance Characterization of Material Identification Systems, 2006). Another way of looking at these tests is that vibrational spectroscopy is *sensitive* to differences in these samples, but that the *selectivity* of the test depends on the setting of the threshold. Note that the algorithm is determining if aspects of the measured spectrum are *different* from those in the reference model for detection of falsified material; this is different from algorithms used in unknown identification, where the criterion is best similarity to a spectrum in the database.

Figure 6.8 outlines the process required to set up an analysis. The mathematical procedure to operate on a spectrum and generate a result is termed a "model."

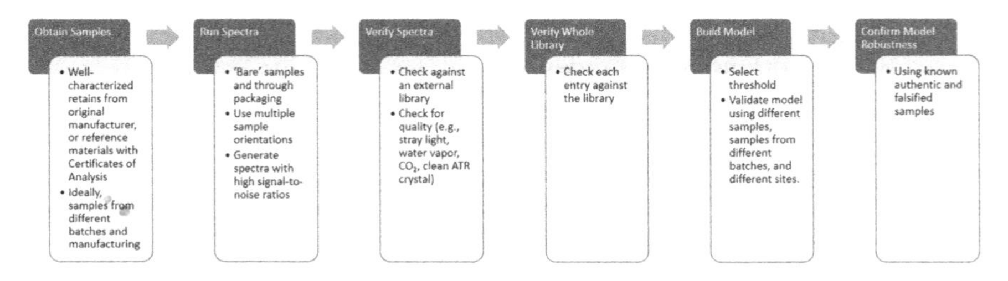

FIGURE 6.8 The process of setting up a counterfeit detection analytical method.

Note that inputs to the model are the spectra of authentic materials; these can be "manufacturing retains" from the legitimate producer or, in the case of pure materials, obtained from a reputable chemical company with a Certificate of Analysis. Because there is some variability in drug product manufacturing from different sites, and from functionally identical but chemically slightly different materials, it is important to include a number of authentic samples of each product in the library, from different lots and different sites, if appropriate. It should also be noted that NIR spectra are sensitive to particle size within the product, and to water, which could be absorbed during storage or transportation.

Library generation is ideally performed on the instrument model on which it will be deployed. This "builds into" the library, the characteristics of that instrument, and in the case of Raman spectrometers that can be a combination of the spectral range and resolution. These characteristics vary for instruments from different vendors, as they use a wavelength-dispersive approach, and some effort is required to transfer libraries from one instrument type to another. In principle, FT-IR spectra are much more "standard," as these instruments use an internal frequency standard (the reference laser), and have even data point spacing in frequency (cm^{-1}) space. However, the use of ATR sampling in portable FT-IR spectrometers introduces some complications. First, the depth of penetration into the sample varies linearly across the spectrum, increasing with shorter wavelengths. This effect can be "corrected" so as to make the spectra resemble transmission spectra, using a simple linear ramp on the absorbance spectrum. Second, the refractive index of the sample undergoes a significant excursion in the vicinity of absorption bands—an effect known as anomalous dispersion, which then affects the depth of penetration (Chantry, 1984; Griffiths & de Haseth, 2007). This can cause band maxima to shift and band shapes to become asymmetrical (Wilks, 1972). For those reasons, it is also better for FT-IR spectral libraries to be built on the instrument model on which they will be deployed and especially using the same ATR crystal material.

Library and method development for portable instruments has recently been reviewed (Schreyer, 2021). As noted above, generation of a method is not necessarily a one-off process; it tends to be cyclical, as the model is refined to generate the desired results. This is illustrated in Figure 6.9.

6.4.2.3 Mass Spectrometry (MS)

Mass spectrometry (MS) is a valuable method of identification used routinely to identify APIs and other pharmaceutically relevant chemicals. MS provides qualitative and quantitative information about the atomic and molecular composition of a sample. It produces charged particles that consist of the parent ion and ionic fragments of the original molecule. It sorts these ions according to their mass-to-charge (*m/z*) ratio. The mass spectrum is a record of the relative numbers of different kinds of ions and is characteristic of every compound, including isomers (Williard, Merritt Jr., Dean, & Settle Jr., 1988). Although MS systems are available as stand-alone analytical instruments, interpretation of data from multicomponent samples can be complicated. As a result, it is common for MS methods to be coupled with front-end separation methods to enable not only separation of

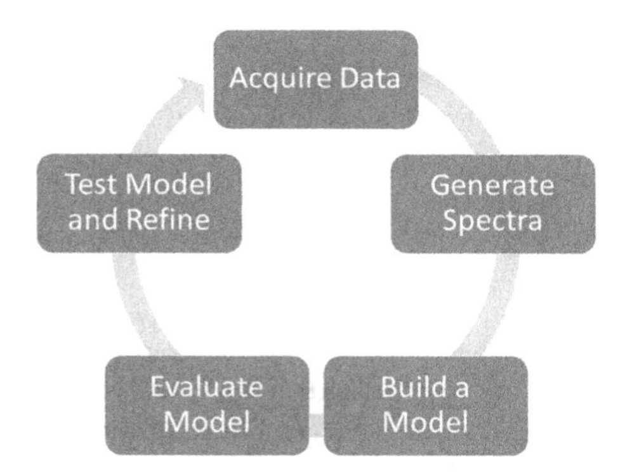

FIGURE 6.9 Model generation for vibrational spectroscopy. This may require several refinement cycles. (Adapted from Schreyer, 2021 with permission.)

components in a mixture sample but also concentration and purification of each analyte. Liquid chromatography (LC) and gas chromatography (GC) are separation methods frequently coupled with MS systems to analyze counterfeit drugs. A primary advantage LC–MS is that it can analyze a wider range of chemicals than GC–MS, including compounds that are thermally labile, exhibit high polarity, or have a high molecular weight. While LC–MS methods are more extensively used for pharmaceutical analysis, GC–MS provides benefits to some niche applications within this industry, including for the analysis of residual solvents and other volatile organic compounds (VOCs).

6.4.2.3.1 Liquid Chromatography–Mass Spectrometry (LC–MS) and LC–MS/MS

As previously mentioned, the use of LC in the pharmaceutical industry is extensive, and a significant portion of analytical testing of pharmaceuticals uses HPLC. The difference between traditional LC and HPLC is that the solvent in LC travels by the force of gravity, whereas the solvent travels under high pressure during HPLC. The pressure required to overcome the pressure drop in the packed column is achieved by means of a pump, which reduces the time of separation (Chemyx Inc., 2021). In spite of this distinction, when HPLC systems are coupled with MS detectors, the method is commonly referred to as LC–MS. LC–MS systems are used throughout a variety of stages within the pharmaceutical development lifecycle, including during drug discovery, metabolism studies (*in vitro* and *in vivo*), and to identify impurities and degradation products (Aliouche, 2019). LC–MS is also commonly used to analyze drugs of abuse within forensic laboratories (Scientific Working Group for the Analysis of Seized Drugs, 2019).

In addition, LC–MS/MS also has value in the identification of chemical constituents within drug samples. LC–MS/MS operates with a combination of chromatography and multiple quadrupole mass spectrometers. The chromatographic system first separates the different components, concentrating the amount of each single component reaching the mass spectrometer. The first quadrupole of a triple quadrupole configuration ionizes the molecules of the analyte. Selected molecular ions are then fragmented in the second quadrupole, and selectively isolated by the third and final quadrupole for measurement by a detector. This series of separation, ionization, and selective fragmentation provides highly sensitive detection. Detection levels using LC–MS/MS can be as sensitive as several parts per billion and are consistently in the part per million range (Champagne, n.d.). Figure 6.10 shows separation and identification of different cutting agents and drug substances that can be identified using a single LC–MS/MS method. This approach is useful in forensic laboratories for the analysis of drugs and cutting agents. Illicit drugs are known to have different adulterants and diluents, known as cutting agents. The cutting agents play an important role in the identification of trafficking routes, and they can also modify or intensify signs and symptoms of drug intoxication increasing the risk to the user's health (Fiorentin, Fogarty,

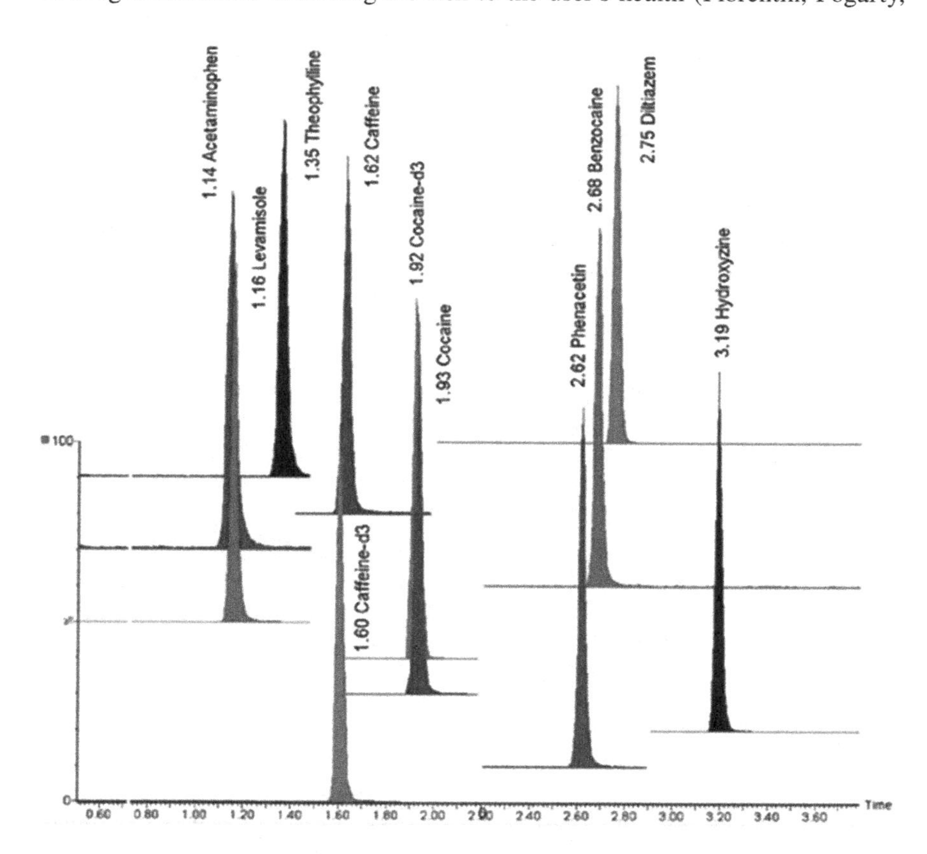

FIGURE 6.10 Representative chromatogram obtained by LC–MS/MS of the analytes acetaminophen (1.14 min), levimasole (1.16 min), theophylline (1.35 min), caffeine-d3 (1.60 min), caffeine (1.62 min), cocaine-d3 (1.92 min), Cocaine (1.93 min), phenacetin (2.62 min), benzocaine (2.68 min), diltiazem (2.75 min), and hydroxyzine (3.19 min) at 50 ng/mL (Fiorentin, Fogarty, Limberger, & Logan, 2019).

Limberger, & Logan, 2019). In this instance, the analytical results can be used to identify the chemical composition of the sample and even compare it to results from authentic tablets, and can also to help establish provenance with the intention of identifying and prosecuting criminal counterfeiting networks.

Although HPLC systems are not exclusively coupled with MS detectors, LC–MS offers some distinct advantages over other detection methods. Unlike the more classical HPLC using UV detection, chromatographic detection of a drug using MS relies on monitoring the mass of the molecular or fragment ion of the drug rather than monitoring its UV absorbing properties. In this respect, determination of weak UV absorbing compounds with conventional UV detectors is always a problem; these compounds are simply determined using mass spectrometric analysis. Moreover, due to higher specificity of LC–MS, a compound can be easily recognized by its molecular ion, and any other interference from co-existing compounds can be identified, even if they co-elute (Abdel-Hamid, 2000). Nevertheless, the pharmaceutical industry relies heavily upon UV-visible (UV-vis) detectors to achieve performance results required. In the pharmaceutical and chemical industries, the normalized area-under-the-curve (AUC) values with UV detection are often equated with purity percentages by weight (Dong, 2019). The International Council of Harmonization (ICH) guidelines (International Council for Harmonisation of Technical Requirements for Pharmaceuticals for Human Use, n.d.), followed by all pharmaceutical laboratories in production and late-stage development, require sensitivity in the range of 0.05%–0.10% for the stability-indicating HPLC methods of drug substances and drug products. The use of UV detection is implicitly assumed in the ICH Q3A guidelines for these methods. In addition, for pharmaceutical testing, the higher precision achievable with UV detection (<0.2% relative standard deviation) is pivotal and necessary in this regulatory testing, because a typical potency specification for drug substances is 98.0%–102.0% (Dong, 2019). Since the use of HPLC with UV-vis detectors is so common within pharmaceutical analysis, these methods may serve as valuable methods that can easily differentiate counterfeits from authentic versions of pharmaceutical products. As an example, Figure 6.11 shows the HPLC-UV chromatograms used to clearly distinguish authentic pharmaceuticals used to treat erectile dysfunction from the counterfeit and imitation versions.

6.4.2.3.2 Gas Chromatography–Mass Spectrometry (GC–MS)

The use of GC–MS for pharmaceutical analysis is more limited than the use of LC–MS, but GC–MS is critical for some niche pharmaceutical applications, including the analysis of residual solvents and other dangerous chemicals including volatile mutagenic impurities and leachables (Kumar, Zhang, & Wigman, 2015; Teasdale & Elder, 2018; Gooty et al., 2016; Reddy, et al., 2019). In addition, GC–MS is useful for the analysis of seized drugs analyzed in forensic laboratories (Scientific Working Group for the Analysis of Seized Drugs, 2019). The ICH Q3C guideline recommends acceptable amounts for residual solvents in pharmaceuticals for the safety of the patient, specifically the use of less-toxic solvents

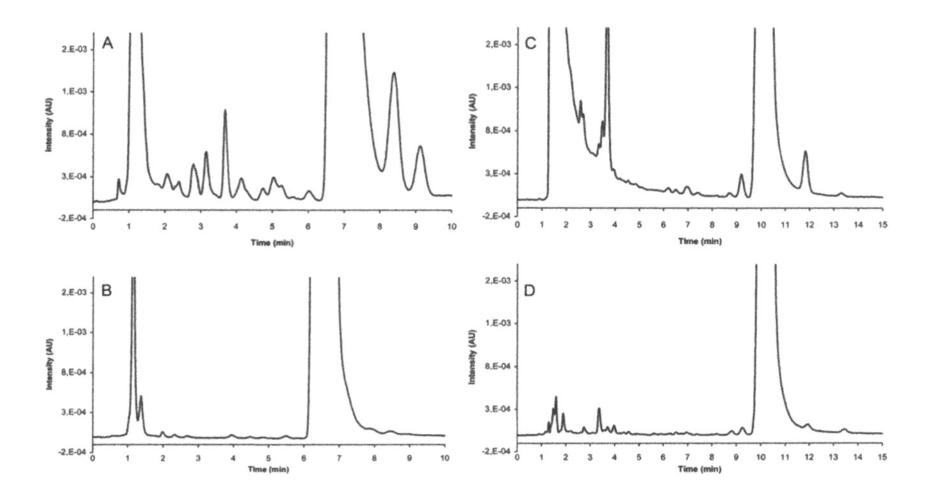

FIGURE 6.11 Impurity profiles: (a) Counterfeit Viagra, (b) Authentic Viagra, (c) Imitation Cialis, and (d) Authentic Cialis. (Reprinted with permission from Sacré, et al., 2011.).

and describes levels considered to be toxicologically acceptable for some residual solvents (International Council for Harmonisation of Technical Requirements for Pharmaceuticals for Human Use, 2021). The ICH M7 guideline describes a consistent approach to identify, categorize, and control DNA (deoxyribonucleic acid) reactive, mutagenic, impurities in pharmaceutical products to limit the potential carcinogenic risk related to such impurities (International Council for Harmonisation of Technical Requirements for Pharmaceuticals for Human Use, 2017). The ICH Q3E is intended to provide guidelines for the assessment and control of extractables and leachables. Extractables are any chemical entities that will extract from components of a manufacturing or packaging system into a solvent under forced conditions. This provides an effective worst-case scenario in terms of what can migrate from a component. Knowledge of these extractables is important in identifying potential "leachables" that can migrate via contact with manufacturing systems, container-closure systems, and drug delivery device components (International Council for Harmonisation of Technical Requirements for Pharmaceuticals for Human Use, 2020). Analysis of suspect dosage forms can be analyzed using established methods of the authentic manufacturer to determine whether or not the suspect dosage form has solvent and impurity profiles that are consistent with the profiles from authentic dosage forms.

In addition, portable GC–MS can also be used to quickly detect and identify dangerous substances contained within lower-concentration street drugs that might not alarm when analyzed in the field using methods that are less sensitive than GC–MS, such as mid-IR and Raman. When samples of drugs like cocaine that have been diluted extensively with cutting agents are subject to analysis, the concentration of the illicit substance may be below the detection limit of the mid-IR or Raman methods. In these cases, no illicit substance is identified. However,

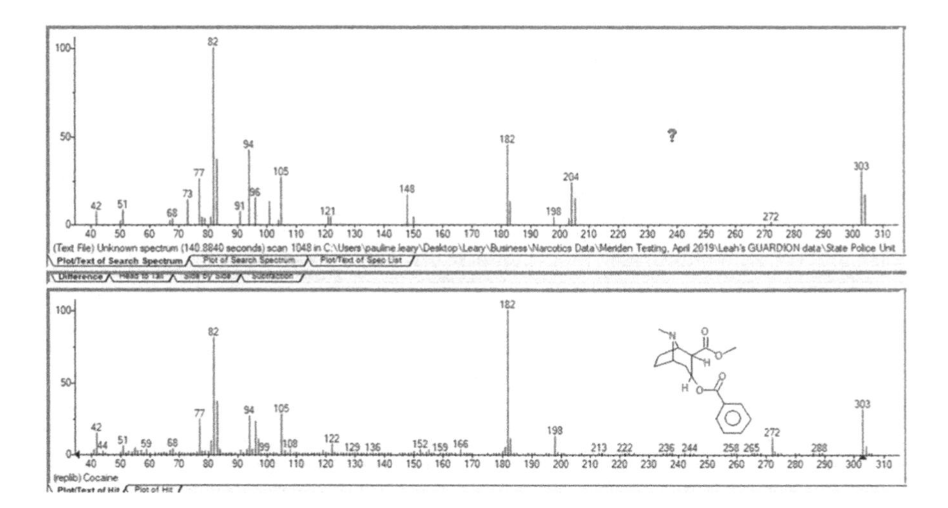

FIGURE 6.12 Spectrum collected from a dilute street-drug sample of cocaine, which tested negatively for cocaine using both Raman and mid-IR spectroscopies. The top shows the mass spectrum from the sample, and the bottom shows the mass spectrum of cocaine from the NIST-MS database. Source: https://webbook.nist.gov/cgi/cbook.cgi?ID=C50362&Mask=200#Mass-Spec

portable GC–MS may be useful in these instances to detect and identify the illicit substance. Figure 6.12, shows the mass spectrum of a cocaine street sample that was identified using a toroidal ion-trap portable GC–MS. The illicit substance was diluted in a solvent and introduced to the instrument using solid-phase micro-extraction (SPME). Total analysis time including sample dilution and analysis was approximately 10 minutes (Rynearson et al., 2020).

It is important to recognize that the use of portable GC–MS in the field tends to be more challenging than deployment of more streamlined methods such as mid-IR and Raman. Some of these challenges are due to the vacuum and other logistics requirements for these systems. In addition, interpretation of data from these systems is more complicated. Nevertheless, these systems have been deployed to environmental scientists for the detection of hazardous air pollutants since 1996 (Crume, 2009) and to military users since about 2000. They have also recently been deployed to conventional US military forces (Leary, Kammrath, Lattman, & Beals, 2019).

6.4.2.3.3 High-Pressure Mass Spectrometry (HPMS)

In 2014, 908 Devices, Inc. (Boston, MA) introduced the M908 portable HPMS system, which was the first commercially available mass spectrometer that operated at 1 Torr (Blakeman & Miller, 2021). Unlike the other technologies that have previously been described in this chapter, HPMS started out as a portable (rather than laboratory-based) system. The M908 weighs approximately 2 kg, uses a glow-discharge ionization source, and a Faraday cup detector (Brown, Knopp, Krylov, & Miller, United States of America Patent No. US 9,099,286 B2, 2015). Pumping is performed by a small, custom, mechanical scroll pump to achieve an operating pressure in the 1 Torr range in under 10 seconds. A notable feature of the M908 is the absence of an expensive, fragile, turbomolecular pump. The use of a mechanical pump as the sole source of vacuum increases the overall ruggedness of the device. The M908 is compliant with military (MIL-STD-810G) standards for shock/drop/vibration (Blakeman & Miller, 2021). These features, especially the absence of a turbomolecular pump and compliance with aggressive standards for shock/drop/vibration, provide significant benefits when deploying a portable system to the field.

A more recent version of this product—the MX908—was introduced in 2017 (908 Devices, Inc., 2017). The MX908 uses a dual-polarity atmospheric pressure chemical ionization (APCI) source driven by fast-switching electronics capable of scanning both positive- and negative-mode polarities in approximately 0.1 seconds. APCI is a soft ionization technique and, therefore, fragmentation is minimized (McLafferty & Turecek, 1993). To induce additional fragmentation and enable better chemical specificity, the MX908 also uses the process of collision-induced dissociation (CID) (Blakeman & Miller, 2021). Figure 6.13 shows an example of the improvement in fragmentation that occurs as a result of the use of CID.

As mentioned, HPMS systems are primarily available as portable instruments and only recently entered the commercial market. With regard to pharmaceutical analysis, these authors are not aware of any applications that use commercial HPMS for analysis during pharmaceutical development or as part of regulatory submissions within the pharmaceutical industry. However, HPMS systems can still play a critical role in the analysis of counterfeit drugs, specifically regarding the rapid detection of potent drugs of abuse such as fentanyl. The goal of this type of analysis is not necessarily to support legal prosecutions but rather to protect from the threats these deadly drugs pose to drug abusers, emergency responders, and the public at large. As previously mentioned, the lethal ingested dose of fentanyl is reported by the DEA to be just over 2 mg (United States Drug Enforcement Administration

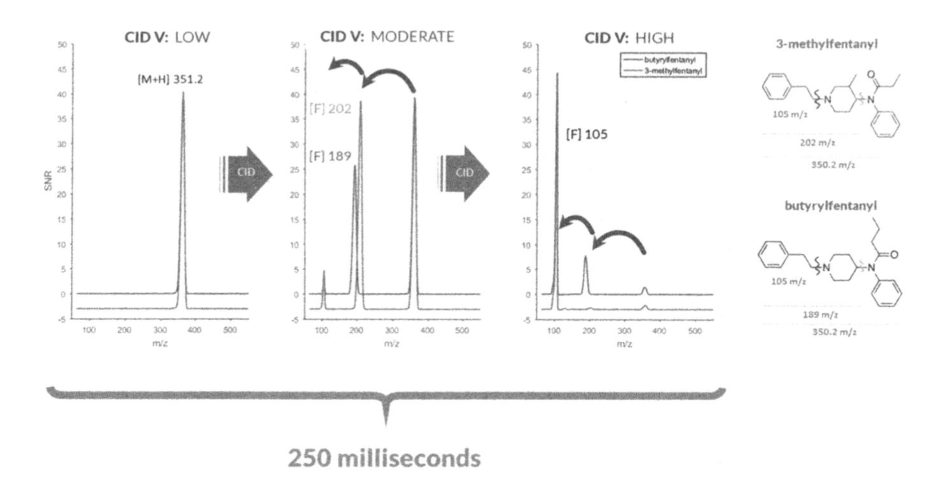

FIGURE 6.13 Improvement in fragmentation that occurs due to use of CID (Blakeman & Miller, 2021).

Strategic Intelligence Section, 2019). This is a very small amount (see Figure 6.1). With no sample preparation required, HPMS systems can quickly detect 100 ng (0.0001 mg) or less of fentanyl and other dangerous drugs. To perform an analysis, the operator, donning an appropriate level of personal protective equipment (PPE), swipes trace amounts of the drug product or its packaging with a specialized sampling swab, then introduces the swab to the instrument. No other sample preparation is required. After a few seconds, if fentanyl or other dangerous substances programmed into the HPMS system are present, the system will alarm for that substance. An additional benefit of using this analysis at the scene is that it can help establish appropriate levels of PPE to be used by law-enforcement personnel when processing the scene. Figure 6.14 shows a tablet-press machine covered in powder raw materials recovered at the site of clandestine drug laboratory. The level of PPE worn by the emergency responder in this image was critical, because this substance tested positive for fentanyl within seconds using the MX908. Figure 6.15 shows the alarm result screen on the screen of the MX908.

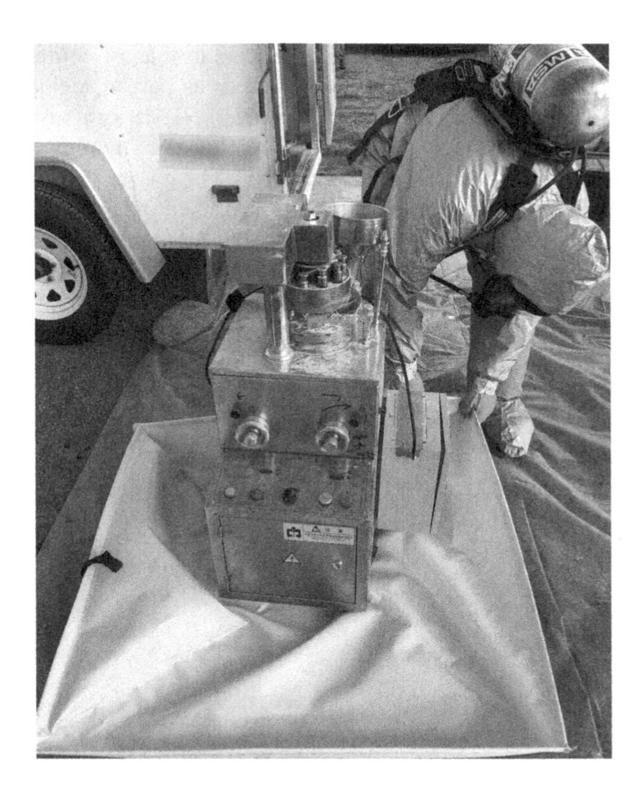

FIGURE 6.14 Tablet press recovered by law enforcement from a clandestine drug lab contaminated with a significant amount of powder residue that tested positive for fentanyl in seconds using the MX908 (see Figure 6.15). (Photo courtesy of Michael Cashman.)

FIGURE 6.15 Display of fentanyl alarm on the MX908 from the analysis of powder residue on the tablet press shown in Figure 6.14. (Photo courtesy of Michael Cashman.)

6.4.2.4 Ion Mobility Spectrometry (IMS)

IMS is a gas-phase separation technique. It works by predictably and reproducibly creating ions from neutral analyte molecules, and separating these ions based on their size and shape. This separation is performed under the influence of an electric field in the presence of a buffer gas. IMS is likely the most widely used trace detection method deployed in the field for testing of dangerous substances. It is selective, offers low detection limits, provides reliable identification, and fast analysis times. IMS systems are SWaP friendly and perform at ambient pressure. These features are important for portability and amenable to miniaturization with a robust deployment of portable and handheld systems in a variety of different markets (DeBono & Leary, 2021). They have been extensively used by the military, as well as in aviation security, border inspection, and by first responders to detect chemical warfare agents (CWAs), explosives, toxic industrial chemicals (TICs), and narcotics. They are also used at drug interdictions, as well as at correctional facilities to screen for narcotics at entry points and in mailrooms (DeBono & Leary, 2021; Leary & Joshi, 2021).

IMS is a valuable method but is quite sensitive to signal interference from sample matrix components (DeBono & Leary, 2021). This sensitivity is one of the reasons that their use within the pharmaceutical industry is limited. It has been proposed as a solution for cleaning validation—an application where the features of fast analysis with sensitive detection limits are quite beneficial (U.S. Food & Drug Administration, 2014; O'Donnell, Sun, & de B. Harrington, 2008; Armenta, Alcalà, Blanco, & González, 2013). Nevertheless, as with HPMS, IMS can play a critical role when analyzing counterfeit drugs, specifically to protect from the threats deadly drugs like fentanyl pose to drug abusers, emergency responders, and the public at large. IMS has been shown to detect fentanyl in an acetone solution at levels of 0.1% fentanyl/acetone (Leary, et al., 2020).

In these applications, IMS systems work in a similar manner to HPMS in that the operator, donned in appropriate PPE, swipes trace amounts of the drug product or its packaging with a specialized sampling swab, then introduces the swab to the instrument. If programmed threats like fentanyl are detected, the instrument will alarm and display the detected substance within seconds. Practically, the greatest challenge to the use of IMS (and HPMS) is the potential overload of the system, which subsequently requires downtime for instrument recovery. For this reason, as well as because the systems tend to be easy to operate, training for these systems is generally focused on optimizing sampling. Regardless of this challenge, the use of both IMS and HPMS to detect potent drugs in the field is an important and critical capability.

6.4.2.5 X-ray Fluorescence (XRF) Spectroscopy for Elemental Analysis

X-ray fluorescence spectroscopy (Jenkins, 1999) is an elemental analysis (not molecular) technique. These systems are available in both laboratory-based and handheld/portable formats (Piorek, 2021). They are sensitive to elements with an atomic number larger than that of silicon, and, therefore, XRF does not detect hydrogen, carbon, nitrogen, and oxygen. Its value in pharmaceutical analysis may appear to be limited, but for drug product it will detect elements like potassium, titanium, iron, zinc, platinum, etc. It can be used to analyze trace metals (e.g., Zn, Mn, Al and Si) in the aluminum foil in blister packs and on the caps on vials, as these trace elements will vary depending on the source of the foil or cap. It is "point-and-shoot" technique, being non-invasive and non-destructive, and typically has short measurement times from a few seconds to about a minute. As an open-beam x-ray device, it will require licensing, and those regulations differ from country to country and from state to state in the United States of America.

One of the most common product safety-related analytical tests within the pharmaceutical industry is the quantification of inorganic impurities within a pharmaceutical product. This includes toxic heavy metals, such as As, Cd, Hg, and Pb. Other metals, such as Fe, Cr, Ni and Zn, are also of interest due to health risks. In addition, many APIs may contain residual metal catalysts, such as Ru, Pt, and Pd. Since there are many potential sources of contamination, it may be of interest to measure raw materials, intermediates, as well as final products (Campbell, et al., 2012).

In addition, portable XRF has been used to analyze vitamins, nutraceuticals, and supplements for their elemental inorganic content ("minerals"). This approach works best if the solid dosage form is powdered and placed in an XRF sample cup with a thin Mylar or polypropylene window and at a sufficient depth that it is "infinitely thick" to the x-ray beam. This has been shown to work for elements like Ca, K, Fe, Mg, etc. The method may not be precise for trace and ultra-trace elements (e.g., Se). XRF methodology also only indicates the presence or concentration of those specific elements and not how they are bound (i.e., what anions are associated). This is important as the binding has a big impact in terms of human uptake.

6.4.2.6 Nuclear Magnetic Resonance Spectroscopy

NMR instruments can be divided into imagers (magnetic resonance imaging or MRI), spectrometers for chemical-shift measurements (conventional NMR), and relaxometers [time domain nuclear magnetic resonance (TD-NMR)]. TD-NMR determines the spin–spin (T_2) relaxation times of excited ^1H nuclei in the sample.

The use of NMR spectroscopy is common within the pharmaceutical industry for a variety of different purposes (United States Pharmacopeia, n.d.; United States Pharmacopeia, n.d.; Holzbrabe, Wawer, & Diehl, 2008; Everett, Harris, Lindon, & Wilson, 2015; Vogt, 2010; Maggio, Calvo, Vignaduzzo, & Kaufman, 2014), and it is possible to use NMR methods for the analysis of counterfeit drugs (Keizers, et al., 2020). Similar to the use of methods for GC–MS and LC–MS, NMR methods developed by authentic pharmaceutical manufacturers to analyze their drugs, and drug products can be applied to suspect drugs to potentially differentiate them from authentic samples. In addition, TD-NMR spectroscopy has recently been suggested as a screening method to detect counterfeit biologic pharmaceuticals (Akhunzada, et al., 2021). The authors of this paper report that TD-NMR is very sensitive to differences in the amount of antibody in solution, with the ability to detect variations in as low as 2 mg/mL. It is therefore capable, by comparison with data from known formulations, of determining whether a particular sample is likely to be of an authentic biologic formulation. For the experiments they describe, they propose that TD-NMR is a fingerprinting measurement, one that is very sensitive to differences between samples. The tradeoff is that the determination of the origin of these differences requires the use of other analytical techniques, as TD-NMR does not provide chemical-shift information.

NMR systems are frequently lab based, but relaxometers have been commercially available, in compact benchtop formats, for decades. With advances in magnet and RF technologies, this type of instrument can be miniaturized (Blümich, Perlo, & Casanova, 2008), and the current state of portable NMR has recently been summarized (Kizzier & Cassata, 2021). However, although they are available, portable NMR systems have not yet been widely adopted.

6.5 CONCLUSIONS

In conclusion, laboratory methods are very useful for confirmed identifications of falsified drugs. Field screening methods for detection of counterfeit pharmaceuticals at the sample site have been extensively deployed, and range from visual inspection of both the packing and its contents, through "track-and-trace" methods, to physical, spectroscopic, and chromatographic testing.

Portable analytical instruments that can provide more specific chemical information on the drug product have been developed over the past twenty years, with rapid and ongoing improvements. Portable vibrational spectroscopy analyzes the drug product directly ("point-and-shoot"), and this can be done very rapidly, and non-invasively, via both Raman and NIR spectroscopy. Mid-IR is as specific as Raman spectroscopy but has the disadvantage in the field that it requires sample handling in order to present the sample to the instrument's ATR interface. However, in a laboratory setting, FT-IR may be superior to Raman spectroscopy, because it does not suffer from fluorescence interference.

Portable vibrational spectroscopy instruments from major vendors are equipped with onboard processing that can turn a spectrum into actionable information, given the appropriate spectroscopic databases and calibrations. The major vendors may supply spectroscopic libraries of common pharmaceutical precursors and products, but in most cases the end user will have to add spectra of genuine materials and products to those libraries.

Field portable GC–MS instruments are available, and can be used to detect and identify dangerous materials in street drugs (e.g., fentanyl and its derivatives) at low concentrations. However, these instruments are more challenging to use in the field, and the data is more difficult to interpret, as compared to vibrational spectroscopy.

A recent technical development is the introduction of portable HPMS. This has proven to be useful to detect low levels (100ng) of fentanyl, protecting emergency responders and the public. IMS, commonly used for detection of explosives and chemical warfare agents, has also been employed for fentanyl detection. Finally, NMR is a proven method, especially in lab format, for the analysis of pharmaceuticals.

Portable spectroscopic methods are "market surveillance," and suspect materials will be brought back to a laboratory for further analysis. Field analysis enables rapid response, removing counterfeit and\or dangerous materials immediately, and also reduces the burden on laboratories, where analyses (e.g., by HPLC) may require considerable time and expertise. A barrier to wider implementation is the cost of these instruments, but very recent developments are bringing that cost down considerably. Each technique has its own strengths and weaknesses, and no one technique is a "catch-all," so that a combination of approaches will continue to be required.

ACKNOWLEDGMENTS

The authors thank Michael Gallagher (Thermo Fisher Scientific), Debbie Schatzlein Griggs (Rigaku), and David Jenkins (FHI360) for helpful discussions. The authors would also like to thank Michael Cashman for helpful discussions as well as providing a number of images used within this chapter.

ACRONYMS AND ABBREVIATIONS

AIDS	Acquired Immune Deficiency Syndrome
ANDA	Abbreviated New Drug Application
API	Active Pharmaceutical Ingredient

APCI	Atmospheric Pressure Chemical Ionization
ATR	Attenuated Total Reflection
AUC	Area Under the Curve
FR	Code of Federal Regulations
CID	Collision-Induced Dissociation
CWA	Chemical Warfare Agent
DEA	Drug Enforcement Administration
DNA	Deoxyribonucleic Acid
FDA	Food and Drug Administration
FT	Fourier Transform
FT-IR	Fourier Transform Mid-Infrared
GC	Gas Chromatography
GC–MS	Gas Chromatography–Mass Spectrometry
GPHF	Global Public Health Fund
GPS	Global Positioning System
HPLC	High-Performance Liquid Chromatography
HPLC-UV	High-Performance Liquid Chromatography-Ultraviolet
HPMS	High-Pressure Mass Spectrometry
ICH	International Council for Harmonization of Technical Requirements for Pharmaceuticals for Human Use
IMS	Ion Mobility Spectrometry
IP	Intellectual Property
IR	Infrared
LC	Liquid Chromatography
LC–MS	Liquid Chromatography–Mass Spectrometry
LC–MS/MS	Liquid Chromatography–Mass Spectrometry/Mass Spectrometry
LED	Light-Emitting Diode
LMIC	Low- and Middle-Income Countries
mid-IR	Mid-Infrared
MRI	Magnetic Resonance Imaging
MS	Mass Spectrometry
NAFDAC	National Agency for Food and Drug Administration and Control
NDA	New Drug Application
NICPBP	National Institute for the Control of Pharmaceutical and Biological Products of China
NIR	Near-Infrared
NIST	National Institute of Standards and Technology
NMR	Nuclear Magnetic Resonance
PAD	Paper Analytical Device
PCN	Pharmaceutical Council of Nigeria
PPE	Personal Protective Equipment
QA/QC	Quality Assurance/Quality Control
RF	Radio Frequency
RFID	Radio Frequency Identification

ROC	Receiver Operator Curves
SNR	Signal-to-Noise Ratio
SORS	Spatially Offset Raman Spectroscopy
SPME	Solid-Phase Microextraction
SWaP	Size, Weight, and Power
TB	Tuberculosis
TD-NMR	Time Domain Nuclear Magnetic Resonance
TIC	Toxic Industrial Chemical
TLC	Thin-Layer Chromatography
USD	United States Dollar
UV	Ultraviolet
UV-vis	Ultraviolet-visible
VNIR	Visible-Near Infrared
VOC	Volatile Organic Compound
WHO	World Health Organization

REFERENCES

908 Devices, Inc. (2017, June 17). *Business Wire.* Retrieved October 21, 2021, from 908 Devices Introduces Next-Generation HPMS Device for the FIrst Responser Community – https://www.businesswire.com/news/home/20170614005086/en/908-Devices-Introduces-Next-Generation-HPMS-Device-for-the-First-Responder-Community.

Abdel-Hamid, M.E. (2000). Comparative LC-MS and HPLC Analyses of Selected Antiepileptics and Beta-Blocking Drugs. *Il Farmaco, 55*, 136–145.

Abe, K., Suzuki, K., & Citterio, D. (2008, August 13). Inkjet-Printed Microfluidic Multianalyte Chemical Sensing Paper. *Analytical Chemistry, 80*(18), 6928–6934. doi:10.1021/ac800604v.

Adegoke, J.A., Kochan, K., Heraud, P., & Wood, B.R. (2021). A Near-Infrared "Matchbox Size" Spectrometer to Detect and Quantify Malaria Parasitemia. *Analytical Chemistry, 93*(13), 5451–5458. doi:10.1021/acs.analchem.0c05103.

Adeyeye, M.C. (2019, November 11). *Curbing Standard, Falsified (SFS) and Counterfeit Medicines*. Retrieved October 27, 2021, from National Agency for Food and Drug Administration annd Control (NAFDAC): https://www.nafdac.gov.ng/curbing-substandard-falsified-sfs-and-counterfeit-medicines/.

Ahmad, F., Rossen, L., & Sutton, P. (2021, July 4). *Vital Statistics Rapid Release: Provisional Drug Overdose Death Counts*. Retrieved July 23, 2021, from Centers for Disease Control and Prevention: https://www.cdc.gov/nchs/nvss/vsrr/drug-overdose-data.htm.

Akhunzada, Z., Wu, Y., Haby, T., Jayawickrama, D., McGeorge, G., La Colla, M.,... Abraham, A. (2021). Analysis of Biopharmaceutical Formulations by Time Domain Nuclear Magnetic Resonance (TD-NMR) Spectroscopy: A Potential Method for Detection of Counterfeit Biologic Pharmaceuticals. *110*(7). doi:10.1016/j.xphs.2021.03.011.

Alcalà, M., Blanco, M., Moyano, D., Broad, N.W., O'Brien, N., Friedrich, D.,... Siesler, H.W. (2013). Qualitative and Quantitative Pharmaceutical Analysis with a Novel Hand-Held Miniature Near Infrared Spectrometer. *Journal of Near Infrared Spectroscopy, 21*(6), 445–457. doi:10.1255/jnirs.1084.

Aliouche, H. (2019, February 5). *LC-MS Analysis of Pharmaceutical Drugs*. Retrieved October 19, 2021, from News Medical: https://www.news-medical.net/life-sciences/LC-MS-Analysis-of-Pharmaceutical-Drugs.aspx.

Amin, A., Snow, R., & Kokwaro, G. (2005, December). The Quality of Sulphadoxine-Pyrimethamine and Amodiaquine Products in the Kenyan Retail Sector. *Journal of Clinical Pharmacy and Therrapeutics, 30*(6), 559–565. doi:10.1111/j.1365-2710.2005.00685.x.

Andria, S.E., Fulcher, M., Witkowski, M.R., & Platek, F.S. (2012, April 1). The Use of SEM/EDS and FT-IR Analysis in the Identification of Counterfeit Pharmaceutical Packaging. *American Pharmaceutical Review*. Retrieved October 27, 2021, from https://www.americanpharmaceuticalreview.com/Featured-Articles/111769-The-Use-of-SEM-EDS-and-FT-IR-Analyses-in-the-Identification-of-Counterfeit-Pharmaceutical-Packaging/.

Apilux, A., Ukita, Y., Chikae, M., Chailapakul, O., & Takamura, Y. (2013, October 22). Development of Automated Paper-Based Devices for Sequential Multistep Sandwich Enzyme-Linked Immunosorbent Assays Using Inkjet Printing. *Lab on a Chip, 13*, 126–135. doi:10.1039/c2lc40690j.

Armenta, S., Alcalà, M., Blanco, M., & González, J. (2013). Ion Mobility Spectrometry for the Simultaneous Determination of Diacetyl Midecamycin and Detergents in Cleaning Validation. *Journal of Pharmaceutical and Biomedical Analysis, 83*, 265–272. doi:10.1016/j.jpba.2013.05.016.

Assi, S. (2016, April 30). Evaluating Handheld Spectroscopic Techniques for Identifying Counterfeit Branded and Generic Medicines Worldwide. *American Pharmaceutical Review, 19*(3).

Assi, S., & Fortunato, L. (2017, May/June). Optimizing the Method for Authenticating Pharmaceutical Products of Different Dosage Forms Using Dual Laser Handheld Raman Spectroscopy. *American Pharmaceutical Review, 20*(4).

Assi, S., Arafat, B., Lawson-Wood, K., & Robertson, I. (2021). Authentication of Antibiotics Using Portable Near-Infrared Spectroscopy and Multivariate Data Analysis. *Applied Spectroscopy, 75*(4), 434–444. doi:10.1177/0003702820958081.

Assi, S., Robertson, I., Coombs, T., McEachran, & Evans, K. (2019, May 1). The Use of Near-Infrared Spectroscopy for Authenticating Cardiovascular Medicines. *Spectroscopy, 34*(5), pp. 46–54.

Bate, R. (2014). *PHAKE. The Deadly World of Falsified and Substandard Medicines*. Washington, D.C.: The AEI Press.

Bate, R., & Mathur, A. (2011). *The Impact of Improved Detection Technology on Drug Quality: A Case Study of Lagos, Nigeria*. American Enterprise Institute for Public Policy Research. https://papers.ssrn.com/sol3/papers.cfm?abstract_id=2212974

Bate, R., Ayodele, T., Tren, R., Hess, K., & Sotola, O. (2009). *Drug Use in Nigeria*. American Enterprise Institute, Initiative for Public Policy Analysis, and Africa Fighting Malaria.

Bate, R., Tren, R., Hess, K., Mooney, L., & Porter, K. (2009, April). Pilot Study Comparing Technologies to Test for Substandard Drugs in Field Settings. *African Journal of Pharmacy and Pharmacology, 3*(4), 165–170. doi:10.5897/AJPP.9000228.

BaySpec, Inc. (n.d.). *Breeze™ – Smart Palm Spectrometer*. Retrieved October 27, 2021, from BaySpec: https://www.bayspec.com/spectroscopy/breeze-smart-palm-spectrometer/.

Beć, K.B., Grabska, J., & Huck, C.W. (2021). Principles and Applications of Miniaturized Near-Infrared (NIR) Spectrometers. *Chemistry – A European Journal, 27*, 1514–1532.

Blakeman, K.H., & Miller, S.E. (2021). Development of High-Pressure Mass Spectrometry for Handheld and Benchtop Analyzers. In R.A. Crocombe, P.E. Leary, & B.W. Kammrath (Eds.), *Portable Spectroscopy and Spectrometry 1: Technologies and Instrumentation* (1st ed., Vol. 1, pp. 391–413). Chichester: John Wiley & Sons Ltd.

Blümich, B., Perlo, J., & Casanova, F. (2008). Mobile Single-Sided NMR. *Progress in Nuclear Magnetic Resonance Spectroscopy, 52*(4), 197–269. doi:10.1016/j.pnmrs.2007.10.002.

Bonvallet, J., Auz, B., Rodriguez, J., & Olmstead, T. (2018). Miniature Raman Spectrometer Development. *Proceedings of SPIE BiOS. 10490*, p. 104900W. San Francisco: Society of Photo-Optical Instrumentation Engineers (SPIE). doi:10.1117/12.2315636.

Bottoni, P., & Caroli, S. (2019, September). Fake Pharmaceuticals: A Review of Current Analytical Approaches. *Microchemical Journal, 149*. doi:10.1016/j.microc.2019.104053.

Brown, C.D., & Davis, H.T. (2006, January 20). Receiver Operating Characteristics Curves and Related Decision Measures: A Tutorial. *Chemometrics and Intelligent Laboratory Systems, 80*(1), 24–38. doi:10.1016/j.chemolab.2005.05.004.

Brown, C.D., & Green, R.L. (2006). Performance Characterization of Material Identification Systems. *Chemical and Biological Sensors for Industrial Environmental Monitoring II. 637809*, p. 6378. Society of Photo-Optical Instrumentation Engineers (SPIE). doi:10.1117/12.686103.

Brown, C.D., & Rhodes, G.H. (2007, August 7). *United States of America Patent No. US 7,254,501 B1.*

Brown, C.D., Knopp, K.J., Krylov, E., & Miller, S. (2015, August 4). *United States of America Patent No. US 9,099,286 B2.*

Burggraaff, O., Schmidt, N., Zamorano, J., Pauly, K., Pascual, S., Tapie, C.,... Snik, F. (2019, July 8). Standardized Spectral and Radiometric Calibration of Consumer Cameras. *Optics Express, 27*(14), 19075–19101. doi:10.1364/OE.27.019075.

Campbell, I., Xiao, Y., Vrebos, B., Kempenaers, L., Coler, D., & Macciarola, K. (2012, November 9). The Use of EDXRF for Pharmaceutical Material Elemental Analysis. *American Pharmaceutical Review*. Retrieved October 26, 2021, from https://www.americanpharmaceuticalreview.com/1504-White-Papers-Application-Notes/124874-The-Use-of-EDXRF-for-Pharmaceutical-Material-Elemental-Analysis/.

Centers for Disease Control and Prevention. (2021, March 17). *Centers for Disease Control and Prevention*. Retrieved July 24, 2021, from Opioid Basics: Understanding the Epidemic: https://www.cdc.gov/opioids/basics/epidemic.html.

Champagne, A. (n.d.). *LC-MS/MS: What is It? Why Use It?* Retrieved October 22, 2021, from The McCrone Group: https://www.mccrone.com/mm/lc-msms-chemical-analysis/.

Chantry, G.W. (1984). Long-Wave Optics – The Science and Technology of Infrared and Near-Millimetre Waves: Applications. In G. W. Chantry, *Long-Wave Optics – The Science and Technology of Infrared and Near-Millimetre Waves: Applications* (Vol. 1, pp. 222–225). London: Academic Press.

Chemyx Inc. (2021, October 11). *Basic Principles of LC, HPLC, MS, & MS*. Retrieved October 20, 2021, from Chemyx Inc. North America: https://www.chemyx.com/support/knowledge-base/applications/basic-principles-hplc-ms-lc-ms/.

Chen, Y., van Berkel, N., Luo, C., Sarsenbayeva, Z., & Kostakos, V. (2020). Application of Miniaturized Near-Infrared Spectroscopy in Pharmacuetical Identification. *Smart Health, 18*. doi:10.1016/j.smhl.2020.100126.

Christesen, S.D., Guicheteau, J.A., Curtiss, J.M., & Fountain, A. (2016, July 11). Handheld Dual-Wavelength Raman Instrument for the Detection of Chemical Agents and Explosives. *Optical Engineering, 55*(7). doi:10.1117/1.OE.55.7.074103.

Chu, X. (2015). Near Infrared Spectroscopy in China. *Journal of Near Infrared Spectroscopy, 23*(Special Issue), v-vii. doi:10.1255/jnirs.1183.

Ciurczak, E., & Igne, B. (2015). *Pharmaceutical and Medical Applications of Near-Infrared Spectroscopy,* (2nd ed.). Boca Raton: CRC Press.

Ciza, P., Sacre, P.-Y., Waffo, C., Coïc, L., Avohou, H., Mbinze, J.,... Ziemons, E.J. (2019, September 1). Comparing the Qualitative Performances of Handheld NIR and Raman Spectrophotometers for the Detection of Falsified Pharmaceutical Products. *Talanta, 202*, 469–478. doi:10.1016/j.talanta.2019.04.049.

Compton, D.A., & Compton, S.V. (1991). Examination of Packaged Consumer Goods by Using FT-Raman Spectrometry. *Applied Spectroscopy, 45*(10), 1587–1589.

Counterfeit. (2014). In *Collins English Dictionary* (12th ed.). Collins. Retrieved July 7, 2021, from https://search.credoreference.com/content/entry/hcengdict/counterfeit/0?institutionId=4100.

Creasey, D., Sullivan, M., Paul, C., & Rathmell, C. (2018). Extending Raman's Reach: Enabling Applications Via Greater Sensitivity and Speed. *Proceedings of the SPIE BiOS. 10490*, p. 104900V. San Francisco: Society of Photo-Optical Instrumentation Engineers (SPIE). doi:10.1117/12.2315634.

Crocombe, R.A. (2018). Portable Spectroscopy. *Applied Spectroscopy, 72*(12), 1701–1751. doi:10.1177/0003702818809719.

Crocombe, R.A. (2021). The Future of Portable Spectroscopy. In R. A. Crocombe, P. E. Leary, & B. W. Kammrath (Eds.), *Portable Spectroscopy and Spectrometry 2: Applications* (Vol. 2, pp. 547–575). Chichester: John WIley & Sons, Ltd.

Crocombe, R.A., Leary, P.E., & Kammrath, B.W. (2021, January). Packing Light: Spectroscopy Goes Mobile. *Photonics Spectra, 55*(1), pp. 68–71.

Crocombe, R.A., Leary, P.E., & Kammrath, B.W. (Eds.). (2021). *Portable Spectroscopy and Spectrometry 1: Technologies and Instrumentation* (1st ed., Vol. 1). Chichester: John Wiley & Sons, Ltd.

Crocombe, R.A., Leary, P.E., & Kammrath, B.W. (Eds.). (2021). *Portable Spectroscopy and Spectrometry 2: Applications* (1st ed., Vol. 2). Chichester: John WIley & Sons, Ltd.

Crume, C. (2009). *History of Inficon HAPSITE From Classic to ER: A Brief Report on the Differences between HAPSITE® Iterations*. Retrieved June 5, 2016, from KD Analytical: http://www.kdanalytical.com/instruments/inficon-hapsite-history.aspx.

Cui, P., Zhao, J., Liu, M., Qi, M., Wang, Q., Li, Z., Suo, T., Li, G. (2021, June). *Infrared Physics and Technology, 115*. doi:10.1016/j.infrared.2021.103687.

Cummings, J. (2018, January). Taking Down a Goliath. *The Analytical Scientist, 37*. https://theanalyticalscientist.com/techniques-tools/taking-down-a-goliath

DeBono, R.F., & Leary, P.E. (2021). Key Instrumentation Developments That Have Led to Portable Ion Mobility Spectrometer Systems. In R. A. Crocombe, P. E. Leary, & B. W. Kammrath (Eds.), *Portable Spectroscopy and Specctrometry 1: Technologies and Instrumentation* (1st ed., Vol. 1, pp. 415–448). Chichester: John WIley & Sons, Ltd.

Deconinck, E.A., Bothy, J., Courselle, P., & De Beer, J. (2014, August). Comparative Dissolution Study on Counterfeit Medicines of PDE-5 Inhibitors. *Journal of Pharmaceutical Analysis, 4*(4), 250–257. doi:10.1016/j.jpha.2014.03.002.

Dégardin, K., Desponds, A., & Roggo, Y. (2017, September). Protein-Based Medicines Analyiss by Raman Spectroscopy for the Detection of Counterfeits. *Forensic Science International, 278*, 313–325. doi:10.1016/j.forsciint.2017.07.012.

Dégardin, K., Guillemain, A., & Roggo, Y. (2017, January). Comprehensive Study of a Handheld Raman Spectrometer for the Analysis of Counterfeits of Solid-Dosage Form Medicines. *Journal of Spectroscopy, 4*, 1–13. doi:10.1155/2017/3154035.

Dégardin, K., Guillemain, A., Guerreiro, N.V., & Roggo, Y. (2016, September 5). Near Infrared Spectroscopy for Counterfeit Detection Using a Large Database of Pharmaceutical Tablets. *Journal of Pharmaceutical and Biomedical Analysis, 128*, 89–97. doi:10.1016/j.jpba.2016.05.004.

Dégardin, K., Guillemain, A., Klespe, P., Hindelang, F., Zurbach, R., & Roggo, Y. (2018). Packaging Analysis is Counterfeit Medicines. *Forensic Science International, 291*, 144–157. doi:10.1016/j.forsciint.2018.08.023.

Dégardin, K., Roggo, Y., Been, F., & Margot, P. (2011, October 31). Detection and Chemical Profiling of Medicine Counterfeits by Raman Spectroscopy and Chemometrics. *Analytica Chimica Acta, 705*(102), 334–341. doi:10.1016/j.aca.2011.07.043.

Deidda, R., Sacre, P.-Y., Clavaud, M., Coïc, L., Avohou, H., Hubert, P., & Ziemons, E. (2019, May). Vibrational Spectrosocpy in Analysis of Pharmaceuticals: A Critical Review of Innovative Portable and Handheld NIR and Raman Spectrometers. *TrAC Trends in Analytical Chemistry, 114*, 251–259. doi:10.1016/j.trac.2019.02.035.

Dempster, M.A., MacDonald, B.F., Gemperline, P.J., & Boyer, N.R. (1995, June). Subsurface Probing in Diffusely Scattering Media Using Spatially Offset Raman Spectroscopy. *Analytica Chimica Acta, 310*(1), 43–51. doi:10.1016/0003-2670(95)00117-I.

Diehl, B., Grout, B., Hernández, J., O'Neil, S., McSweeney, C., Alvarado, J., & Smith, M. (2012). An Implementation Perspective on Handheld Raman Spectrometers for the Verification of Material Identity. *European Pharmaceutical Review, 17*(5), 3–8.

Dong, M.W. (2019). *HPLC and UHPLC for Practicing Scientists*. Hoboken: John Wiley & Sons, Inc.

Dong, M.W. (2019). Ultraviolet Detecotrs: Perspectives, Principles, and Practices. *LCGC North America, 37*(10), pp. 750–759.

Eady, M., Payne, M., Sortijas, S., Bethea, E., & Jenkins, D. (2021, October 4). A Low-Cost and Portable Near-Infrared Spectrometer Using Open-Source Multivariate Data Analysis Software for Rapid Discriminatory Quality Assessment of Medroxyprogesterone Acetate Injectables. *Spectrochimical Acta Part A: Molecular and Biomolecular Spectroscopy, 259*. doi:10.1016/j.saa.2021.119917.

Emmanuel, N., Nair, R.B., Abraham, B., & Yoosaf, K. (2021). Fabricating a Low-Cost Raman Spectrometer to Introduce Students to Spectroscopy Basics and Applied Instrument Design. *Journal of Chemical Education, 98*(6), 2109–2116. doi:10.1021/acs.jchemed.0c01028.

Everett, J.R., Harris, R.K., Lindon, J.C., & Wilson, I.D. (Eds.). (2015). *NMR in Pharmaceutical Sciences*. Chichester: John Wiley & Sons, Ltd.

Felter, C. (2021, July 20). *Council on Foreign Relations*. Retrieved July 324, 2021, from The U.S. Opioid Epidemic: https://www.cfr.org/backgrounder/us-opioid-epidemic.

Feng, Y.-C., & Hu, C.-Q. (2006, May 3). Construction of Universal Quantitative Models for Determination of Roxithromycin and Erythromycin Ethylsuccinate in Tablets from Different Manufacturers Using Near Infrared Reflectance Spectroscopy. *Journal of Pharmaceutical and Biomedical Analysis, 41*(2), 373–384. doi:10.1016/j.jpba.2005.11.027.

Fernandez, F.M., Green, M.D., & Newton, P.N. (2007, July 13). Prevalence and Detection of Counterfeit Pharmaceuticals: A Mini Review. *Industrial and Engineering Chemistry Research, 47*(3), 585–590. doi:10.1021/ie0703787.

Fiorentin, T.R., Fogarty, M., Limberger, R.P., & Logan, B.K. (2019). Determination of Cutting Agents in Seized Cocaine Sample Using GC-MS, GC-TMS and LC-MS/MS. *Forensic Science International, 295*, 199–206. doi:10.1016/j.forsciint.2018.12.016.

Frank, C.J. (1999). Review of Pharmaceutical Applications of Raman Spectroscopy. In M.J. Pelletier (Ed.), *Analytical Applications of Raman Spectroscopy* (pp. 224–275). Oxford: Blackwell Sciences Ltd.

Frosch, T., Wyrwich, E., Yan, D., Domes, C., Domes, R., Popp, J., & Frosch, T. (2019, September). Counterfeit and Substandard Test of the Antimalarial Tablet Riamet ® by Means of Raman Hyperspectral Multicomponent Analysis. *Molecules, 24*(8), 1–14. doi:10.3390/molecules24183229.

Fu, E., Liang, T., Spicar-Mihalic, P., Houghtaling, J., Ramachandran, S., & Yager, P. (2012, April 26). Two-Dimensional Paper Network Format that Enables Simple Multistep Assays for Use in Low-Resource Settings in the Context of Malaria Antigen Detection. *Analytical Chemistry, 84*(10), 4574–4579. doi:10.1021/ac300689s.

Galia, E., Horton, J., & Dressman, B. (1999). Albendazole Generics – A Comparative In Vitro Study. *Pharmaceutical Research, 16*, 1871–1875. doi:10.1023/A:1018907527253.

García Sánchez, J., García Sánchez, E., & Merino Marcos, M. (2006). Antibacterial Agents in the Cinema. *Revista Española de Quimioterapia, 19*(4), 397–402.

Gardner, C., & Green, R.L. (2014). Identification and Confirmation Algorithms for Handheld Spectrometers. In R. Meyers (Ed.), *Encyclopedia of Analytical Chemistry: Applications, Theory and Instrumentation* (pp. 1–18). Chichester: John Wiley & Sons, Ltd. doi:10.1002/9780470027318.a9381.

Gaudiano, M.C., Di Maggio, A., Cocchieri, E., Antoniella, E., Bertocchii, P., Alimonti, S., & Valvo, L. (2007). Medicines Informal Market In Congo, Burundi and Angola: Counterfeit and Sub-Standard Antimalarials. *Malaria Journal, 6*(22). doi:10.1186/1475-2875-6-22

Global Pharma Health Fund. (2021). *Global Pharma Health Fund (GPFH) Objectives.* Retrieved July 20, 2021, from Global Pharma Health Fund: https://www.gphf.org/en/gphf/index.htm.

Global Pharma Health Fund. (n.d.). *The GPHF-Minilab™ – Protection Against Counterfeit Medicines.* Retrieved July 20, 2021, from Global Pharma Health Fund: https://www.gphf.org/en/minilab/index.htm.

Global Public Health Fund. (n.d.). Retrieved July 20, 2021, from The GPHF-Minilab™ – Focusing on Prevalent Medicines against Infectious Diseases: https://www.gphf.org/en/minilab/wirkstoffe.htm.

Global Public Health Fund. (n.d.). *GPHF-Minilab™ – Fact Sheet.* Retrieved July 20, 2021, from Global Public Health Fund: https://www.gphf.org/en/minilab/factsheet.htm.

Gooty, A.R., Katreddi, H.R., Reddy S, R., Hunnur, R.K., Sharma, H.K., & Masani, N.K. (2016). Simultaneous Determination of Genotoxic Impurities in Fudosteine Drugs by GC-MS. *Journal of Chromatographic Science*, 1–5. doi:doi: 10.1093/chromsci/bmw070.

Green, M.D., Hostetler, D.M., Nettey, H., Swamidoss, I., Ranieri, N., & Newton, P.N. (2015, June 3). Integration of Novel Low-Cost Colorimetric, Laser Photometric, and Visual Fluorescent Techniques for Rapid Identification of Falsified Medicines in Resource-Poor Areas: Application to Artemether–Lumefantrine. *American Journal of Tropical Medicine and Hygiene, 92*(Suppl 6), 8–16. doi:10.4269/ajtmh.14-0832

Green, M., Mount, D., Wirtz, R., & White, N. (2000, December). A Colorimetric Field Method to Assess the Authenticity of Drugs Sold as the Antimalarial Artesunate. *Journal of Pharmaceutical and Biomedical Analysis, 24*(1), 65–70. doi:10.1016/s0731-7085(00)00360-5.

Green, R.L., Hargreaves, M.D., & Garnder, C.M. (2013). Performance Characterization of a Combined Material Identification and Screening Algorithm. *Proceedings Volume 8726, Next-Generation Spectroscopic Technologies VI. 87260F*, p. 87260F. Baltimore: Society of Photo-Optical Instrumentation Engineers (SPIE). doi:10.1117/12.2015953.

Greene, G. (Writer), & Reed, C. (Director). (1949). *The Third Man* [Motion Picture].

Griffiths, P.R., & de Haseth, J.A. (2007). *Fourier Transform Infrared Spectrophotometry* (2nd ed.). Hoboken: John Wiley & Sons, Inc.

Guillemain, A., Dégardin, K., & Roggo, Y. (2017, April 1). Performance of NIR Handheld Spectrometers for the Detection of Counterfeit Tablets. *Talanta, 165*, 632–640. doi:10.1016/j.talanta.2016.12.063.

Gwaziwa, N., Dzomba P., & Mupa, M. (2017, April 29). A Quality Control Study of Ibuprofen Tablets Available in the Formal and Informal Market in Harare, Zimbabwe. *African Journal of Pharmacy and Pharmacology, 11*(16), 195–203. doi:10.5897/AJPP2017.4746.

Hajjou, M., Qin, Y., Bradby, S., Bempong, D., & Lukulay, P. (2013, February 23). Assessment of the Performance of a Handheld Raman Device for Potential Use as a Screening Tool in Evaluating Medicines Quality. *Journal of Pharmaceutical and Biomedical Analysis, 74*, 47–55. doi:10.1016/j.jpba.2012.09.016.

Handzo, B., Luczak, A., Huffman, S., Peters, J., & Kalyanaraman, R. (2020, April 1–6). Benchtop and Portable Raman Spectrometers to Screen Counterfeit Drugs. *American Pharmaceutical Review, 23*(3).

Hargreaves, M. (2021). Handheld Raman, SERS, and SORS. In R.A. Crocombe, P.E. Leary, & B.W. Kammrath (Eds.), *Portable Spectroscopy and Spectrometry 2: Applications* (1st ed., Vol. 2, pp. 347–376). Chichester: John Wiley & Sons, Ltd.

Holt, T., Millroy, L., & Mmopi, M. (2017, May 12). *Winning in Nigeria: Pharma's Next Frontier*. Retrieved October 27, 2021, from McKinsey & Company: https://www.mckinsey.com/industries/life-sciences/our-insights/winning-in-nigeria-pharmas-next-frontier.

Holzbrabe, I., Wawer, I., & Diehl, B. (Eds.). (2008). *NMR Spectroscopy in Pharmaceutical Analysis*. Oxford: Elsevier.

Hu, C., Feng, Y., & Yin, L. (2015). Review of the Characteristics and Prospects of Near Infrared Spectroscopy for Rapid Drug-Screening Systems in China. *Journal of Near Infrared Spectroscopy, 23*(5), 272–283. doi:10.1255/jnirs.1154.

Igne, B., & Ciurczak, E.W. (2021). Pharmaceutical Analysis. In E.W. Ciurczak, B. Igne, J. Workman Jr., & D.A. Burns (Eds.), *Handbook of Near-Infrared Analysis* (4th ed., Chapter 40, pp. 817–828). Boca Raton: CRC Press. https://www.taylorfrancis.com/chapters/edit/10.1201/b22513-45/pharmaceutical-analysis-benoît-igne-emil-ciurczak

InnoSpectra Corporation. (n.d.). *InnoSpectra Corporation Standard Wavelength NIR Spectrometer*. Retrieved October 27, 2021, from InnoSpectra Corporation: http://www.inno-spectra.com/en/product.

International Council for Harmonisation of Technical Requirements for Pharmaceuticals for Human Use. (2017). *ICH Harmonised Guideline: Assessment and Control of DNA Reactive (Mutagenic) Impurities in Pharmaceuticals to Limit Potential Carcinogenic Risk M7(R1)*. Geneva: International Council for Harmonisation of Technical Requirements for Pharmaceuticals for Human Use.

International Council for Harmonisation of Technical Requirements for Pharmaceuticals for Human Use. (2020). *Final Concept Paper ICH Q3E: Guideline for Extractables and Leachables (E&L)*. Geneva: International Council for Harmonisation.

International Council for Harmonisation of Technical Requirements for Pharmaceuticals for Human Use. (2021). *ICH Harmonised Guideline – Impurities: Guideline for Residual Solvents Q3C(R8)*. Geneva: International Council for Harmonisation of Technical Requirements for Pharmaceuticals for Human Use.

International Council for Harmonisation of Technical Requirements for Pharmaceuticals for Human Use. (n.d.). *ICH Guidelines*. Retrieved October 26, 2021, from International Council for Harmonisation of Technical Requirements for Pharmaceuticals for Human Use: https://www.ich.org/page/ich-guidelines.

Jähnke, R.W., & Dwornik, K. (2020). *A Concise Quality Control Guide on Essential Drugs and Other Medicines Review and Extensions 2020: Manual Accompanying the GPHF-Minilab™, Physical Testing & Thin-Layer Chromatography*. Retrieved from Global Public Health Fund: https://www.gphf.org/images/downloads/Manual_2020_English_Demo.pdf.

Jenkins, R. (1999). *X-Ray Fluorescence Spectrometry, Volume 152* (2nd ed.). Chichester: John Wiley & Sons, Inc. doi:10.1002/9781118521014.

Kaale, E., Hope, S.M., Jenkins, D., & Layloff, T. (2016, January). Implementation of 350–2500 nm Diffuse Reflectance Spectroscopy and High-Performance Thin-Layer Chromatography to Rapidly Assess Manufacturing Consistency and Quality of Cotrimoxazole Tablets in Tanzania. *Tropical Medicine and International Health, 21*(1), 61–69. doi:10.1111/tmi.12621.

Kakio, T., Yoshida, N., Macha, S., Moriguchi, K., Hiroshima, T., Ikeda, Y.,... Kimura, K. (2017). Classification and Visualization of Physical and Chemical Properties of Falsified Medicines with Handheld Raman Spectroscopy and X-Ray Computed Tomography. *American Journal of Tropical Medicine and Hygiene, 97*(3), 684–689. doi:10.4269/ajtmh.16-0971.

Kalyanaraman, R., Dobler, G., & Ribick, M. (2010). Portable Spectrometers for Pharmaceutical Counterfeit Detection. *American Pharmaceutical Review, 13*, pp. 38–45.

Kalyanaraman, R., Dobler, G., & Ribick, M. (2011, May/June). Near-Infrared (NIR) Spectral Signature Development and Validation for Counterfeit Drug Detection Using Portable Spectrometer. *American Pharmaceutical Review*, pp. 98–104.

Keizers, P.H., Bakker, F., Ferreira, J., Wackers, P.F., van Kollenburg, D., Aa, v. d.,... van Beers, A. (2020, January 30). Benchtop NMR Spectroscopy in the Analysis of Substandard and Falsified Medicines as well as Illegal Drugs. *Journal of Pharmaceutical and Biomedical Analysis, 178*. doi:10.1016/j.jpba.2019.112939.

Khuluza, F., Kigera, S., Jähnke, R.W., & Heide, L. (2016). Use of Thin-Layer Chromatography to Detect Counterfeit Sulfadoxine/Pyrimethamine Tablets with the Wrong Active Ingredient in Malawi. *Malaria Journal, 15*(215), 1–7. doi:10.1186/s12936-016-1259-9

Kizzier, K.L., & Cassata, G. (2021). Field-Deployable Utility of Benchtop Nuclear Magnetic Resonance Spectrometers. In R.A. Crocombe, P.E. Leary, & B.W. Kammrath (Eds.), *Portable Spectroscopy and Spectrometry 1: Technologies and Instrumentation* (1st ed., Vol. 1, pp. 501–513). Chichester: John Wiley & Sons, Ltd.

Koesdjojo, M., Wu, Y., Boonloed, A., Dunfield, E., & Remcho, V. (2014, December). Low-Cost, High-Speed Ddentification of Counterfeit Antimalarial Drugs on Paper. *Talanta, 30*, 122–127. doi:10.1016/j.talanta.2014.05.050.

Kovacs, S., Hawes, S.E., Maley, S.N., Mosites, E., Wong, L., & Stergachis. (2014). Technologies for Detecting Falsified and Substandard Drugs in Low and Middle-Income Countries. *PLOS ONE, 9*(3), 1–11. doi:10.1371/journal.pone.0090601.

Kumar, A., Zhang, K., & Wigman, L. (2015). Analytical Technologies for Genotoxic Impurities in Pharmaceutical Compounds. *LCGC North America, 33*(5), pp. 344–359. Retrieved from https://www.chromatographyonline.com/view/analytical-technologies-genotoxic-impurities-pharmaceutical-compounds.

Lanzarotta, A., Kern, S., Batson, J., Falconer, T., Fulcher, M., Gaston, K.,... Witkowski, M. (2021, September 5). Evaluation of "Toolkit" Consisting of Handheld and Portable Analytical Devices for Detecting Active Pharmaceutical Ingredients in Drug Products Collected During a Simultaneous Nation-Wide Mail Blitz. *Journal of Pharmaceutical and Biomedical Analysis, 203*. doi:10.1016/j.jpba.2021.114183.

Lanzarotta, A., Kimani, M.M., Thatcher, M.D., Lynch, J., Fulcher, M., & Witkowski, M.R. (2020, July). Evaluation of Suspected Counterfeit Pharmaceutical Tablets Declared to Contain Controlled Substances Using Handhedl Raman Spectrometers. *Journal of Forensic Sciences, 65*(4), 1274–1279. doi:10.1111/1556-4029.14287.

Lanzarotta, A., Lorenz, L., Batson, J.S., & Flurer, C. (2017, 30). Development and Implementation of a Pass/Fail Field-Friendly Method for Detecting Sildenafil in Suspect Pharmaceutical Tablets Using a Handheld Raman Spectrometer and Silver Colloids. *Journal of Pharmaceutical and Biomedical Analysis, 146*, 420–425. doi:10.1016/j.jpba.2017.09.005.

Lanzarotta, A., Ranieri, N., Albright, D.C., Witkowski, M.R., Batson, J.S., & Fulcher, M. (2015, May 20). Analysis of Counterfeit FDA-Regulated Products at the Forensic Chemistry Center: Rapid Visual and Chemical Screening Procedures Inside and Outside of the Laboratory. *American Pharmaceutical Review*, pp. 24–29.

Lawson, G., Ogwu, J., & Tanna, S. (2018, August 10). Quantitative Screening of the Pharmaceutical Ingredient for the Rapid Identification of Substandard and Falsified Medicines Using Reflectance Spectroscopy. *PLoS One, 13*(8), 1–17. doi:10.1371/journal.pone.0202059.

Leary, P.E. (2014). Counterfeiting: A Challenge to Forensic Science, the Criminal Justice System, and Its Impact on Pharmaceutical Innovation. New York: Ph.D. Dissertation.

Leary, P.E., & Joshi, M. (2021). Applications of Ion Mobility Spectrometry. In R.A. Crocombe, P.E. Leary, & B.W. Kammrath (Eds.), *Portable Spectroscopy and Spectrometry 2: Applications* (1st ed., Vol. 2, pp. 159–178). Chichester: John Wiley & Sons Ltd.

Leary, P.E., Crocombe, R.A., & Kammrath, B.W. (2021). Introduction to Portable Spectroscopy. In R.A. Crocombe, P.E. Leary, & B.W. Kammrath (Eds.), *Portable Spectroscopy and Spectrometery 1: Technology and Instrumentation* (1st ed., Vol. 1, pp. 1–13). Chichester: John Wiley & Sons, Ltd.

Leary, P.E., Davis, S.M., Sanchez-Melo, M.I., Rynearson, L.R., Vallee, L., Langlois, E.,... Kammrath, B.W. (2020). On Scene Detection of Low-Dose Fentanyl Tablets. *Annual Meeting of the American Academy of Forensic Sciences*. Anaheim.

Leary, P.E., Kammrath, B.W., & Reffner, J.A. (2018). Field-portable Gas Chromatography-Mass Spectrometry. In R. A. Meyers (Ed.), *Encyclopedia of Analytical Chemistry: Applications, Theory and Instrumentation* (pp. 1–13). Bognor Regis: John Wiley & Sons, Ltd. doi:10.1002/9780470027318.a9583.

Leary, P.E., Kammrath, B.W., & Reffner, J.A. (2021). Portable Gas Chromatography-Mass Spectrometry: Instrumentation and Applications. In R.A. Crocombe, P.E. Leary, & B.W. Kammrath (Eds.), *Portable Spectroscopy and Spectrometery 1: Technologies and Instrumentation* (1st ed., Vol. 1, pp. 367–389). Chichester: John WIley & Sons, Ltd.

Leary, P.E., Kammrath, B.W., Lattman, K.J., & Beals, G.L. (2019). Deploying Portable Gas Chromatography-Mass Spectrometry (GC-MS) to Military Users for the Identification of Toxic Chemical Agents in Theater. *Applied Spectroscopy, 73*(8), 841–858. doi:10.1177/0003702819849499.

Liu, L., Oza, S., Hogan, D., Perin, J., Rudan, I., Lawn, J. E.,... Black, R.E. (2015). Global, Regional, and National Causes of Child Mortality in 2000–2013, With Projections to Inform Post-2015 Priorities: An Updated Systematic Analysis. *Lancet, 385*, 430–440. doi:http://dx.doi.org/10.1016/S0140-6736(14)61698-6.

Luczak, A., & Kalyanaraman, R. (2014, September/October 27). Portable and Benchtop Raman Technologies for Product Authentication and Counterfeit Detection. *American Pharmaceutical Review, 17*(6), pp. 1–5.

Luczak, A., & Kalyanaraman, R. (2015, March/April). Analysing the Analytical Techniques for Counterfeit Drug Detection. *BioPharma Asia*, pp. 24–29.

Lutz, B., Trinh, P., Ball, C., Fu, E., & Yagera, P. (2011, December 21). Two-Dimensional Paper Networks: Programmable Fluidic Disconnects for Multi-Step Processes in Shaped Paper. *Lab on a Chip, 11*(24), 4274–4278. doi:10.1039/c1lc20758j.

Maggio, R.M., Calvo, N.L., Vignaduzzo, S., & Kaufman, T.S. (2014, December). Pharmaceutical Impurities and Degradation Products: Uses and Applications of NMR Techniques. *Journal of Pharmaceutical and Biomedical Analysis, 101*, 102–122. doi:10.1016/j.jpba.2014.04.016.

Markl, D., & Zeitler, J.A. (2017). A Review of Disintegration Mechanisms and Measurement Techniques. *Pharmaceutical Research, 34*(5), 890–917. doi:10.1007/s11095-017-2129–z.

Martinez, A., Phillips, S., & Whitesides, G. (2008). Three-Dimensional Microfluidic Devices Fabricated in Layered Paper and Tape., *105*, pp. 19606–19611. doi:10.1073/pnas.0810903105.

Martinez, A., Phillips, S., & Whitesides, G. (2010, January 1). Diagnostics for the Developing World: Microfluidic Paper-Based Analytical Devices. *Analytical Chemistry, 82*(1), 3–10.

Martinez, A., Phillips, S., Nie, Z., Cheng, C., Carrilho, E., Wileya, B., & Whitesides, G. (2010, October 7). Programmable Diagnostic Devices Made from Paper and Tape. *Lab on a Chip,* (10), 2499–2504. doi:10.1039/C0LC00021C.

Martino, R., Malet-Martino, M., Gilard, V., & Balayssac, S. (2010). Counterfeit Drugs: Analytical Techniques for Their Identification. *Analytical and Bioanalytical Chemistry, 398*(1), 77–92. doi:10.1007/s00216-010-3748-y.

Matousek, P., Clark, I., Draper, E., Morris, M., Goodship, A., Everall, N.,... Parker, A. (2005, April). Subsurface Probing in Diffusely Scattering Media Using Spatially Offset Raman Spectroscopy. *Applied Spectroscopy, 59*(4), 393–400. doi:10.1366/0003702053641450.

McLafferty, F., & Turecek, F. (1993). *Interpretation of Mass Spectra,* (4th ed.). Sausalito: University Science Books.

McVey, C., Gordon, U., Haughey, S.A., & Elliott, C.T. (n.d.). Assessment of the Analytical Performance of Three Near-Infrared Spectroscopy Instruments (Benchtop, Handheld and Portable) through the Investigation of Coriander Seed Authenticity. *Foods, 10*(5), 956. doi:10.3390/foods10050956.

Moffatt, A.C., Assi, S., & Watt, R.A. (2010). Identifying Counterfeit Medicines Using Near Infrared Spectroscopy. *Journal of Near Infrared Spectroscopy, 18*(1), 1–15. doi:10.1255/jnirs.856.

Mukhopadhyay, R. (2007, April 1). The Hunt for Counterfeit Medicine. *Analytical Chemistry, 79*(7), 2623–2637. doi:10.1021/ac071892p.

Nair, A., Strauch, S., Lauwo, J., Jähnke, R., & Dressman, J. (2011, November). Are Counterfeit or Substandard Anti-Infective Products the Cause of Treatment Failure in Papua New Guinea? *Journal of Pharmaceutical Sciences, 100*(11), 5059–5068. doi:10.1002/jps.22691.

Newton, P.N., & Timmermann, B. (2016). Fake Penicillin, The Third Man, and Operation Claptrap. *BMJ, 355*, i6494. doi:10.1136/bmj.i6494.

Newton, P.N., McGready, R., Fernandez, F., Green, M.D., Sunjio, M., Bruneton, C., Phanouvong, S., Millet, P., Whitty, C.J.M., Talisuna, A.O., Proux, S., Christophel, E.M., Malenga, G., Singhasivanon, P., Bojang, K., Kaur, H., Palmer, K., Day, N.P.J., Greenwood, B.M., Nosten, F., White, N.J. (2006, June). Manslaughter by Fake Artesunate in Asia – Will Africa Be Next? *Public LIbrary of Service Medicine, 3*(6). doi:10.1371/journal.pmed.0030197.

Nynomic. (2021, June 21). *Corporate News/Ad Hoc/Director Dealings – Nynomic AG: Strategic Technology Partnership with Novartis AG for Fighting Against Counterfeit Tablets.* Retrieved June 27, 2021, from Nynomic: https://www.nynomic.com/en/corporate-news-ad-hoc-directors-dealings/.

O'Donnell, R.M., Sun, X., & de B. Harrington, P. (2008, January). Pharmaceutical Applications of Ion Mobility Spectrometry. *Trends in Analytical Chemistry, 27*(1), 44–53. doi:10.1016/j.trac.2007.10.014.

Olds, W.J., Jaatinen, E., Fredericks, P., Cletus, B., Panayiotou, H., & Izake, E.L. (2011, October 10). Spatially Offset Raman Spectroscopy (SORS) for the Analysis and Detection of Packaged Pharmaceuticals and Concealed Drugs. *Forensic Science International, 212*(1–3), 69–77. doi:10.1016/j.forsciint.2011.05.016.

Olsen, B.A., Borer, M.W., Perry, F.M., & Forbes, R.A. (2002, June). Screening for Counterfeit Drugs Using Near-Infrared Spectroscopy. *Pharmaceutical Technology, 26*(6), p. 62.

O'Neill, A.J., Lee, R.D., Lee, G., Charvill, A., & Moffat, A.C. (2008). Use of a Portable Near Infrared Spectrometer for the Authentication of Tablets and the Detection of Counterfeit Versions. *Journal of Near Infrared Spectroscopy, 16*(3), 327–333. doi:10.1255/jnirs.796.

Osborn, J., Lutz, B., Fu, E., Kauffman, P., Stevens, D., & Yager, P. (2010, October 21). Microfluidics Without Pumps: Reinventing the T-Sensor and H-Filter in Paper Networks. *Lab on a Chip, 10*(20), 2659–2665. doi:10.1039/c004821f.

Owen, H. (2006). Volume Phase Holographic Optical Elements. In J. Chalmers, & P. Griffiths (Eds.), *Handbook of Vibrational Spectroscopy: Theory and Instrumentation* (Vol. 1). Chichester: John Wiley & Sons, Ltd.

Patil, D., Pandit, V., Pore, S., & Chavan, C. (2012). Fighting Counterfeit and Substandard Drugs at Periphery: The Utility of Basic Quality Control Tests. *Pharmacie Globale, 3*(3), 1–3.

Peters, J., Luczak, A., Ganesh, V., Park, E., & Kalyanaraman, R. (2016, March). Raman Spectral Fingerprinting for Biologics Counterfeit Drug Detection. *American Pharmaceutical Review, 19*(2), pp. 1–6.

Peveler, W.J., & Algar, W.R. (2021). Toward Clinical Applications of Smartphone Spectroscopy and Imaging. In R. A. Crocombe, P. E. Leary, & B.W. Kammrath (Eds.), *Portable Spectroscopy and Spectrometry 2: Applications* (1st ed., Vol. 2, pp. 199–226). Chichester: John Wiley & Sons, Ltd.

Pfizer. (2013, May 5). *Facing Off Against Counterfeit Online Pharmacies: Pfizer Launches New Purchasing Website to Alleviate the Guesswork Around Buying Legitimate Viagra® (Sildenafil Citrate) Online*. Retrieved July 7, 2021, from Pfizer: https://www.pfizer.com/news/press-release/press-release-detail/facing_off_against_counterfeit_online_pharmacies_pfizer_launches_new_purchasing_website_to_help_alleviate_the_guesswork_around_buying_legitimate_viagra_sildenafil_citrate_online.

Piorek, S. (2021). Handheld X-Ray Fluorescence (HHXRF). In R.A. Crocombe, P.E. Leary, & B.W. Kammrath (Eds.), *Portable Spectroscopy and Spectrometery 2: Applications* (1st ed., Vol. 2, pp. 423–455). Chichester: John WIley & Sons, Ltd.

Platek, S.F., Ranieri, N., & Batson, J. (2016). Applications of the FDA's Counterfeit Detection Device (CD3+) to the Examination of Suspect Counterfeit Pharmaceutical Tablets and Packaging. *Microscopy and Microanalysis, 22*(Suppl 3). doi:10.1017/S1431927616006206.

Potts, P.J., & West, M. (Eds.). (2008). *Portable X-ray Fluorescence Spectrometry Capabilities for In Situ Analysis*. London: Royal Society of Chemistry.

Ranieri, N., Tabarnero, P., Green, M.D., Verbois, L., Herrington, J., Sampson, E.,... Mitkowski, M. R. (2014, November 5). Evaluation of a New Handheld Instrument for the Detection of Counterfeit Artesunate by Visual Fluorescence Comparison. *American Journal of Tropical Medicine and Hygiene, 91*(5), 920–924. doi:10.4269/ajtmh.13-0644.

Rather, D. (2010, September 14). The Mysterious Case of Kevin Xu. *Dan Rather Reports*. AXS TV.

Rathmell, C., Bingemann, D., Zieg, M., & Creasey, D. (2021). Portable Raman Spectroscopy: Instrumentation and Technology. In R.A. Crocombe, P.E. Leary, & B.W. Kammrath (Eds.), *Portable Spectroscopy and Spectrometry 1: Technologies and Instrumentation* (Vol. 1, pp. 115–145). Chichester: John Wiley & Sons, Ltd.

Reddy, S.R., Reddy, K.H., Kumar, M.N., Reddy, P.M., Reddy, J.V., & Sharma, H.K. (2019). A Validated GC-MS Method for the Determination of Genotoxic Impurities in Divalproex Sodium Drug Substance. *Journal of Chromatographic Science, 57*(2), 101–107. doi:10.1093/chromsci/bmy089.

Riegler, T. (2020). The Spy Story Behind The Third Man. *Journal of Austrian-American History, 4*, 1–37.

Risha, P.G., Msuya, Z., Clark, M., Johnson, K., Ndomondo-Sigonda, M., & Layloff, T. (2008). The Use of Minimlabs® to Improve the Testing Capacity of Regulatory Authorities in Resource Limited Settings: Tanzanian Experience. *Health Policy, 87*, 217–222. doi:10.1016/j.healthpol.2007.12.010.

Rodionova, O.Y., Houmøller, L.P., Pomerantseva, A.L., Geladi, P., Burger, J., Dorofeyev, V.L., & Arzamastsev, A.P. (2005, September). NIR Spectrometry for Counterfeit Drug Detection: A Feasibility Study. *Analytical Chimica Acta, 549*(1–2), 151–158. doi:10.1016/j.aca.2005.06.018.

Rodionova, O., Titova, A., Balyklova, K., & Pomerantsev, A. (2019, December 2019). Detection of Counterfeit and Substandard Tablets Using Non-Invasive NIR and Chemometrics – A Conceptual Framework for a Big Screening System. *Talanta, 205*. doi:10.1016/j.talanta.2019.120150.

Rodionova, O., Titova, A., Demkin, N., Balyklova, K., & Pomerantsev, A. (2019, April 1). Qualitative and Quantitative Analysis of Counterfeit Fluconazole Capsules: A Non-Invasive Approach Using NIR Spectroscopy and Chemometrics. *Talanta, 195*, 662–667. doi:10.1016/j.talanta.2018.11.088

Roth, L., Biggs, K.B., & Bempong, D.K. (2019). Substandard and Falsified Medicine Screening Technologies. *AAPS Open, 5*(2), 1–12. doi:10.1186/s41120-019-0031-y.

Roth, L., Nalim, A., Turesson, B., & Krech, L. (2018). Global Landscape Assessment of Screening Technologies for Medicine Quality Assurance: Stakeholder Perceptions and Practices from Ten Countries. *Globalization and Health, 14*(43), 1–19. doi:10.1186/s12992-018-0360-y.

Rotinwa, A. (2018, July 10). *How Nigeria Partners with Tech Companies to Outwit Drug Counterfeiters*. Retrieved October 27, 2021, from Devex: https://www.devex.com/news/how-nigeria-partners-with-tech-companies-to-outwit-drug-counterfeiters-92953.

Rynearson, L.R., Lawton, Z., McMahon, M., Leary, P.E., Kizzire, K., & Kammrath, B.W. (2020). An Analysis of Illicit Drugs by Portable Ion-Trap Gas Chromatography–Mass Spectrometry (GC-MS). *Pittsburgh Conference*. Chicago.

Sacré, P.-Y., Doconinck, E., Daszykowski, M., Courselle, P., Vancauwenberghe, R., Chiap, P., & De Beer, J. O. (2011, September 9). Impurity Fingerprints for the Identificaiton of Counterfeit Medicines – A Feasibility Study. *Analytica Chimica Acta, 70*(2), 224–331. doi:10.1016/j.aca.2011.05.041.

Sanada, T., Yoshida, N., Kimura, K., & Tsuboi, H. (2021). Discrimination of Falsified Erectile Dysfunction Medicines by Use of an Ultra-Compact Raman Scattering Spectrometer. *Pharmacy, 9*(1), 1–16. doi:10.3390/pharmacy9010003.

Scafi, S., & Pasquini, C. (2001, December). Identification of Counterfeit Drugs Using Near-Infrared Spectroscopy. *Analyst, 126*(12), 2218–2224. doi:10.1039/b106744n.

Scheeline, A. (2021). Smartphone Technology – Instrumentation and Applications. In R.A. Crocombe, P.E. Leary, & B.W. Kammrath (Eds.), *Portable Spectroscopy and Spectrometry 1: Technologies and Instrumentation* (1st ed., Vol. 1, pp. 209–235). Chichester: John Wiley & Sons, Ltd.

Schiering, D., & Stein, J. (2021). Design Considerations for Portable Mid-Infrared FTIR Spectrometers Used for In-Field Identification of Threat Materials. In R.A. Crocombe, P.E. Leary, & B.W. Kammrath (Eds.), *Portable Spectroscopy and Spectrometery 1: Technologies and Instrumentation* (1st ed., Vol. 1, pp. 41–65). Chichester: John Wiley & Sons, Ltd.

Schreyer, S.K. (2021). Library and Method Development for Portable Instrumentation. In R.A. Crocombe, P.E. Leary, & B.W. Kammrath (Eds.), *Portable Spectroscopy and Spectrometry 2: Applications* (Vol. 2, pp. 43–63). Chichester: John WIley & Sons, Ltd.

Scientific Working Group for the Analysis of Seized Drugs. (2019). *Recommendations of the Scientific Working Group for the Analysis of Seized Drugs, Version 8.0.* Washington, D.C.: United Stateed Department of Justice Drug Enforcement Administration; Executive Office of the President Office of National Drug Control Policy Counterdrug Technology Assessmeent Center.

Sherma, J. (2007). Analysis of counterfeit drugs by thin layer chromatography. *Acta Chromatographica, 19*, 5–20.

Sherma, J., & Rabel, F. (2019). Advances in the thin layer chromatographic analysis of pharmaceutical products: 2008–2019. *Journal of Liquid Chromatography & Related Technologies*, 1–13. doi:10.1080/10826076.2019.1610772.

Shin, M., Jee, R., Chae, K., & Moffat, A. (2005, September). Identification of Counterfeit Cialis, Levitra and Viagra Tablets by Near-Infrared Spectroscopy in Short Talks on Pharmaceuticals. *Journal of Pharmaceutical and Pharmacology, 57*(S1), pp. S-11. doi:10.1211/002235705778248398.

Simpson, M.B. (2010). Near-Infrared Spectroscopy for Process Analytical Technology: Theory, Technology and Implementation. In K. A. Bakeevk (Ed.), *Process Analytical Technology: Spectroscopic Tools and Implementation Strategies for the Chemical and Pharmaceutical Industries*. Chichester: John Wiley & Son, Inc. doi:10.1002/9780470689592.ch5.

Singh, B., Parwate, D., & Shukla, S. (2009). Rapid Color Test Identification System for Screening of Counterfeit Fluoroquinolone. *E-Journal of Chemistry, 6*(2), 377–384.

Sroka, A., Ishizaki, K., & Barańczuk, Z. (2020). Detecting Falsified Viagra Using Miniaturized Consumer Near-Infrared Spectroscopy. *Pharmaceutial Regulatory Affairs, 9*(1), pp. 1–3. doi:10.37421/Pharmaceut Reg Affairs.2020.9.226.

Swets, J.A. (1988, June 3). Measuring the Accuracy of Diagnostic Systems. *Science, 240*(4857), 1285–1293. doi:10.1126/science.3287615.

Teasdale, A., & Elder, D.P. (2018). Analytical Control Strategies for Mutagenic Impurities: Current Challenges and Future Opportunities? *Trends in Analytical Chemistry, 101*, 66–84. doi:https://doi.org/10.1016/j.trac.2017.10.027.

U.S. Attorney's Office, District of Idaho. (2021, March 30). Illicitly Manufactured PIlls Sold as Oxycodone Can Lead To Overdose and Death. *Acting U.S. Attorney Addresses Increasing Danger of Counterfeit Prescription Opioids [Press Release]*. Boise, Idaho, United States of America. Retrieved July 9, 2021, from https://www.justice.gov/usao-id/pr/acting-us-attorney-addresses-increasing-danger-counterfeit-prescription-opioids-0.

U.S. Food & Drug Admininstration. (2019, May 22). *Abbreviated New Drug Application (ANDA)*. Retrieved October 25, 2021, from U.S. Food & Drug Administration: https://www.fda.gov/drugs/types-applications/abbreviated-new-drug-application-anda.

U.S. Food & Drug Admininstration. (2019, June 10). *New Drug Application (NDA)*. Retrieved October 25, 2021, from U.S Food & Drug Administration: https://www.fda.gov/drugs/types-applications/new-drug-application-nda.

U.S. Food & Drug Administration. (2014, August 26). *Validation of Cleaning Processes (7/93)*. Retrieved October 22, 2021, from U.S. Food & Drug Administration: https://www.fda.gov/validation-cleaning-processes-793.

U.S. Food & Drug Administration. (2018, June 1). *Generic Drug Facts*. Retrieved July 8, 2021, from U.S. Food & Drug Administration: https://www.fda.gov/drugs/generic-drugs/generic-drug-facts.

U.S. Food & Drug Administration. (2021, January 5). *Counterfeit Medicine*. Retrieved July 9, 2021, from U.S. Food & Drug Administration: https://www.fda.gov/drugs/buying-using-medicine-safely/counterfeit-medicine.

U.S. Food & Drug Administration. (n.d.). *A Hand Held Portable Device Based on LEDs for Use in the Detection of Counterfeit Pharmaceutical Drugs and Packaging*. Retrieved October 27, 2021, from U.S. Food & Drug Admininstration: https://www.fda.gov/media/130322/download.

Ulmschneider, M., & Pénigault, E. (2000). Non-Invasive Confirmation of the Identity of Tablets by Near-Infrared Spectroscopy. *Analusis, 28*, 336–346. doi:10.1051/analusis:2000124.

United States Customs and Border Protection. (2020, December 17). *Counterfeit Viagra Pills, Footwear, Belts, Car Emblems and Headphones Worth Over $32 Million Seized at the LA/Long Beach Seaport [Press Release]*. Los Angeles, California, United States of America. Retrieved July 9, 2021, from https://www.cbp.gov/newsroom/local-media-release/counterfeit-viagra-pills-footwear-belts-car-emblems-and-headphones.

United States Drug Enforcement Administration. (n.d.). *Image Gallery*. Retrieved July 23, 2021, from United States Drug Enforcement Administration: https://www.dea.gov/galleries/drug-images/fentanyl.

United States Drug Enforcement Administration Strategic Intelligence Section. (2019). *2019 National Drug Threat Assessment*. U.S. Department of Justice, Drug Enforcement Administration. Retrieved March 8, 2020, from https://www.dea.gov/sites/default/files/2020-01/2019-NDTA-final-01-14-2020_Low_Web-DIR-007-20_2019.pdf.

United States Drug Enforcement Administration. (n.d.). *United States Drug Enforcement Administration*. Retrieved July 24, 2021, from Facts about Fentanyl: https://www.dea.gov/resources/facts-about-fentanyl.

United States Pharmacopeia. (n.d.). USP. *General Chapter Nuclear Magnetic Resonance Spectrosocpy*.

United States Pharmacopeia. (n.d.). USP. *General Chapter Applications of Nuclear Magnetic Resonance Spectroscopy*.

Vickers, S., Bernier, M., Zambrzycki, S., Fernandez, F.M., Newton, P.N., & Caillet, C. (2018). Field Detection Devices for Screening the Quality of Medicines: A Systematic Review. *BMJ Global Health, 3*(e000725), 1–16. doi:10.1136/.

Voelker, S.E., Kern, S.E., Falconer, T.M., Thatcher, M.D., Skelton, D.M., Gaston, K.W., & Litzau, J.J. (2021, July 15). Evaluation of Four Field Portable Devices for the Rapid Detection of Mitragynine in Suspected Kratom Products. *Journal of Pharmaceutical and Biomedical Analysis, 201*. doi:10.1016/j.jpba.2021.114104.

Vogt, F. (2010, June 14). Evolution of Solid-State NMR in Pharmaceutical Analysis. *Future Medicinal Chemistry, 2*(6), pp. 915–921. doi:10.4155/fmc.10.200.

Wang, W., Keller, M.D., Baughman, T., & Wilson, B.K. (2020). Evaluating Low-Cost Optical Spectrometers for the Detection of Simulated Substandard and Falsified Medicines. *Applied Spectroscopy, 74*(3), 323–333. doi:10.1177/0003702819877422.

Weaver, A.A., Reiser, H., Barstis, T., Benvenuti, M., Ghosh, D., Hunckler, M., Lieberman, M. (2013). Paper Analytical Devices for Fast Field Screening of Beta Lactam Antibiotics and Antituberculosis Pharmaceuticals. *Analytical Chemistry, 85*(13), 6453–6460. doi:10.1021/ac400989p.

Wilks, J.P. (1972). A Practical Approach to Internal Reflection Spectroscopy. In R. Miller, & B. Stace (Eds.), *Laboratory Methods in Infrared Spectrscopy*. London: Heyden.

Williard, H., Merritt Jr., L., Dean, J., & Settle Jr., F. (1988). *Instrumental Methods of Analysis* (2nd ed.). Belmont: Wadsworth Publishing Company.

Wilson, B., Kaur, H., Allan, E.L., Lozama, A., & Bell, D. (2017, May). A New Handheld Device for the Detection of Falsified Medicines: Demonstration on Falsified Artemisinin-Based Therapies from the Field. *American Journal of Tropical Medicine and Hygiene, 96*(5), 1117–1123. doi:10.4269/ajtmh.16-0904.

World Health Organization. (2017). *A Study on the Public Health and Socioeconomic Impact of Substandard and Falsified Medical Products*. Geneva: World Health Organization. Retrieved from https://www.who.int/publications/i/item/study-on-public-health-socioeconomic-impact-substandard-falsified-medical-products-978-92-4-151343-2.

World Health Organization. (2018, January 31). *Substandard and Falsified Medical Products*. Retrieved July 9, 2019, from World Health Organization: https://www.who.int/news-room/fact-sheets/detail/substandard-and-falsified-medical-products.

Yaghoobi, M., Wu, D., Clewes, R.J., & Davies, M.E. (2016). *Fast Sparse Raman Spectral Unmixing for Chemical Fingerprinting and Quantification* (Vol. 9995). Edinburgh: Society of Photo-Optical Instrumentation Engineers (SPIE). doi:10.1117/12.2241834.

Yoon, W. (2005, January). Near-Infrared Spectroscopy: A Novel Tool to Detect Pharmaceutical Counterfeits. *American Pharmaceutical Review, 8*(5), pp. 115–118.

Zhang, H., Hua, D., Huang, C., Samal, S.K., Xiong, R., Sauvage, F., & De Smedt, S.C. (2020, February 3). Materials and Technologies to Combat Counterfeiting of Pharmaceuticals: Current and Future Problem Tackling. *Advanced Materials, 32*(11). doi:10.1002/adma.201905486.

Zhang, L., Lee, L.M., & Schreyer, S. (2021). Identification and Confirmation Algorithms for Handheld Analyzers. In R.A. Crocombe, P. E. Leary, & B. W. Kammrath (Eds.), *Portable Spectroscopy and Spectrometry 2: Applications* (1st ed., Vol. 2, pp. 19–42). Chichester: John Wiley & Sons, Ltd.

7 Trends in Counterfeit Drugs and Toxicology

Kelly M. Elkins
Towson University

CONTENTS

7.1 INTRODUCTION

Consumers may unknowingly purchase counterfeit drug products to fill their prescriptions, especially if they purchase pharmaceuticals online. In 2008, the World Health Organization (WHO) reported that 10% of the products in the global pharmaceutical trade are counterfeits, with the highest concentration in developing countries (Wertheimer and Norris, 2009; Dégardin et al., 2014). Counterfeit pharmaceuticals can cause hospitalization and death.

Counterfeit drug's harm to individuals can be determined by close inspection of the drugs that the patient was taking in conjunction with the review of the toxicology report. Counterfeit drug products may be similar or nearly indistinguishable from the authentic product. For example, in a high-profile case in 2018, the entertainer Prince was determined to have died of a fentanyl overdose after taking a counterfeit Vicodin painkiller pill that was laced with fentanyl (Silva, 2018). Only individuals with a trained eye or sophisticated instrumentation may be able to detect the counterfeit drug. Counterfeit pills and tablets can differ from the authentic medicine in terms of color, size, shape, packaging, ink used to print markings and packaging, font used for markings and packaging, and lack of barcoding or authentication tags.

Also of serious concern is that counterfeit products with insufficient active pharmaceutical ingredient(s) or an alternate ingredient may not suppress disease or its progression, while a product with a larger than indicated quantity of an active pharmaceutical ingredient (API) or substituted opioid can lead to toxic effects (Bolla et al., 2020).

DOI 10.1201/9781003183327-7

As noted in the other chapters, physical product is essential to enable testing to determine if the product is authentic or counterfeit. Connecting counterfeit drugs to toxicology cases is difficult. Oftentimes, law enforcement does not find the vial containing pills or tablets, and the packaging has been discarded so that there is no substance or material to compare. Law enforcement also may not be able to identify subtle counterfeits, and agencies may not submit the evidence for further testing by a state or federal lab. In some cases, however, unused suspect drug substance or product can be compared to the toxicology report and detect that a counterfeit product caused harm or even death. Coroners and medical examiners occasionally can identify that counterfeits were the cause of death in cases in which the toxicology results for the deceased do not match their medication list or pills in their possession, or indicate that some other substance(s) were detected in the samples' body fluids. The focus of this chapter will be on those toxicology case reports from the primary literature, although there are dozens more in the news media. Cases have been reported in Argentina, Australia, Canada, China, Europe, Japan, Myanmar, Niger, Nigeria, Singapore, and the United States.

Instrumental methods including thin-layer chromatography (TLC), gas chromatography–mass spectrometry (GC–MS), high-performance liquid chromatography (HPLC), liquid chromatography–mass spectrometry (LC–MS), ultraviolet–visible spectroscopy (UV–Vis), Raman spectroscopy, infrared (IR) spectroscopy, near-infrared (NIR) spectroscopy, and nuclear magnetic resonance spectroscopy (NMR) can be used to detect counterfeit drugs in toxicology cases (Bolla et al., 2020). The application of these tools is demonstrated in the cases described in this chapter.

7.2 COUNTERFEIT PHARMACEUTICALS—FAKE DRUGS

Cases of counterfeit medicines causing hospitalization and death were reported in Niger, Argentina, Myanmar, Singapore, and China from 1995 to 2009 (Dégardin et al., 2014). The counterfeit products implicated include those to treat diabetes, malaria, impotence, and anemia as well as vaccines used to prevent infection (Dégardin et al., 2014).

Nigerian consumers became acutely aware of the threat of counterfeit pharmaceuticals in 1990 when paracetamol (Tylenol)-containing counterfeited raw ingredients led to acute renal failure in 102 children in the "killer syrup" case. Specifically, in Nigeria's Plateau state, 24 of 26 children that received the counterfeit paracetamol died. Laboratory analysis showed that the starting material propylene glycol used to make the Tylenol contained the toxic chemical ethylene glycol (Alubo, 1994).

In 1995, Niger experienced a meningitis epidemic. To thwart the epidemic, people were vaccinated. However, 50,000 people were treated with fake vaccine. The event led to a reported 2,000 deaths (Wertheimer and Norris, 2009).

HIV/AIDS drugs are prone to counterfeiting. In another study in Nigeria, 12.8% of patients that took counterfeit drugs resulted in adverse reactions and fatalities. Other noted complications that arose from the fake medicines included resistance to drug therapy (52.9%), increased severity (42.2%) and complications from the counterfeit drugs (34.2%) (Wertheimer and Norris, 2009).

Illicit fentanyl and its analogs have been detected in counterfeit oxycodone, hydrocodone, and alprazolam tablets (Pichini et al., 2018). In a U.S. case reported by Martucci et al. (2018), a 23-year-old consumed blue pills imitating oxycodone. Toxicology screening did not detect the presence of oxycodone (Martucci et al., 2018). An immunoassay detected the presence of the fentanyl analog furanyl fentanyl (Martucci et al., 2018). The concentration of furanyl fentanyl in postmortem samples was recorded as follows: "1.9 ng/mL in peripheral blood, 2.8 ng/mL in cardiac blood, and ~55,000 ng in gastric contents" (Martucci et al., 2018). The metabolite (and precursor) 4-anilino-*N*-phenethyl-piperidine (4-ANPP) was detected and confirmed by the test to be present at levels 4.3 ng/mL in peripheral blood, 5.8 ng/mL in cardiac blood, and >40 ng/g in the liver (Martucci et al., 2018). Trace amounts of furanyl fentanyl and 4-ANPP were detected in the urine and vitreous humor samples (Martucci et al., 2018).

In a U.S. case, GC–MS, liquid chromatography quadrupole time-of-flight mass spectrometry, and liquid chromatography tandem mass spectrometry were used to analyze blood and urine samples collected from a young adult who reported taking crushed Xanax (alprazolam). GC–MS testing of the urine only detected caffeine. The LC methods determined the presence of U-47700 and its metabolites in the blood and urine. Naloxone reversed his symptoms, which included altered mental status and respiratory depression (Chapman et al., 2021).

In a study of several hundreds of cases, fentanyl analogs were identified in the blood/urine of fentanyl-analog-associated deaths with drugs from several common classes of pharmaceuticals (Rauf et al., 2021). Fentanyl analogs were detected in 2.22% ($n=24$) of cases with antidepressant/antipsychotic drugs (clozapine, sertraline, mirtazapine, amitriptyline, olanzapine, trazodone, aripiprazole, duloxetine, nortriptyline, paroxetine, quetiapine), 2.13% ($n=23$) of cases with antiepileptic drugs (gabapentin, pregabalin, topiramate, phenytoin), 0.28% ($n=3$) of cases with antihypertensive drugs (clonidine, lisinopril, loperamide), 9.92% ($n=107$) of cases with benzodiazepine drugs (alprazolam, clonazepam, diazepam, nitrazepam, oxazepam, temazepam), 10.47% ($n = 113$) of cases with cannabinoid drugs, 38.00% ($n=410$) of cases with stimulant drugs (cocaine, methamphetamine, and amphetamine), 0.65% ($n=7$) of cases with hypnotic/tranquilizer drugs (ketamine, Zolpidem, Zopiclone, Zolmitriptan), and 0.37% ($n = 4$) of cases with antihistamine drugs (promethazine, chlorpheniramine, and diphenhydramine) (Rauf et al., 2021).

7.3 ILLICIT OPIOIDS

Separately from consumers seeking to find an alternative supplier for their prescription drugs, drug users may encounter another type of counterfeiting. Dealers may proffer illicit opioids—substances containing novel synthetic or designer opioids, because they are cheaper and more accessible than the API. Consuming these substances can be deadly and detected in toxicology screens. An example is a set of cases reported by Papsun et al. (2017) in which ultra-high-performance liquid chromatography tandem mass spectrometry (UPLC–MS/MS) identified U-47700 and furanyl fentanyl in whole blood samples of heroin users. Consumption of the designer opioids was fatal in the 11 reported cases (Papsun et al., 2017). In another case, a 62-year-old male overdosed on acrylfentanyl (Raheemullah and Andruska, 2019). Treatment included two doses of 2 mg of naloxone and ventilator with adverse effects, including ventilator-associated pneumonia followed by a pulmonary embolism (Raheemullah and Andruska, 2019). In a set of 14 case reports reviewed by Rauf et al. (2021), fentanyl analogs including cyclopropylfentanyl, acetylfentanyl, acrylfentanyl, furanylfentanyl, carfentanil, norfentanyl, butyryl fentanyl, 4F-butyryl fentanyl, and tetrahydrofuranyl fentanyl were associated with mostly accidental overdose deaths.

Illicit opioids have been found as adulterants in street cocaine and methamphetamine (Pichini et al., 2018; Han et al., 2019). Fentanyl and fentanyl analogs have been found as adulterants or substituted in street heroin (Pichini et al., 2018; Jannetto et al., 2019). Synthetic opioids (not including methadone) resulted in 9,580 deaths in 2015, as reported by the CDC (Prekupec et al., 2017; Jannetto et al., 2019). This reflects a 72% increase from 2014 (Jannetto et al., 2019). Most of the cases resulted from the use of illicitly manufactured fentanyl or its analogs (Prekupec et al., 2017; Jannetto et al., 2019). Deaths from the use of fentanyl analogs resulted in 17% of the reported cases. The fentanyl analogs included acetylfentanyl, butyrylfentanyl, and furanylfentanyl. Unknown substances also were identified in the toxicology analysis. Jannetto et al. (2019) included a compilation of structural and mass spectral data for fentanyl analogs in their report. In a review of toxicology reports and drugs identified in the blood/urine of the deceased, fentanyl analogs were detected in 42.53% ($n=459$) of cases with opioids (Rauf et al., 2021). The opioids included heroin, morphine, oxycodone, methadone, codeine, hydrocodone, tramadol, hydromorphone, buprenorphine, oxymorphone, dihydrocodeine, and U-47700 (Rauf et al., 2021). The reports can be used to determine cause of illness or death.

7.4 NEW PSYCHOACTIVE SUBSTANCES IN THE COUNTERFEIT DRUG MARKET

Flualprazolam—a benzodiazepine—is not registered as a medicinal drug but is a new psychoactive substance (NPS) that has been found in the illegal drug market. In June 2019, the Oregon Poison Center detected a patient cluster of six teens in a 1-week period who had consumed flualprazolam. Blood and urine samples from three patients were collected and tested as well as a tablet fragment. The samples were analyzed using liquid chromatography quadrupole time-of-flight mass spectrometry, which led to the identification of flualprazolam. In this case, all of the patients recovered within 6 hours of ingesting the drug (Blumenberg et al., 2020).

7.5 CONCLUSION

The cases reviewed in this chapter demonstrate that counterfeit drugs have been the root cause of death in several cases. The substances have ranged from counterfeit nonprescription Tylenol and vaccines to oxycodone, hydrocodone, and alprazolam tablets to new psychoactive substances such as flualprazolam to illicit opioids, including acrylfentanyl, fentanyl, furanyl fentanyl, and 4-ANPP. An additional case is introduced in Chapter 9. Counterfeit drugs, illicit opioids, and NPSs pose challenges for medical professionals, law enforcement, and policymakers and significant risks for the public. Strategies that screen and detect counterfeits and counterfeit operations before the counterfeit drugs reach the market are essential to keeping the pharmaceutical marketplace safe for consumers. More oversight and international agreements may reduce the access of web markets to counterfeit sales.

REFERENCES

Alubo, S.O. "Death for sale: A study of drug poisoning and deaths in Nigeria." *Social Science and Medicine* 38 (January 1994): 97–103. doi: 10.1016/0277-9536(94)90304-2.

Blumenberg, A., Hughes, A., Reckers, A., Ellison, R., and R. Gerona. Flualprazolam: Report of an outbreak of a new psychoactive substance in adolescents. *Pediatrics* 146 (June 2020): e20192953. doi: 10.1542/peds.2019-2953.

Bolla, A.S., Patel, A.R., and R. Priefer. "The silent development of counterfeit medications in developing countries – A systematic review of detection technologies." *International Journal of Pharmaceutics* 587 (September 25, 2020): 119702. doi: 10.1016/j.ijpharm.2020.119702.

Chapman, B.P., Lai, J.T., Krotulski, A.J., Fogarty, M.F., Griswold, M.K., Logan, B.K., and K.M. Babu. "A Case of Unintentional Opioid (U-47700) Overdose in a young adult after counterfeit Xanax use." *Pediatric Emergency Care* 37 (September 1, 2021): e579–e580. doi: 10.1097/PEC.0000000000001775.

Dégardin K., Roggo, Y., and P. Margot. "Understanding and fighting the medicine counterfeit market." *Journal of Pharmaceutical and Biomedical Analysis* 87 (January 18, 2014): 167–175. doi: 10.1016/j.jpba.2013.01.009.

Han, Y., Yan, W., Zheng, Y., Khan, M.Z., Yuan, K., and L. Lu. "The rising crisis of illicit fentanyl use, overdose, and potential therapeutic strategies." *Translational Psychiatry* 9 (November 11, 2019): 282. doi: 10.1038/s41398-019-0625-0.

Jannetto, P.J., Helander, A., Garg, U., Janis, G.C., Goldberger, B., and H. Ketha. "The fentanyl epidemic and evolution of fentanyl analogs in the United States and the European Union." *Clinical Chemistry* 65 (February 2019): 242–253. doi: 10.1373/clinchem.2017.281626.

Martucci, H., Ingle, E.A., Hunter, M.D., and L.N. Rodda. "Distribution of furanyl fentanyl and 4-ANPP in an accidental acute death: A case report." *Forensic Science International* 283 (February 2018): e13–e17. doi: 10.1016/j.forsciint.2017.12.005.

Papsun, D., Hawes, A., Mohr, A., Friscia, M., and B.K. Logan. "Case series of novel illicit opioid-related deaths." *Academic Forensic Pathology* 7 (September 2017): 477–486. doi: 10.23907/2017.040.

Pichini, S., Solimini, R., Berretta, P., Pacifici, R., and F.P. Busardò. "Acute intoxications and fatalities from illicit fentanyl and analogues: An update." *Therapeutic Drug Monitoring* 40 (February 2018): 38–51. doi: 10.1097/FTD.0000000000000465.

Prekupec, M.P., Mansky, P.A., and M.H. Baumann. "Misuse of novel synthetic opioids: A deadly new trend." *Journal of Addiction Medicine* 11 (July/August 2017): 256–265. doi: 10.1097/ADM.0000000000000324.

Raheemullah, A., and N. Andruska. "Fentanyl analogue overdose: Key lessons in management in the synthetic opioid age." *Journal of Opioid Management* 15 (September/October 2019): 428–432. doi: 10.5055/jom.2019.0531.

Rauf, U., Ali, M., Dehele, I., Paudyal, V., Elnaem, M.H., and E. Cheema. "Causes, nature and toxicology of fentanyl-analogues associated fatalities: A systematic review of case reports and case series." *Journal of Pain Research* 14 (August 24, 2021):2601–2614.

Silva, D. Prince died after taking fake Vicodin laced with fentanyl, prosecutor says. NBC News. April 19, 2018. Accessed October 21, 2021. https://www.nbcnews.com/news/us-news/no-criminal-charges-prince-s-overdose-death-prosecutor-announces-n867491.

Wertheimer, A.I. and J. Norris. "Safeguarding against substandard/counterfeit drugs: Mitigating a macroeconomic pandemic." *Research in Social and Administrative Pharmacy* 5 (March 2009): 4–16. doi: 10.1016/j.sapharm.2008.05.002.

8 Analytical Tools for Examining Counterfeit Drug Products

Adam Lanzarotta, Mark Witkowski,
Nicola Ranieri, Douglas Albright,
Lianji Jin, and Martin Kimani
U.S. Food & Drug Administration

CONTENTS

DOI 10.1201/9781003183327-8

Disclaimer: The contents of this book chapter are the authors' opinions and should not be considered as opinions or policy of the US FDA. The mention of trade names and manufacturers is for technical accuracy, and should not be considered as endorsement of a specific product or manufacturer. Cases discussed in this chapter have been adjudicated.

8.1 OVERVIEW

At the request of the World Health Assembly (WHA), the Director General of the World Health Organization (WHO) first initiated programs to combat counterfeit drug products in 1988 (WHO, 1999). This resolution put into motion a decades-long initiative that continues to urge participation from countless individuals and organizations all over the world. WHA and WHO continue to pass resolutions and issue newsletters that regularly update background information, consequences, statistics, and causes related to counterfeit drug products (WHO, 1999, 2018). These organizations even recommended the establishment of working groups and other committees to help combat falsified drug products. For example, the WHA created the Member State Mechanism in 2012 to establish a unified global effort to combat falsified (including counterfeit) drug products (WHO, 2018). At this meeting, the group developed a three-pronged approach for this endeavor, which includes objectives dedicated to preventing, detecting, and responding to counterfeit drug products. The group indicated that prevention can be achieved by securing the supply chain via multi-collaborative stakeholder engagement and preventative measures, detection can be achieved by improving current analytical capabilities, and response can be achieved through governmental and regulatory action. While this chapter will briefly discuss commonly employed preventative measures used to curtail drug counterfeiting, it will primarily focus on analytical tools used to detect and source suspected counterfeit drug products. Response actions will not be discussed here, since these typically vary significantly by country and regulatory agency.

8.2 ANTI-COUNTERFEITING TECHNOLOGY AND PREVENTATIVE MEASURES

Several reviews and articles describe challenges posed by counterfeit drugs and the history of anti-counterfeiting technology/preventative measures undertaken by international organizations, government agencies, drug manufacturers, and many others (Bansal et al., 2013, Kumar et al., 2013, Kumar and Baldi, 2016, Mackey and Nayyar, 2017, Zhang et al., 2020, Xie and Tan, 2021). These articles highlight several different initiatives to prevent counterfeiting, which include the establishment of verified internet pharmacy practice sites (VIPPS) and public–private partnerships (PPP). They also describe commonly used anti-counterfeiting measures and their pitfalls, including holograms, security threads, color-shifting ink, barcodes, quick response (QR) codes, laser marking, watermarks, and radio-frequency identification devices (RFID). Kumar et al. and Rasheed et al. describe mass encryption and track and trace authentication methods to prevent counterfeit drug products (Kumar and Baldi, 2016, Rasheed et al., 2018). These technologies involve tagging each authentic product with a unique digital code so that the manufacturer can monitor the precise location of their products, and the end user can validate the products using a web-based application. Zhang et al. reviewed advantages and limitations of several in-drug/on-drug labeling technologies that have been employed to protect the drug supply chain, which include using encryption, polymers, isotope-labeled excipients, microscopic tags, and nanoscopic tags for tablet/capsule formulations (Zhang et al., 2020). Sharief et al. review advantages and challenges of using deoxyribonucleic acid (DNA)-based anti-counterfeiting measures with a wide variety of detection techniques, including polymerase chain reaction (PCR), surface plasmon resonance (SPR), electrochemical, and colorimetric biosensor-based methods (Sharief et al., 2021). While a significant amount of research has been dedicated to developing methods such as these to prevent counterfeit pharmaceuticals and secure the supply chain, there have been just as many studies dedicated to reactive analytical methods for detecting/sourcing counterfeit drug products using a wide variety of analytical instrumentation; these methods are discussed in the following sections of this chapter.

8.3 COUNTERFEIT DETECTION METHODS

8.3.1 Reviews

Several reviews, minireviews, and book chapters that describe analytical methods for counterfeit drug detection have been published over the last 10–15 years. Some review articles are broad, some are narrowly focused by technique(s), and some are focused by target analyte(s). Many authors have reviewed several different analytical instruments that have been commonly used to compare a wide range of authentic and counterfeit drug products (Fernandez et al., 2008, Fernandez et al., 2011, Martino et al., 2010, Talati et al., 2011, Rebiere et al., 2017, Bottoni and Caroli, 2019, Bakker-'t Hart et al., 2021) These authors described how analytical capabilities for detecting counterfeit drug products have increased and improved with time. Singh et al. (2009) reviewed analytical techniques used to differentiate authentic and counterfeit products labeled to contain some of the most counterfeited drug products, phosphodiesterase type-5 (PDE-5) inhibitors (e.g., Viagra, Cialis and Levitra), and Bonsu et al. (2021) reviewed analytical techniques used to discriminate authentic and counterfeit anorectic-containing products. Deconinck et al. described the role of chromatography for detecting and characterizing illegal pharmaceutical products, including counterfeits (Deconinck et al., 2013). Vickers et al. (2018) and Caillet et al. (2018) directly compared the performance of 41 portable devices covering 19 technologies for the analysis of counterfeit drug products, and Leary et al. (2021) later wrote a book chapter describing the advantages and limitations of portable spectrometers for comparing authentic and suspect counterfeit drug products. These and other analytical methods that have been routinely used for counterfeit drug analysis are described in more detail below in the order of least to most destructive.

8.3.2 Printing Analysis

Printing examinations are an often underutilized technique that can screen for and detect counterfeit packaging. Most chemists within regulatory or forensic laboratories understandably perceive that only chemical analysis of a suspect counterfeit drug is useful. However, it should be noted that in the Food, Drug, and Cosmetic (FD&C) Act Section 201(g)(2), a counterfeit drug is defined as the drug "or the container or labeling of which" does not have authorization from the authentic manufacturer. Thus, examinations of cartons, boxes, bottle labels, pouches, vials labels, etc., can be a reliable indicator regarding whether counterfeit drugs may be contained within the packaging (Lanzarotta et al., 2015, Albright 2016, Degardin et al., 2018). Packaging examinations can also be useful for screening bulk shipments of suspected products to isolate the counterfeits prior to more time-consuming, laboratory-based chemical analyses. This is often the case when large shipments of packaged drugs are detected at international mail facilities (IMF) or other ports of entry. Packaging analyses often consist of alternate light source (ALS) exams as well as visual examinations of the printing processes using magnification devices such as stereo microscopes, digital microscopes, or even small, inexpensive, handheld magnifiers.

Printing examinations conducted at the FCC primarily focus on the printing of the graphics (artwork) and the variable data (lot/batch numbers, expiration dates). Pocket Pal is a useful and easy-to-read reference for anyone performing printing examinations (Romano and Riordan, 2003). Expertise in the identification of printing processes was developed over many years by taking courses on printing technologies, attending printing conferences and trade shows, and by analyzing numerous sample types. For large authentic manufacturers, the printing of these two regions of interest (graphics and variable data) is performed at distinctly different times using different printing processes, as lots/batches are filled. The graphics for cartons, boxes, inserts, foil pouches, blister packs, vial labels, syringe labels, etc. are usually printed in bulk using traditional impact printing technologies like offset lithography, gravure, flexography, and screen printing. Once the finished product is packaged, the variable data is added later using a variety of non-impact printing technologies such as drop on demand (DOD)/continuous inkjet printing (CIJ), thermal transfer, and laser marking (ablation). Using magnifiers, all these technologies have distinct microscopic features that are easily discernable to the trained eye. Counterfeit packaging products are often printed using the same technology for both the graphics and variable data, and may be used as an indication of counterfeit packaging. With offset lithography, in particular, defects that occur on the printing plate can be reproducibly transferred to the substrate and may allow for linking of counterfeits of the same origin. A few examples of these types of printing processes will be discussed.

8.3.2.1 Graphics Printing

Offset lithography has been widely used for decades and can be divided into two types based upon the color reproduction or halftone process implemented. Halftone is a technique that attempts to recreate color by printing dots of varying size or spacing using the four process colors of ink (i.e., cyan, magenta, yellow, and black, which can be viewed in the digital copy). Amplitude modulated (AM) and frequency modulated (FM) halftones have very distinctive characteristics under magnification. AM halftones are easily distinguished by the varying diameters of the halftones, while the centers of the halftones are equally spaced, as shown in Figure 8.1a. FM halftones are comprised of very small dots of the same diameters, but they appear randomly spaced, as shown in Figure 8.1b. The

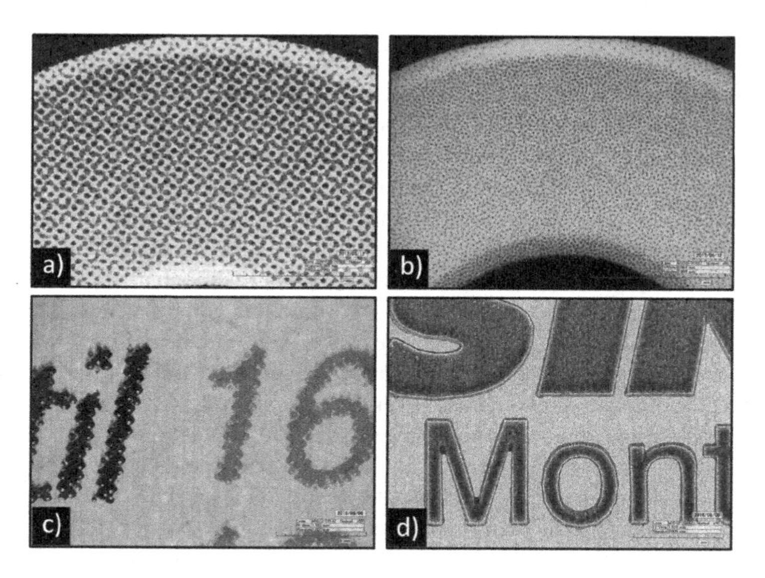

FIGURE 8.1 AM offset printing (a) and FM offset printing (b) on product packaging material. Gravure printing (c) and flexography printing (d) on blister foils.

diameter of the AM halftones and the density of the FM halftones dictate the saturation/intensity of the printed color. In our experience, many authentic manufacturers have used AM offset lithography for decades but are transitioning to FM to increase the complexity of their graphics printing.

Gravure printing is an impact printing process that is used more frequently on the second layer of packaging such as foil pouches and blister packs that directly contain finished dosages. Gravure printing is a process in which the printing cylinders are engraved with a series of recessed cells corresponding to the graphics being reproduced. These cells are commonly shaped as ovals or diamonds. Unlike offset halftones that use varying dot diameters or densities, gravure cells create color saturation based upon the depth of the cells. The deeper the cells, the more ink that is printed onto the substrate. The shallower the cells, the less saturated the color. These cells leave distinctive characteristics of the cell shapes on the printed substrate and are easily seen using magnification. Another characteristic is the appearance of void areas (fisheyes) or small dots without ink (Figure 8.1c). Flexography printing is another impact printing process used on a variety of packaging especially bottle and vial labels. Flexography is a more versatile version of the letterpress printing process invented by Johannes Gutenberg in 1440. It uses flexible polymer printing plates that have a raised image area that transfers ink to the substrate using pressure. The flexibility of the plate allows the ink to spread somewhat unevenly onto the substrate, creating a haloed appearance at the edges of the graphics (Figure 8.1d).

8.3.2.2 Variable Data Printing

Variable data printing processes were determined using Pocket Pal (Romano and Riordan, 2003). Inkjet printing is used extensively in variable data printing to apply the lot/batch numbers and expiration dates, and is comprised of either DOD or CIJ printing. DOD inkjet ejects minute picoliter-sized droplets of liquid ink through tiny nozzles on a print head. This can be done using either thermal means to create a vaporized bubble (bubble jet) or by an oscillating piezoelectric crystal to force the ink out of the nozzles. DOD is the same technology found in home office inkjet printers. CIJ utilizes a continuous flow of ink that separates into drops as it travels through the air with the flight path being directed using static electric charges into the shape of the desired alphanumeric characters. Both processes have distinct characteristics on the substrate that can be seen using magnification. For DOD inkjet, the droplets can have small tails (satellites), while CIJ droplets are much larger and more translucent (Figure 8.2a and b, respectively). In the past, FCC has also come across several suspect counterfeits that have used a combination of non-impact printing processes, such as DOD inkjet and/or xerography (laser printers), to copy entire graphics and variable data packaging. This is most common on lower volume counterfeits of vial labels that can be easily printed on sheets of labels then transferred to a vial.

Thermal transfer printing is used on a variety of packaging substrates, including paperboard, vial labels, bottle labels, and syringe labels. This non-impact process involves the use of a ribbon embedded with pigmented material that is

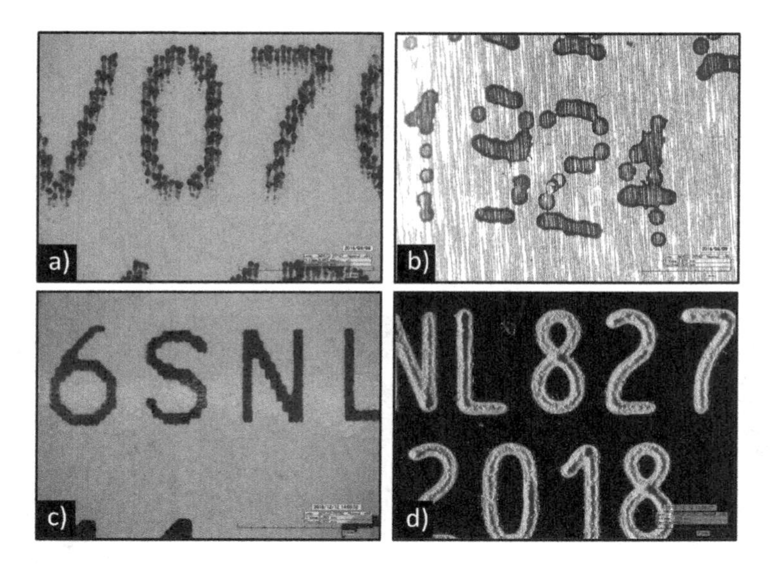

FIGURE 8.2 DOD inkjet printing on paperboard (a), CIJ printing on metal (b), thermal transfer printing on bottle label (c), and laser marking on carton (d).

transferred to the substrate using a thermal print head. The print head heats the opposing side of the ribbon and melts the pigmented material, thus transferring it to the substrate. The shapes of the print head heating elements leave a signature stair-step pattern on the substrate that can be easily seen under magnification (Figure 8.2c).

Unlike other non-impact printing technologies, laser marking (laser coding) does not transfer any ink or pigmented material to the substrate. Rather, it is a laser ablation technique that uses a high energy infrared laser to vaporize or ablate the substrate into the desired alphanumeric characters. Typically, a pre-printed region of the graphics is removed using ablation to expose the underlying substrate. Under magnification, the ablated material can be easily detected as a recessed area beneath the substrate surface (Figure 8.2d). This can even be seen with the unaided eye by simply using specular reflectance to observe the disturbed surface of the substrate, which can be accomplished by holding the packaging up to a light source at an oblique angle, while tilting it back and forth to observe the difference in the surface gloss from the ablated material. The ablated material appears dark in Figure 8.3a and bright in Figure 8.3b, whereas the text "LOT" and "EXP" are bright in both images.

Additional examples of how counterfeit packaging materials have been differentiated from authentic products using printing analysis are described below. For example, variable data offset printing on a suspect counterfeit carton (Figure 8.4a) was differentiated from DOD inkjet printing on an authentic carton (Figure 8.4b) at 60X magnification.

FIGURE 8.3 Laser marking specular reflection (a) and laser marking variable data (b).

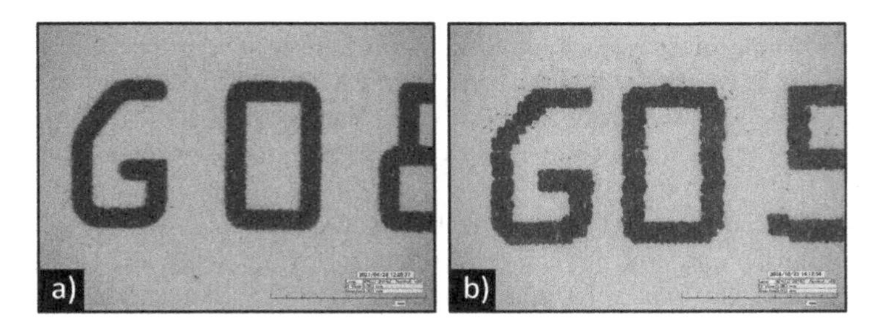

FIGURE 8.4 Offset printing on a suspect counterfeit carton (a) and DOD inkjet printing on an authentic carton (b) at 60X magnification.

Offset printing on a suspect counterfeit carton (Figure 8.5a) was not consistent with gravure printing on an authentic carton (Figure 8.5b) at 120X magnification. Offset printing on a suspect counterfeit carton (Figure 8.5c) was not consistent with laser marking on an authentic carton (Figure 8.5d) at 60X magnification.

AM offset printing on a suspect counterfeit carton (Figure 8.6a) was differentiated from offset printing, since no halftones were present on an authentic carton (Figure 8.6b) at 60X magnification. Xerographic printing on a suspect counterfeit vial label (Figure 8.6c) was not consistent with offset printing (no halftones) on an authentic vial label (Figure 8.6d) at 80X magnification. Using the unaided eye, many of these counterfeit products were not discernable from the authentic product. However, significant differences related to contrasting printing processes are often observed between counterfeit and authentic packaging products at high

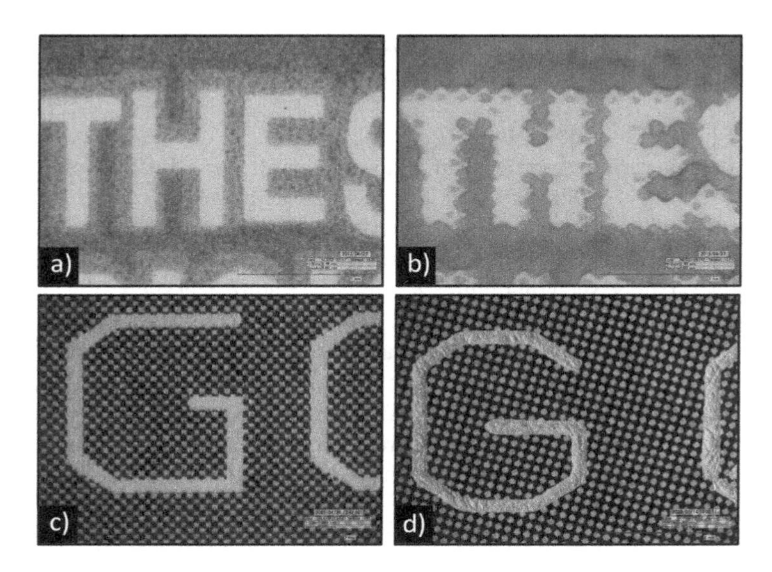

FIGURE 8.5 Offset printing on a suspect counterfeit carton (a) and gravure printing on an authentic carton (b) at 120X magnification. Offset printing on a suspect counterfeit carton (c) and laser marking on an authentic carton (d) at 60X magnification.

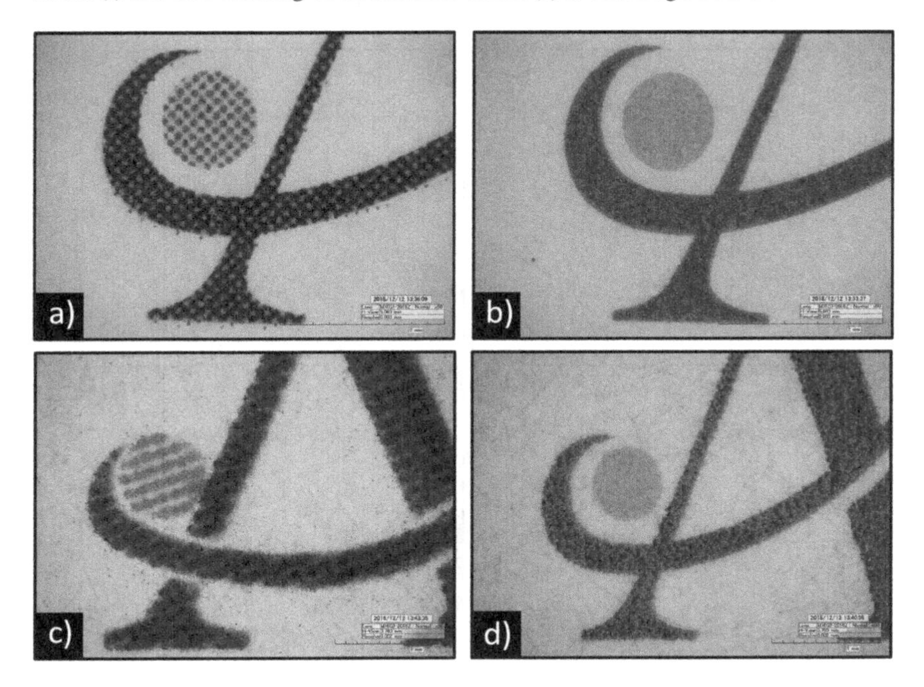

FIGURE 8.6 AM offset printing on a suspect counterfeit carton (a) and offset printing no halftones on an authentic carton (b) at 60X magnification. Xerographic printing on a suspect counterfeit vial label (c) and offset printing no halftones on an authentic vial label (d) at 80X magnification.

magnifications, which is often an effective way to rapidly screen suspect products prior to more time-consuming chemical analyses.

8.3.2.3 Printing Defect Analysis

During the counterfeiting process, it is not unusual for small offset lithographic printing defects to be introduced on the packaging. Images of small bits of dust or debris during the platemaking process can be transferred to the printing plates. The defects will then be printed on every carton printed from the same plate. These defects can be seen on Figures 8.7 and 8.8. Figure 8.7a shows the authentic carton graphics with no printing defects, and Figure 8.7b–d shows three cartons that were determined to be counterfeit based on variable printing process differences using procedures described above (data not shown). Figure 8.7b–d show the same printing defects (dots highlighted with red circles) observed on three separate counterfeits submitted at different times over the course of 6 months. These defects show that these counterfeits were printed with the same plate and share a common origin.

Images from two additional counterfeit cartons that were received separately are shown in Figure 8.8; the cartons were determined to be counterfeit based on variable printing process differences using procedures described above (data not shown). Figure 8.8a and b are from two different faces of one carton, and Figure 8.8c and d are from the corresponding faces of the second carton. The same "dot"

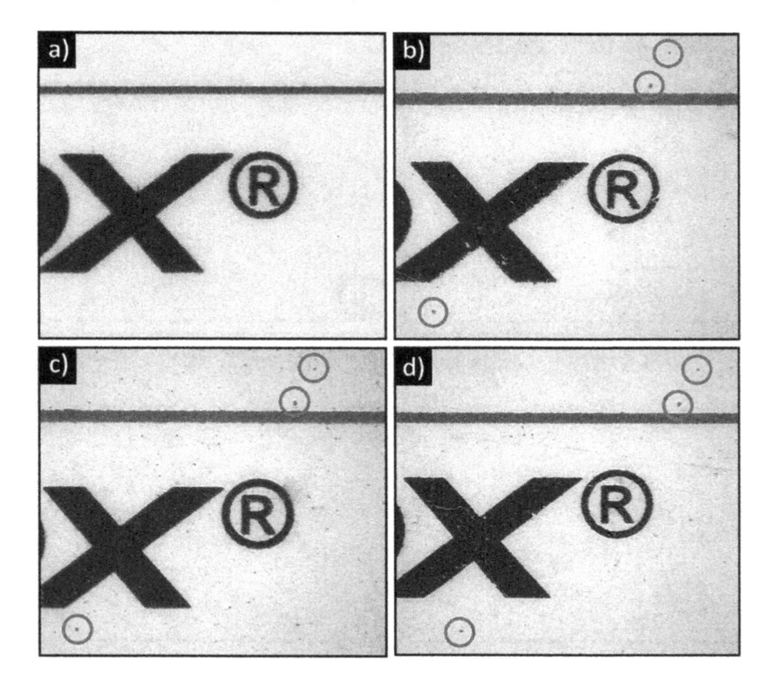

FIGURE 8.7 Images of carton graphics from an authentic product (a) and three counterfeit products (b–d) received by FCC on three separate occasions.

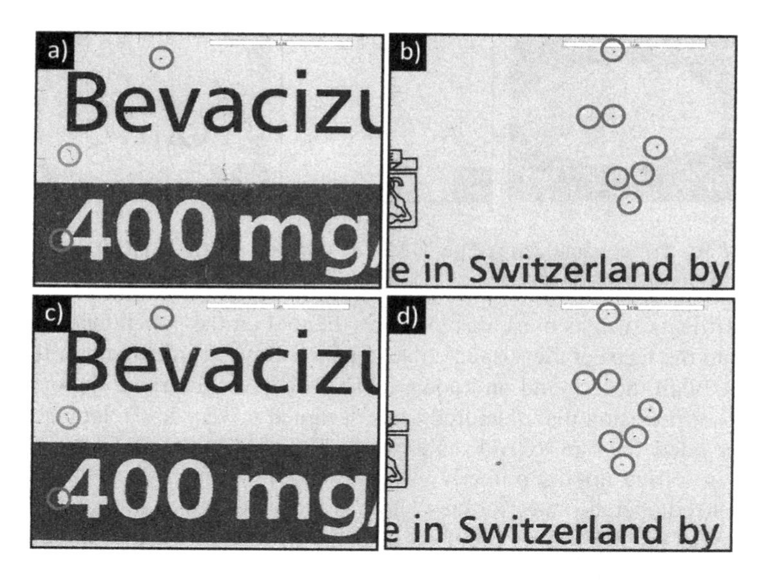

FIGURE 8.8 Two faces from one counterfeit carton (a,b) and corresponding faces from another counterfeit carton (c,d) that was received separately.

defects described above are highlighted with red circles in each image. Additionally, elongated fiber-like defects with more character are also often observed as highlighted by blue circles in Figure 8.8a and c. Again, these defects show that the counterfeit cartons were printed with the same plate and share a common origin.

8.3.3 Forensic Image Analysis (IA)

Forensic image comparison is a process of comparing known and unknown items using digital imagery to determine if they are consistent with each other. One such comparison could be that of an unknown facial image obtained during an investigation from a security surveillance camera with a facial image of an identified suspect. This same procedure can be applied to examine counterfeit pharmaceuticals. It is known that counterfeiting, illegal diversion, and theft incidents of pharmaceuticals over the last multiple decades have been on the rise, and it is also known that an increase of "rogue websites" selling potentially dangerous drugs is one contributing factor (Pharmaceutical Security Institute, 2020 and Bate, 2012). As the income derived from these unlicensed organizations increases, there is a real need to combat this problem, and potentially linking counterfeit tablets from different sources has become an important component for solving criminal investigations. Pharmaceutical tablets are manufactured by the compression of the tablet formulation in a tablet die forming the side of the tablet and the two "faces" of the tablet. They are formed between upper and lower opposing tools, the punches (punch tip/cup), which also apply any debossed character(s) or score lines to the tablet face (Figure 8.9a). Essentially, the debossed surface features as

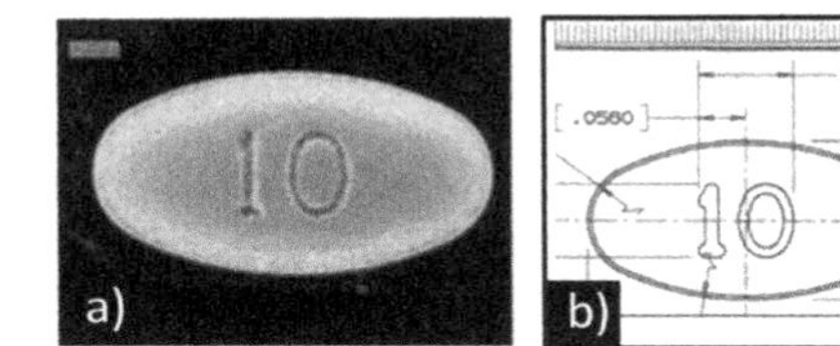

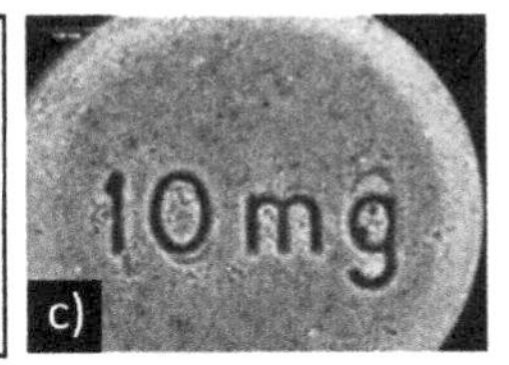

FIGURE 8.9 Tablet manufactured per CAD file diagram (a), CAD file drawing (b) and double press or impression (c).

well as artifacts such as markings made by the tool on the punch face are transferred onto the faces of the suspect tablets, which allows comparison of the measurements with those found on known tablet faces over the same region. Tablet punches are precision manufactured tools designed to very low tolerances using computer aided diagrams (CAD), as shown in Figure 8.9b. Additionally, adequate training in setting up and properly using a tablet press is critical; therefore, an inexperienced operator may not know how to make needed changes to a press to prevent tablet manufacturing problems such as sticking, picking, capping, double press, or impression, as shown in Figure 8.9c (Higgins 2015).

Counterfeit tablets may or may not have been made with high-quality punches or knowledgeable operators, which makes examination of the determining the tablet debossing or the overall tablet shape critical. This examination may include determining the debossed character font, font type and size, overall placement of debossed features, depth of debossed characters, etc. Artifacts such as tool markings and non-conformities with known tablet debossing or original tablet punch CAD files may aid in sourcing a counterfeit tablet. Various microscopic applications, including stereo light microscopy (SLM), can aid in performing two-dimensional (2D) tablet-to-tablet punch face image analysis (2DIA) and tablet surface feature profilometry measurements by 3D image analysis (3DIA) to examine pharmaceutical tablet faces in real color.

8.3.3.1 2D Tablet Face Analysis

SLM 2DIA is performed using an image processing software such as the Image-Pro Plus 2D Image Analysis Software. The program has a comparison analysis feature to perform image overlays (IO). Various digital filters may be used prior to performing an IO, such as the remove noise filter, contrast adjustment, and pseudo-coloring, to aid in identifying counterfeit tablets as well as known tablets manufactured at different facilities. This IO feature is effective for providing an evaluation of a suspect counterfeit tablet based on visual differences with a corresponding authentic tablet. Image capture and analysis of the suspect tablet debossing when overlaid with an authentic tablet CAD file can visually demonstrate differences between the tablet and the original punch face drawing; examples where a suspect tablet debossing is not consistent and is consistent with the CAD diagram are shown in Figure 8.10a and b, respectively.

A similar method is applied to captured images of suspect and authentic tablets, and the debossing comparison image overlay. Adding the color segmentation

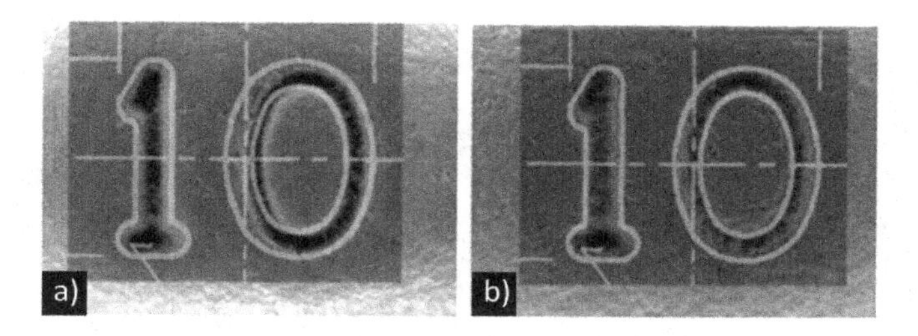

FIGURE 8.10 Debossing not consistent with CAD diagram (a) and consistent with CAD diagram (b).

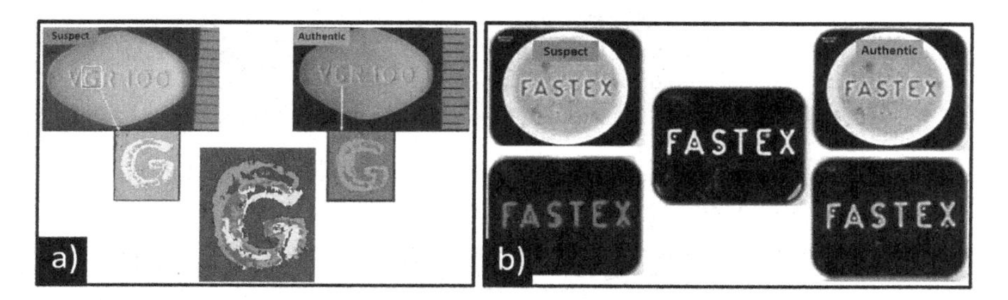

FIGURE 8.11 2DIA overlay exhibiting not consistent debossing (a) and consistent debossing (b).

feature is performed to reveal similarities or differences between the tablets; examples where a suspect tablet debossing is not consistent and is consistent with an authentic tablet debossing are shown in Figure 8.11a and b, respectively. However, the 2D image comparison is unable to help display potential inconsistencies in the third dimension of the suspect tablet(s). This can result in false positives for tablets that have different debossing depth, overall debossing depth shape, and overall tablet surface features characteristics due to the tablet punch tip cup design shape.

8.3.3.2 3D Tablet Face Analysis

Advancements in IA technologies have added the optical third dimension, *z*, to the *x* and *y* spatial dimensions. The improvement in IA by 3D surface profilometry instrumentation such as the InfiniteFocus G4 & G5 shows invaluable ability to successfully determine the consistency between entire surface feature characteristics of two tablets or tablet to punch face—a process referred to as difference measurement (DM) analysis. The same measurement can be used to compare a tablet punch face cup shape CAD-dataset file or of an actual tablet punch face, as shown in Figure 8.12. In this example, imperfections on the tablet can be linked to those on the punch face (indicated with arrows).

The DM allows analysts to examine the overall shape and size of the tablet, as well as the debossing position in relationship to the overall shape of the tablet. Basically, a DM approach is used to numerically compare two different geometries. The technique allows users to measure form deviations to a CAD-dataset

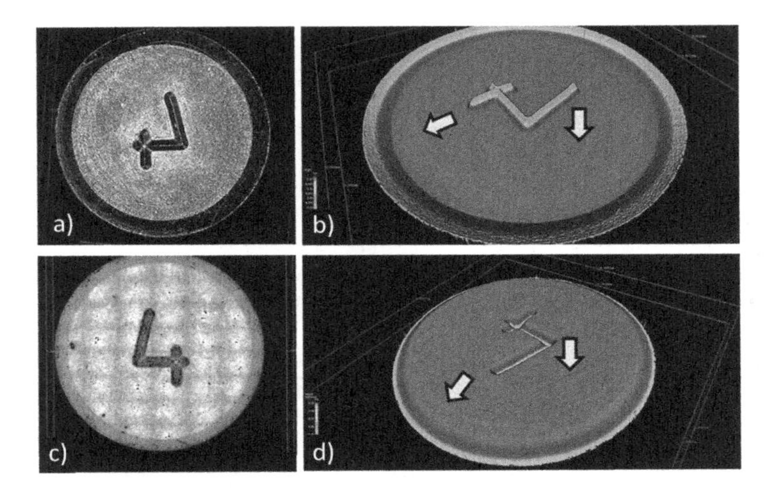

FIGURE 8.12 Tablet punch face (a), tablet punch face by 3DIA (b), tablet face debossing (c), and tablet face debossing by 3DIA (d).

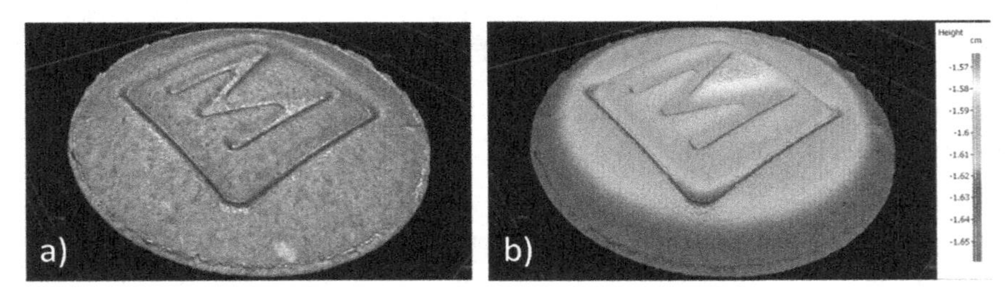

FIGURE 8.13 3D digital elevation model in real color (a) and pseudo-color (b).

or reference component(s). These measurements generate data files referred to as datasets, which can be examined in real color mode as well as pseudo-color as digital elevation mode, as shown in Figure 8.13a and b, respectively.

The DM process measures surfaces from numerous perspectives. Single measurements are then automatically merged into a full 3D dataset, which is essentially in a wireframe format consisting of millions of XYZ data points (Figure 8.14). Examination by DM characteristically allows for the production of a tablet volumetric measurement difference by a diagram or application of a pseudo-color to a 3D model of a tablet volumetric measurement. The DM process also generates quantifiable data.

Surface profilometry, also referred to as profile measurement (PM) by optical microscopy technology, is a different measurement than DM. A PM is a 3D roughness measurement module for high-resolution surface measurements. The technique can calculate the spectrum of a profile by using the Fast Fourier Transform (FFT) of the profile. The PM is created using a transparent red line with consistent width across the 3D image, which results in a spectrum curve or wave form, as shown in Figure 8.15 (top). Depending on the profile width (red line), several profiles are

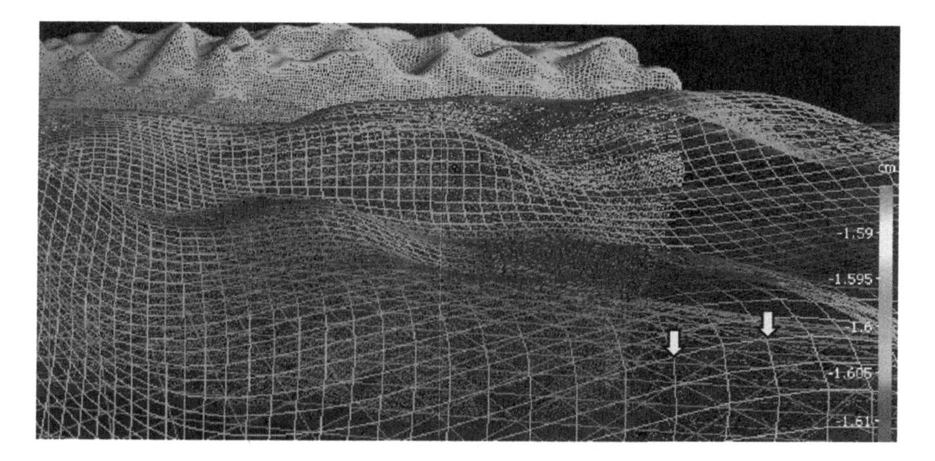

FIGURE 8.14 3D wireframe model view of a tablet surface—each arrow highlights an XYZ data point.

calculated. The statistical parameters of the profiles generate an average profile. This method can measure roughness of flat and curved components. The technique can reveal differences between two items in a visual format by displaying the shape of tablet topographic surface features, which is essentially a surface cross-section displayed in a single line diagram (Figure 8.15 (bottom).

Examination by PM characteristically allows for a production of a tablet surface cross-section in the form of a diagram that can be compared anytime. Examination of tablet(s) by profile measurements provides valuable information at ~0.25 µm *z* step resolution measurements for the entire tablet face or just of a selected region of interest (ROI) such as that of a debossed feature characteristic. The tablet ROI of the measured pixel-by-pixel values can then be plotted over the selected ROI from both the suspect and authentic data. Afterwards, the data files can be compared. Additionally, these single line profile diagrams can be compared among multiple suspect or authentic tablets, and examined anytime during the course of an investigation with a new suspect tablet, as shown in Figure 8.16. In this example, the PM from suspect tablet 1 (Figure 8.16a) is not consistent with the PM from suspect tablet 2 (Figure 8.16b). However, the PM from suspect tablet 2 (Figure 8.16b) is consistent with that of the authentic tablet (Figure 8.16c).

The principle of these two techniques combines the small depth of focus measurement of an optical system with the ~0.25µm *z* step resolution vertical scanning to provide topographical and color information from the variation of focus. A proprietary algorithm reconstructs these measurements into a single *x, y, z* data-set file with highly accurate topographical information (Schroettner et al., 2006). Traceable calibration standards allow the verification of measurement results. The tablet debossing procedure has been validated by comparing known counterfeit tablets to known authentic tablets (received directly from the manufacturer from multiple manufacturing locations and lots) as well as to known authentic

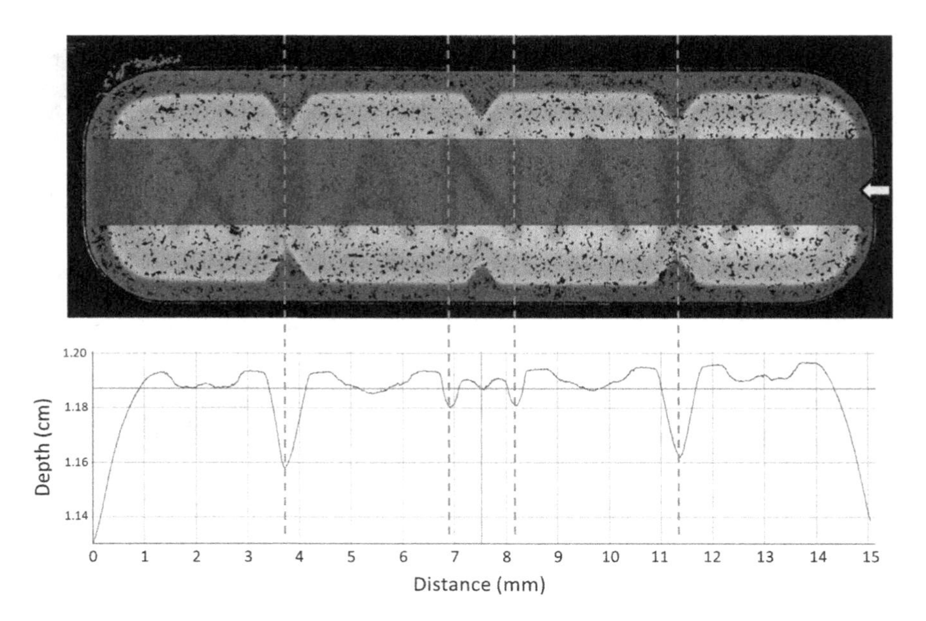

FIGURE 8.15 PM created using a transparent red line (emphasized with an arrow) with a specific consistent width—surface cross-sectioning of a tablet (top) and PM diagram—measure roughness of flat and curved components—surface cross-sectioning of a tablet (bottom). Dashed lines are to show consistency between the top and bottom images.

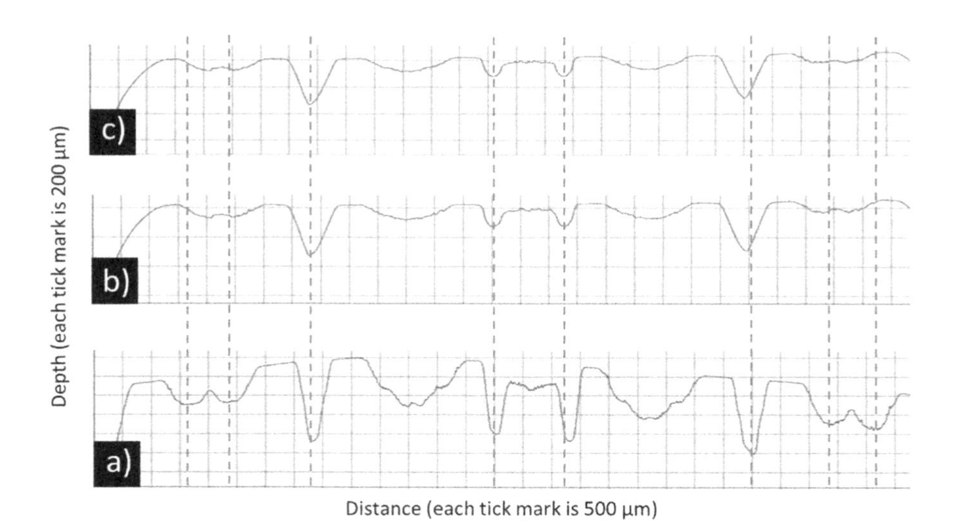

FIGURE 8.16 PM multiple profile comparisons. The PM of suspect tablet 1 (a) is not consistent with the PM of suspect tablet 2 (b) or the PM of the authentic tablet (c). The PM of suspect tablet 2 (b) is consistent with the PM of the authentic tablet (c). Dashed lines have been added for clarity.

tablet engineer drawings/diagrams or actual punches. (Bureau International des Poids et Mesures, 2021) A specific counterfeit tablet case of known origin was used to test the validated method.

The results obtained by the 3D model comparisons can also be used to link counterfeit tablets from two or more investigations. For example, the PM measurements from two different counterfeit alprazolam 2 mg tablets (white, rectangular, quarter-scored tablet debossed "GG 249") shown in Figure 8.17 are consistent with each other, which demonstrates that they share a common origin.

A similar method can be used to link counterfeit tablets with specific tablet punch tips. For example, a reverse image of a tablet punch tip for a generic alprazolam 2 mg tablet (white, rectangular, quarter-scored, debossed "GG 249") is shown in Figure 8.18 (left), a Mikrosil silicone casting of the tablet punch tip is shown in Figure 8.18 (center), and an image of a suspect tablet is shown in Figure 8.18 (right).

Defects in the tablet punch tip, such as those shown in Figure 8.18 (left) with arrows, can be used to link tablet punch tips to suspect tablets. For example, the downward arrow pointing to the manufacturing defect on the tablet punch tip in Figure 8.18 (left) was reproduced on the silicone casting in Figure 8.18 (center) and on the tablet in Figure 8.18 (right). The arrow in Figure 8.18 (left) is pointing to concentric circle lines, which are created by a tooling manufacturing process called Computer Numerical Control (CNC). A CNC tooling apparatus creates these lines as it produces the tool; the lines appear in a concentric circle, indicating

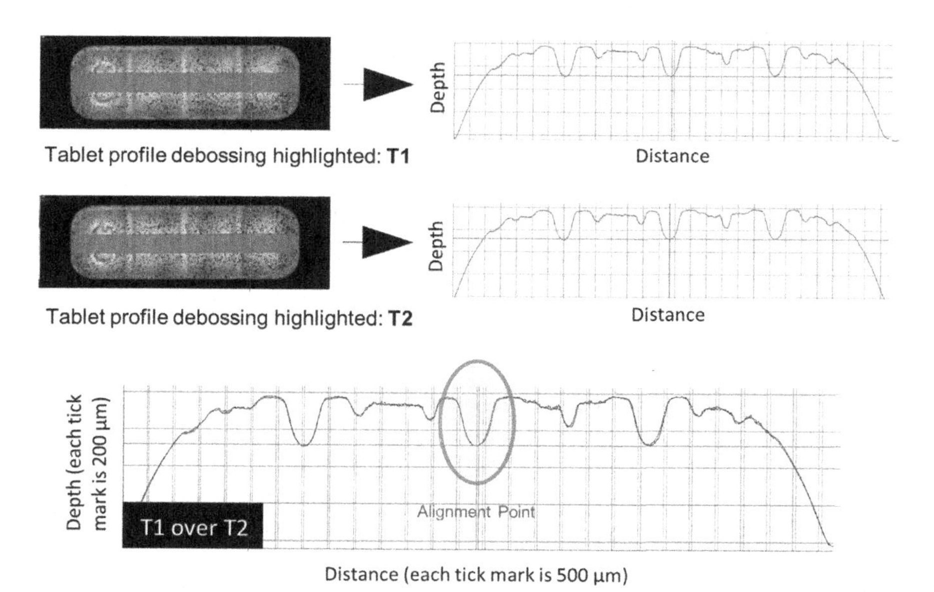

FIGURE 8.17 PM measurement of one counterfeit tablet that is consistent with that of another counterfeit tablet.

FIGURE 8.18 Images of a tablet punch tip (left), silicone casting of the tablet punch tip (center), and a suspect tablet (right).

a specific distance from each other. CNC is essentially a computer-controlled process following specific instructions about the tip without the oversight of a machining operator. These lines on the tablet punch tip in Figure 8.18 (left) were reproduced on the silicone casting in Figure 8.18 (center) and on the tablet in Figure 8.18 (right).

PM measurements from silicone castings and suspect tablets can also be compared to determine if the tablets and tablet punch tips share a common origin as shown in Figure 8.19.

Additionally, imperfections observed on suspect tablets via 3D IA can often be traced back to imperfections on the punches that were used to make the tablet. For example, consider Figure 8.20; zoomed regions are shown below each image. Figure 8.20a is an image of a tablet punch tip that displays little or no sticking and/or picking on the tablet punch tip face surface (e.g., no trapped or adhered tablet formulation material). However, the image of the tablet punch tip in Figure 8.20b exhibits picking, which is a tablet pressing condition when tablet material is trapped/stuck on the tablet punch face and is pulled off the tablet surface. Picking is often specifically noticed in small embossing regions, such as the enclosed region of the “9” (indicated with an arrow). The image of the tablet punch tip in Figure 8.20c displays a high level of tablet formulation material sticking to the surface of the tablet punch tip face surface. Sticking is a tablet pressing condition when tablet granulation material adheres to the tablet punch face surface and is pulled off the tablet surface (indicated with arrows).

Sticking and picking can cause reproducible defects on tablet surfaces that can be used to link two tablets to each other. For example, the PM measurements of

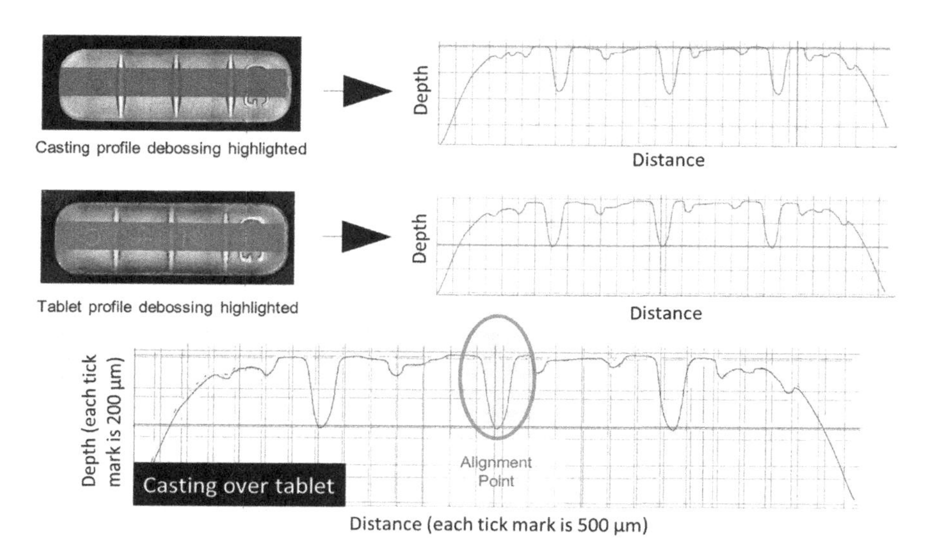

FIGURE 8.19 PM measurement for a silicone casting a tablet punch tip that is consistent with a counterfeit tablet.

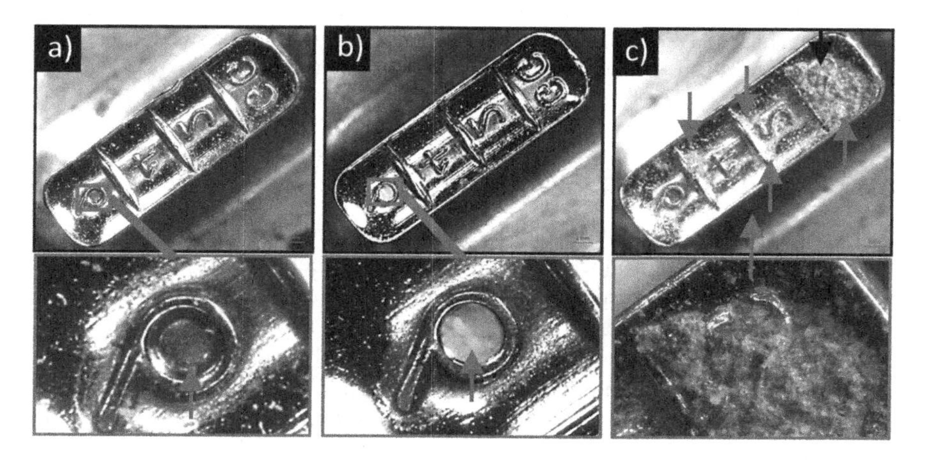

FIGURE 8.20 Images showing little or no sticking/picking on a tablet punch tip (a), picking inside the embossed "9" of a tablet punch tip (b), and a high level of tablet material sticking to the tablet punch tip (c). Images on the bottom row are zoomed in regions of the images on the top row.

two generic alprazolam 2 mg tablets (white, rectangular, quarter-scored, debossed "GG 249") shown in Figure 8.21 are consistent with each other. Furthermore, evidence of sticking and/or picking can be observed just above the score line between the "GG" and "2" debossed on the suspect tablets, which further indicate that the tablets share a common origin.

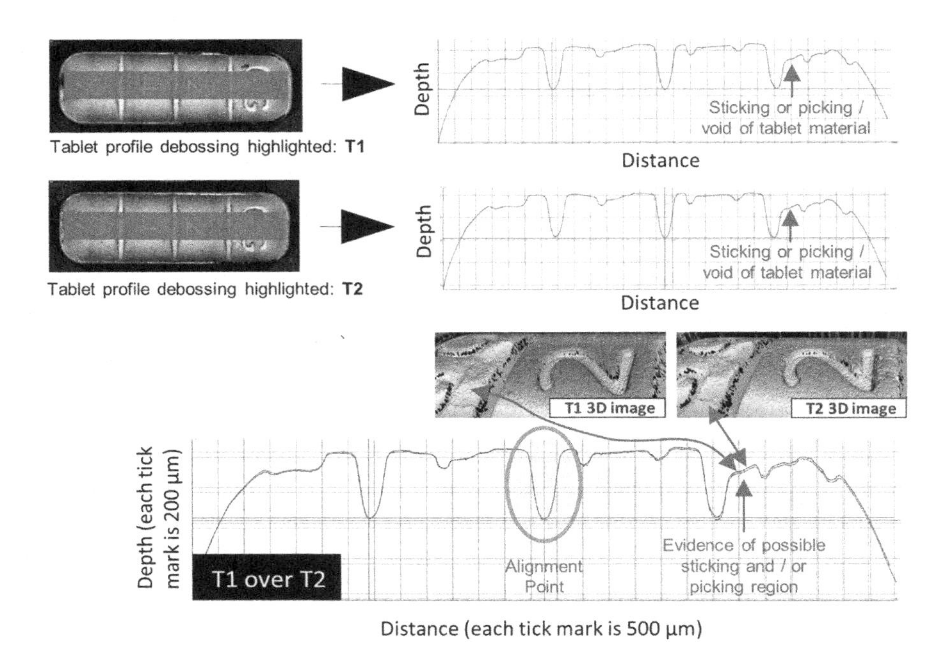

FIGURE 8.21 Consistent PM measurements for two counterfeit tablets that exhibit evidence of sticking and picking on the tablet punch tip. The reproducible sticking and picking further strengthen the conclusion that the two counterfeit tablets share a common source.

8.3.4 Visual Analysis Tools

A wide range of visual analysis tools have been employed for the analysis of counterfeit pharmaceutical products. These devices are effective for providing a rapid pass/fail assessment of suspect counterfeit products, based on visual differences with a corresponding authentic product. However, the devices are unable to determine the composition of suspect samples and can yield false positives for samples that change color over time due to non-ideal storage conditions (e.g., exposure to light, humidity, temperature fluctuations, etc.). Glossmeters (Bawuah et al., 2017), directional reflectance instruments (Wilczynski et al., 2016a), colorimeters (Rodomonte et al., 2010, Dos Santos et al., 2019), laser operated-counterfeit drug identifier (CODI) devices (Green et al., 2015, Opuni et al., 2019), microtomographic imaging instruments (Wilczynski et al., 2019), and alternate light source (ALS) instruments (Jung et al. 2012, Ranieri et al. 2014, Lanzarotta et al. 2015, Green et al. 2015, Batson et al. 2016) have been employed to compare counterfeit and authentic tablets and packaging materials.

The simplest forensic light source is an ultraviolet (UV) lamp that emits a single wavelength of light. Although a UV lamp can be used for examining samples that fluoresce in the UV, it has limited use for more complex samples, since it relies on a single wavelength to generate fluorescence. Multi-wavelength ALS devices such as the CrimeScope Forensic Light Source and the Video Spectral Comparator (VSC) have been used for many years in the forensic science field to examine fibers, fingerprints, body fluids, and documents, because they overcome these limitations of monochromatic light sources. Illuminating a sample over a wavelength range provides the analyst greater flexibility and discriminating power when examining complex samples such as pharmaceuticals. Unlike the UV lamp, multi-wavelength forensic light sources consist of a high-intensity lamp, filters, and a light guide. The lamp is a broad band source that contains both visible and infrared wavelengths. Band pass filters are used to separate individual "bands of light" (wavelengths), and the light guide transmits the light to the sample. The filtered light then interacts with the sample (e.g., absorption, reflection, fluorescence). The analyst uses colored goggles to filter out the transmitted light from the forensic light source and observes the interaction of light with the sample. ALS is a quick, nondestructive technique that has applicability inside and outside of the laboratory (e.g., crime scene investigations, mail blitzes, search warrants, etc.).

The FCC frequently uses ALS devices to examine counterfeit packaging and suspect products for evidence of tampering, but did not apply the technique to counterfeit tablets until 2003. In 2003, suspect counterfeit Lipitor tablets were submitted to the FCC for analysis (Flaherty, 2003). These samples included numerous bottles that contained a mixture of diverted authentic and suspect counterfeit Lipitor tablets. Initially, the tablets in the bottles were visually examined using SLM, since the diverted and suspect counterfeits could be differentiated by their debossing. This was a time-consuming, subjective process, and a more rapid approach was needed to screen larger numbers of tablets. Based on previous experience, ALS devices were employed to separate the diverted authentic products from the suspect counterfeit products. Using this technique, the tablets exhibited different fluorescent responses at different wavelengths (Figure 8.22), thereby allowing mixed tablets in the bottles to be separated into two groups for prioritization using more time-consuming analytical techniques.

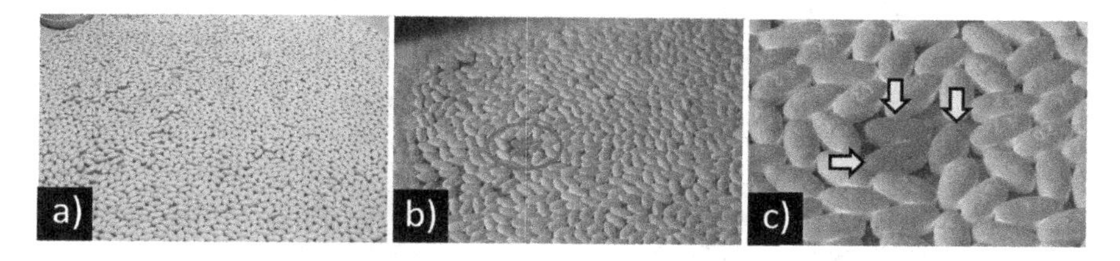

FIGURE 8.22 Approximately 2,500 tablets consisting of diverted authentic and suspect counterfeit tablets using white light (a) and monochromatic light (b). The area marked with an oval in (b) is expanded in (c). The three tablets indicated with arrows (diverted authentic) exhibit a different fluorescent response compared to the other tablets (suspect counterfeit) and were also found to contain a different excipient profile based on chemical analyses.

Based on the experience gained during the 2003 Lipitor case, other authentic products were examined using ALS devices to determine if differences could be visualized. The results were promising, and a small library of images was created of authentic products, which was used to compare to suspect samples submitted to FCC for analysis. With the emergence of online pharmacies in the 2000s, there was a need to screen products entering the US through the IMFs. The ALS technique was determined to be ideal, since it was quick and nondestructive. A forensic light source such as the CrimeScope had shown that this technique could be successfully used to compare authentic and counterfeit products.

Work began at the FCC to develop a handheld battery-operated ALS device that incorporated multiple illumination wavelengths. The result was the CD1 (Counterfeit Detector version 1), as shown in Figure 8.23. The CD1 was the first of several versions of a novel ALS handheld device developed and built by the FCC. The device consists of several single-wavelength light-emitting diode (LED) light sources ranging between UV and the visible (Vis) that are useful for providing visual comparisons between suspect and authentic products (both finished dosage and packaging) (FDA, 2013, Ranieri et al., 2016, 2018). Figure 8.24 shows two examples of suspect counterfeit products compared to their respective authentic

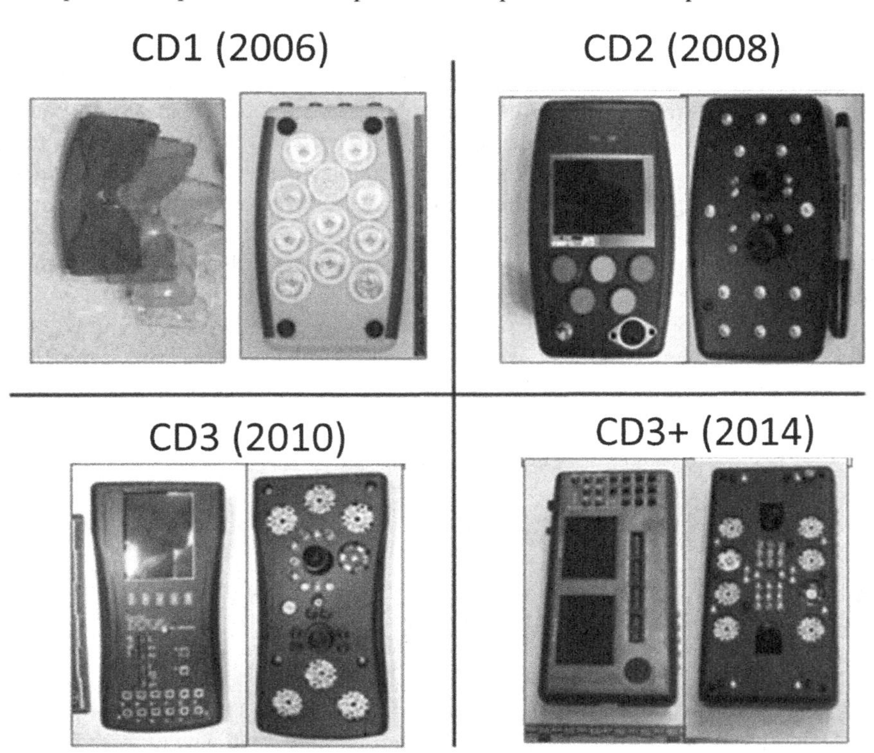

FIGURE 8.23 Comparison of the different versions of the Counterfeit Detector (CD) devices developed and built at the FCC.

products (Figure 8.24a and b). There is a clear color difference between the suspect counterfeit and authentic Viagra 100 mg tablets (Figure 8.24c) and Cialis 20 mg tablets (Figure 8.24d), which makes it easy to instantaneously differentiate suspect counterfeit from authentic. The digital images were captured using a 35 mm digital single lens reflex (SLR) camera fitted with either yellow or orange glass filters to block the incident wavelengths of light from the CD1.

The key to the CD1's success was the ability to use single-wavelength LEDs as illumination sources. Their compact design and low power consumption made them ideal as an ALS device. Each successive version of the CD device was designed and built based on experience gained from the previous versions. For example, the CD1 incorporated illumination wavelengths that were found to be necessary to detect counterfeit drugs and could be operated with a 9-volt battery or an outlet. However, this version still required the user to wear filter goggles. Consequently, the CD2 was outfitted with a charge coupled device (CCD) camera that allowed a filter to be placed over the camera and the user to view the camera

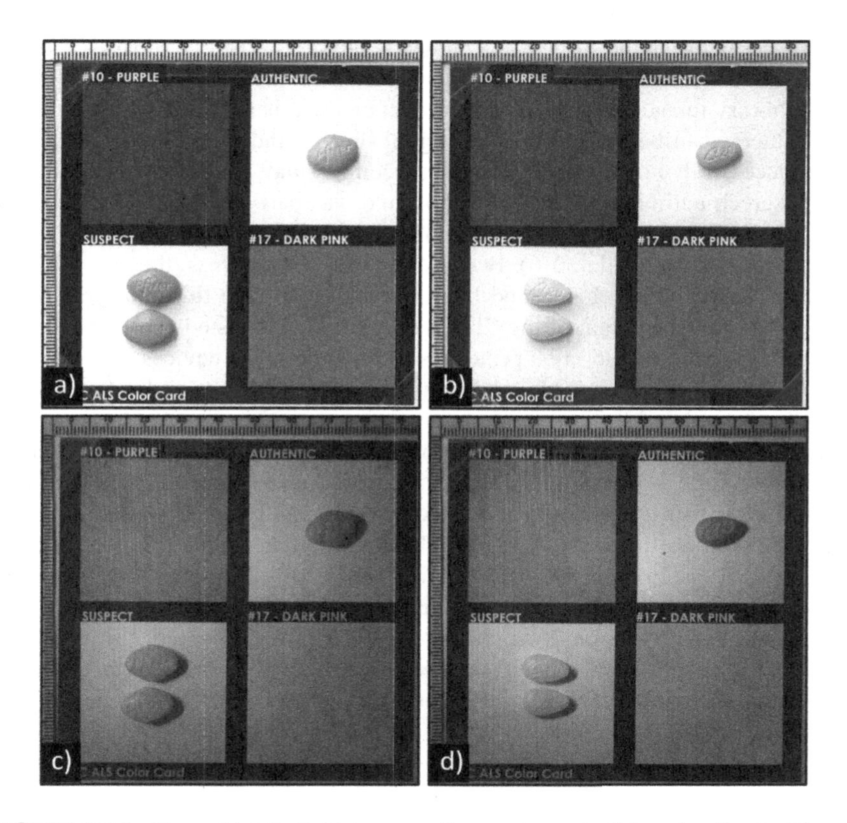

FIGURE 8.24 Normal/white light images of suspect counterfeit and authentic Viagra 100 mg tablets (a) and Cialis 20 mg tablets (b) along with images of the same tablets illuminated with the CD1 device (c and d, respectively).

screen, thus removing the need of the user to wear filter goggles. This version also added an IR LED wavelength. The CD1 and CD2 still required the user to have either a physical authentic product to perform the side-by-side comparison to the suspect tablet or use stored images on a computer to perform the comparisons. The CD3 incorporated additional UV, Vis, and IR LEDs and incorporated a storage device that allowed the authentic product images to be stored on the device. This eliminated the need for physical authentic side-by-side comparisons, since the user was no longer constrained to use the device next to a nearby computer. The CD3 still required the user to switch between live image and stored images to compare the suspect counterfeit to the authentic, and filters had to be continuously changed if the user wanted to examine a product over the full wavelength range, since it had only one camera. The CD3+ was developed and built to address these limitations.

The CD3+device contains two CCD cameras each displaying on a separate screen, which allows for simultaneous UV-Vis and IR mode viewing. Various color or IR cut-off filters can be placed in front of the CCD camera lenses to block the incident LED light wavelength, allowing for better visualization of the differences between the suspect and authentic products. The two separate screens also permit the user to view a live image of the sample on one screen and a stored library image of the authentic product on the other screen for comparison. The cameras can be used to capture digital images and videos. The CD3+ has been successfully used at IMFs to screen incoming mail packages for counterfeit drugs, screen antimalarial drugs in Laos, and was part of a study comparing it other portable techniques, such as the Mini Lab and handheld Raman (Ranieri et al., 2014, Batson et al., 2016). Because of the CD3+ success, the next version, the CD5 has been developed and built with the help of a third-party contractor. The prototype version of the CD5 is currently being tested at the IMFs. The CD5 is still based on the same principles as the other CDx devices, but incorporates many more advances in hardware and software design to make it more user friendly and efficient.

Large sample sizes encountered in the field and/or large sample submissions received by the laboratory are often first screened using normal/white light and an ALS. Whether it is a single tablet, package, a large group of tablets, or packages, ALS allows the analysts to visually examine the sample prior to the start of analysis (Lanzarotta et al., 2015, Ranieri et al., 2014, Batson et al., 2016). The ALS is an extremely versatile tool for screening large quantities of product (finished dosage and packaging) with no sample preparation, but its major limitation is a lack of chemical specificity. However, this allows for intelligent sampling as opposed to random sampling of suspect products prior to conducting further laboratory analysis.

8.3.5 Rapid Low-Cost Analytical Instruments/Methods

Several rapid low-cost methods developed to detect counterfeit drug products have been targeted for use in underdeveloped countries. Colorimetric methods have been used to examine suspect counterfeit drug products for the presence of the labeled active pharmaceutical ingredient (API) in antimalarials, (Green et al., 2000, 2001, 2015, Koesdjojo et al., 2014, Opuni et al., 2019), antibiotics (Hu et al., 2006), and miltefosine (Dorlo et al., 2012). Using these methods, suspect counterfeit finished dosages are treated with reagents that causes a color change if the API is present. Colorimetric analyses can also be combined with image analysis techniques to determine the amount of API present (Green et al., 2000, 2001, 2015, Koesdjojo et al., 2014, Opuni et al., 2019, Sherma and Rabel, 2019). Paper-based microfluidic analytical devices have been employed with and without colorimetric detection to examine suspect counterfeit ceftriaxone, chloroquine, doxycycline, and antimalarial drugs (Bliese et al., 2019, Koesdjojo et al., 2014). Immunoassay strips have been utilized to examine fentanyl- and benzodiazepine-containing counterfeit drug products (Tobias et al., 2021), and lateral flow dipstick assays have been used to examine counterfeit antimalarials (Guo et al., 2016, Ning et al., 2019, He et al., 2014). The Minilab Kit has been used to authenticate antituberculosis, antimalarial, antiretroviral, and antibiotic drug products; this technique involves a physical examination of suspect counterfeit tablets, a disintegration test, and thin layer chromatography (TLC) to determine authenticity (Opuni et al., 2019, Batson et al., 2016, Jähnke, 2004). The PharmaChk device has also been employed to authenticate antimalarial drugs; the device uses a microfluidics/luminescence-based approach for kinetic release tests and API quantitation (Desai, 2014). Vickers et al. provided an excellent review of field-based devices for screening substandard drug products, including the rapid low-cost tools described here and additional more costly tools described in the following sections (Vickers et al., 2018).

8.3.6 Near Infrared (NIR) Spectroscopic Imaging

Several groups have described the use of near infrared (NIR) spectroscopic imaging systems along with multivariate analysis to examine suspect counterfeit drug products (Rodionova et al., 2005, Lopes and Wolff, 2009, Puchert et al., 2010, Guo et al., 2016, Wilczynski et al., 2016b). NIR spectra are typically collected in the 750–2,500 nm (13,000–4,000 cm^{-1}) range, and signals in this region arise from overtones and combination bands of fundamental vibrations in functional groups such as OH, NH, and CH. NIR peaks are typically broad and weak, which is not generally desired for determining the composition of suspect samples but does provide a useful means for assigning a signature (fingerprint) to a given sample. NIR imaging detectors typically consist of focal plane arrays (e.g., InGaAs, InSb, InAs, RT-PBs, MCT) and the output of this instrument is a hyperspectral data cube where the x- and y-axes relate to static spatial positions, and the z-axis relates to variable wavelengths. The x- and y-axis dimensions are typically dictated by a combination of array pixel size and optics of the instrument, and the z-axis dimensions are dictated by the spectral resolution and wavelength range of the instrument. In other words, each pixel on the array relates to an x, y position at the sample plane and contains a NIR spectrum (z-axis) that corresponds to that spatial location. Consequently, this technique can provide the ability to spatially resolve individual ingredients in a given sample if the pixel resolution is smaller than the particle sizes of the formulation ingredients. Resulting three-dimensional hyperspectral images can then undergo multivariate analysis to compare the results of one sample (e.g., a suspect counterfeit tablet) to another sample (e.g., authentic tablet). Advantages of this technique include rapid data analysis, limited sample preparation (if any), and the ability to simultaneously examine multiple samples. Limitations include the need to develop chemometric models, time-consuming data reduction times, and specificity regarding drug ingredients (APIs and excipients). Rodionova et al. employed NIR spectroscopic imaging and principal components analysis (PCA) to authenticate antimicrobial film-coated tablets, antispasmodic un-coated tablets, and un-coated antimicrobial crushed tablets with varying API concentrations (Rodionova et al., 2005, Rodionova and Pomerantsev, 2010). Dubois et al. described the use of NIR spectroscopic imaging to simultaneously examine 30 suspect counterfeit antimalarial tablets, including some that were imaged through the blister package (Dubois et al., 2007). Their analysis determined that ten tablets were consistent with authentic tablets and contained the correct active ingredient, seven tablets were counterfeit and the API was substituted with acetaminophen, and 13 tablets were substituted with another API. Lopes et al. used NIR spectroscopic imaging to authenticate and classify/determine a possible number of sources for 55 counterfeit Heptodin™ 100 mg lamivudine tablets. Eighteen percent of the counterfeit tablets contained the API, and 82% of the counterfeit tablets contained only talc and starch (Lopes and Wolff, 2009). Using PCA and k-means clustering, the authors claim that the counterfeit tablets originated from as many as 15 different sources. Puchert et al. employed NIR spectroscopic imaging to examine suspect Concor 5 mg (bisoprolol fumarate) tablets (Puchert et al., 2010). The suspect tablets were compared to authentic tablets from several batches from two different manufacturing sites to account for variability based on location. Wilczynski et al. described the use of NIR spectroscopic imaging along with Grey-Level Co-Occurrence Matrix (GLCM) analysis and PCA to examine suspect counterfeit Viagra 100 mg (sildenafil citrate) tablets (Wilczynski et al., 2016b). Suspect tablets were determined to be counterfeit, based on significant reflectance differences as well as differences in the distribution of tablet ingredients, even for counterfeit tablets that contain the correct API.

8.3.7 Near Infrared Spectroscopy (NIR)

Several studies have described the use of benchtop and handheld NIR spectrometers for the analysis of counterfeit pharmaceuticals (Kalyanaraman et al., 2010, Sacre et al., 2010, Rodionova et al., 2010, Been et al., 2011, Dorlo et al., 2012, Degardin et al., 2016, Custers et al., 2016b, Guillemain et al., 2017, Rodionova et al., 2019a, b, Wang et al., 2020). Just as with the NIR imaging spectrometers, these devices operate in the 750–2,500 nm region, and signals arise from overtone and combination bands of fundamental vibrations. Unlike NIR imaging spectrometers, however, these devices collect single-point spectra, where the resulting signal is an average of all ingredients probed by the light. These devices are typically effective for rapidly determining if a suspect product is counterfeit and can often examine products through glass and plastic. However, the devices are less useful for determining the composition of the suspect samples and often require chemometric models to be generated on authentic samples prior to analysis of suspect samples.

Kalyanaraman et al. described the use of a handheld NIR spectrometer to examine suspect counterfeit capsules through the blister package and suspect powder through a glass vial; the device determined that both suspect products were counterfeit without having to remove them from their packaging (Kalyanaraman et al., 2010). Sacré et al. employed a benchtop NIR spectrometer along with multivariate analysis to successfully authenticate six suspect Viagra (sildenafil citrate) and five suspect Cialis (tadalafil) tablets (Sacre et al., 2010). Rodionova et al. demonstrated that NIR spectrometers were effective for differentiating high-quality counterfeit dexamethasone ampoules from authentic products (Rodionova et al., 2010). Using chromatographic methods, the only differences found between the counterfeit and authentic products were impurities, which demonstrates the highly selective capabilities of NIR for detecting counterfeits. This same group later applied similar approaches to examine counterfeit furosemide and bisoprolol tablets (Rodionova et al., 2019a), and to analyze counterfeit fluconazole capsules through the blister package and capsule shell (Rodionova et al., 2019b). Degardin et al. (2016) employed a benchtop NIR spectrometer, and Guillemain et al. (2017) employed a handheld NIR spectrometer along with multivariate analysis tools to build models for 29 different pharmaceutical product families with a total of 53 different formulations; the data set represented almost all of Roche's tablet portfolio at the time. Overall, the models were 100% effective for correctly determining if a suspect tablet was counterfeit or authentic.

8.3.8 Benchtop Raman Spectroscopy

Several studies have employed Raman spectrometers for the analysis of counterfeit pharmaceutical tablets. Raman spectra are typically collected in the 10-4000 cm^{-1} range and strong signals using this technique arise from fundamental vibrations of non-polar functional groups such as C=C and C=N. Raman peaks are typically sharp, which is advantageous for authenticity analysis and for identifying APIs. Witkowski demonstrated that Raman spectroscopy is effective for detecting counterfeit drug products based on API and excipient profiles, as well as for tablets containing the incorrect API polymorph (Witkowski, 2005). Trefi et al. (2008), DeVeij et al. (2008), Sacré et al. (2010), Kwok and Taylor (2012) all examined counterfeit Viagra (sildenafil), Cialis (tadalafil) and Levitra (vardenafil) using a benchtop Raman spectrometer and/or Raman microscope. For example, De Veij et al. employed a benchtop Raman spectrometer to examine nine suspect counterfeit and nine authentic Viagra tablets (De Veij et al., 2008). This study demonstrated that even though each of the suspect tablets contained the correct API, this instrument was able to determine that all nine suspect tablets were counterfeit based on spectroscopic differences due to the presence of incorrect excipients. Degardin et al. employed a benchtop Raman spectrometer to generate library spectra for 31 different pharmaceutical product families that consisted of 25 different tablet formulations and six capsule formulations (Degardin et al., 2011). Raman spectra of 27 suspect drug products were collected and compared to corresponding authentic spectra, and were further classified using multivariate analysis. In addition to demonstrating that all 27 drug products were counterfeit, this study determined that the counterfeit drug products could be classified into 15 unique groups based on PCA, which may be able to support results of other analytical tools that are able to determine if two or more counterfeits share a common source. The same group later employed a benchtop Raman spectrometer to generate library spectra for twelve different authentic protein-based medicines by probing the product contents (liquids and lyophilized powders) through their glass vials (Degardin et al., 2017). Results of this study indicated that seven suspect products were counterfeit based on comparison to corresponding authentic products. The authors conclude that it was possible to differentiate counterfeits from authentic products based on contrasting excipient profiles and intensity differences of protein (amide) peaks.

Ricci et al. demonstrated the ability to utilize spatially offset Raman spectroscopy to authenticate counterfeit antimalarial tablets by probing them through the blister package (Ricci et al., 2007a). Neuberger et al. employed a benchtop Raman spectrometer along with multivariate analysis to differentiate five model formulations with different coatings and varying of API and excipient concentrations (Neuberger and Neususs, 2015). The study also monitored how different storage conditions can cause minor spectroscopic changes that may be erroneously attributed to a counterfeit product. Lawson et al. described a novel binary barcode comparison method using Raman spectrometers to identify APIs in finished dosage forms (Lawson and Rodriguez, 2016). The method compares known peaks in the API reference spectrum to peaks present in the finished dosage form. Using this approach, APIs were identified in nine simulated counterfeit drug products and 18 approved drug products with an accuracy of 100%. Degardin et al. employed Raman microspectroscopy to detect 31 suspect counterfeit vials based on analysis of the packaging (Degardin et al., 2019). Along with other techniques, Raman analysis allowed the authors to link eight vials from different seizures to a common source.

A few groups have explored using Raman microspectroscopic mapping/imaging for the analysis of counterfeit pharmaceutical products. The output of this instrument is a hyperspectral data cube where the *x*- and *y*-axis relate to static spatial positions (limited by the diffraction-limited focused beam diameter), and the *z*-axis relates to variable wavelengths; the *x*- and *y*-axis dimensions are typically dictated by the step size, and the *z*-axis dimensions are dictated by wavelength limitations of the detector/instrument optics. In other words, each pixel on the map relates to an *x*, *y* position at the sample plane and contains a Raman spectrum (*z*-axis) that corresponds to that spatial location. Consequently, this technique can provide the ability to spatially resolve individual ingredients in a given sample if the pixel resolution is smaller than the particle sizes of the formulation ingredients. Sacré et al. employed Raman mapping along with multivariate analysis to examine eight authentic and 26 counterfeit Viagra tablets (Sacre et al. 2011b). Counterfeit products were compared to authentic products and, based on PCA results using full spectra, the presence of lactose and the spatial distribution of the API (sildenafil) in the tablets was determined. The authors concluded that the counterfeit products were easily differentiated from the authentic products based on spectral differences in the 200–1,800 cm^{-1} region. Adar et al. employed single-point Raman spectroscopy to differentiate suspect counterfeit packaging components (e.g., papers and inks) from those of authentic products and Raman spectroscopic mapping to differentiate suspect counterfeit and authentic formulation ingredients (Adar et al., 2014). Frosch et al. employed a novel fiber array-based Raman spectroscopic imaging approach to compare authentic Riamet (artemether and lumefantrine) tablets to self-made model counterfeit tablets (Frosch et al., 2019). This study described the ability to identify and quantify artemether and lumefantrine as a means to detect counterfeit products and substandard products, which are typically more difficult to detect than counterfeits using Raman spectroscopy.

8.3.9 Handheld Raman Spectroscopy

Ricci et al. employed a handheld Raman spectrometer to examine suspect counterfeit artensunate tablets both through the packaging and after tablets were removed from the blister packages (Ricci et al., 2008). The results of this analysis demonstrated that the device was effective for differentiating counterfeit and authentic tablets, and demonstrated that fluorescence, which typically hinders the efficacy of Raman spectroscopy, can actually be used as an advantage for counterfeit analysis. Hajjou et al. (2013) and Batson et al. (2016) employed a handheld Raman spectrometer to determine the potential of each device for evaluating the quality of medicines. Both studies determined that the device was effective for differentiating counterfeit and authentic drug products but was not effective for identifying substandard products, specifically products that contain sub-potent API concentrations. These conclusions are expected, since the device is more effective for differentiating spectra based on peak position (qualitative information) rather than peak height (quantitative information). Degardin et al. employed a handheld Raman spectrometer to generate library spectra for 33 different pharmaceutical product families that consisted of 62 different product formulations, including tablets, capsules, and powders (Dégardin et al., 2017). Raman spectra of 44 known counterfeits were collected and compared to those of appropriate authentic drug products. The method was evaluated in terms of accuracy and robustness (e.g., measuring the same sample multiple times, measuring samples multiple times at different locations in the sample holder, measuring through different materials, and measuring the same sample by different analysts). Regarding accuracy, analysis of all authentic products yielded a pass, and analysis of all known counterfeit products yielded a fail, which indicated that the device was highly effective for determining authenticity. Regarding robustness, no significant differences were noted when measuring the same sample multiple times, when measuring samples multiple times at different locations in the sample holder, and when measuring the same sample by different analysts. The authors did, however, determine that the device was effective for sampling through transparent glass vials and through transparent blisters but not effective for sampling through white blister packages. Lanzarotta et al. followed this study by using handheld Raman spectrometers to examine a total of 84 suspect counterfeit drug products from 11 different finished dosage forms labeled to contain controlled substances (Lanzarotta et al., 2020). Authentic spectra were collected on three "parent" devices and then added to a library; the library was transferred to the three "parent" devices and 13 "daughter" devices for a total of 16 devices. High true pass/true fail rates and low false pass/false fail rates were achieved using both parent and daughter devices, which demonstrated that the parent–daughter electronic transfer method was successful, and that methods developed in the laboratory can be seamlessly pushed out to field devices.

While Raman spectroscopy has proven to be an excellent tool for detecting counterfeit finished dosages, it is less effective for identifying APIs in these products for low- concentration formulations. Consequently, recent studies have focused on improving the detection limits of these instruments via surface-enhanced Raman spectroscopy (SERS) (Lanzarotta et al., 2017, Kimani et al., 2021a). Lanzarotta et al. (2017) developed a simple, sensitive, pass/fail field-friendly SERS method to detect sildenafil in falsified sildenafil-containing tablets, including counterfeit Viagra tablets, using handheld spectrometers. The method was 98.3% and 91.7% effective during in-lab and in-field validation studies respectively, and yielded detection limits ranging between 10 and 625 µg/mL (ppm), which is well below concentrations typically encountered with this API. Kimani et al. employed SERS for trace detection of fentanyl, hydrocodone, oxycodone, tramadol, and several benzodiazepines in suspect tablets and capsules, including counterfeits, using two different handheld Raman devices—one equipped with a 785 nm laser and one equipped with a 1,064 nm laser (Kimani et al., 2021a, b). These studies achieved detection limits ranging from 0.5 to 50 µg/mL using the 785 nm device and from 1 to 75 µg/mL using the 1,064 nm device, which were below concentrations typically encountered with these APIs. The same group is currently working on developing methods for detecting additional APIs using SERS, including those listed in Table 8.1. In addition to the previous work, (Lanzarotta et al., 2020) this group demonstrated the ability to both authenticate suspect tablets and identify APIs using a single handheld Raman device. Furthermore, since these devices are handheld, they are transferrable to the field at remote sampling sites, such as satellite laboratories stationed at IMFs and express courier hubs.

8.3.10 Fourier Transform Infrared (FT-IR) Spectroscopy

Fourier transform infrared (FT-IR) spectroscopy is one of the most effective tools for comparing authentic and suspect pharmaceutical products. FT-IR spectra are typically collected in the 400-4000 cm^{-1} range and strong signals using this technique arise from fundamental vibrations of polar functional groups such as OH, C-O, C=O. FT-IR peak widths and intensities vary but the technique is also often effective for identifying APIs. Like other vibrational spectroscopic techniques,

TABLE 8.1
Current Target Analytes for SERS Method Development at FCC

4'-methyl acetyl fentanyl HCI	Diclazepam	Oxycodone
4'-methyl fentanyl HCI	Ephedrine	Papaverine
4-ANPP	Estazolam	Para-chlorofentanyl HCl
Acetyl fentanyl HCI	Etizolam	Para-chloroisobutyryl fentanyl HCl
Acetyl norfentanyl	Fentanyl	Para-fluoro acrylfentanyl salt
Acrylfentanyl HCl	FIBF	Para-fluoro cyclopropyl fentanyl salt
Adinazolam	Finasteride	Para-fluoro tetrahydrofuran fentanyl salt
Adrafinil	Flualprazolam	Para-fluorobutyryl fentanyl HCl
Alprazolam	Flubromazepam	Para-fluorofentanyl HCl
Avanafil	Flubromazolam	Para-methoxy-butyryl fentanyl
Benzodioxole fentanyl	Fluoxetine	Para-methoxyfentanyl
Benzyl acrylfentanyl HCI	Furanyl fentanyl 3-furancarboxamide HCI	Para-methyl fentanyl HCI
Benzyl carfentanil HCI	Furanyl norfentanyl HCI	Phenazepam
Benzyl fentanyl	Furanylethyl fentanyl HCl	Phentermine
Bromazepam	Galanthamine HBr	Phenyl fentanyl HCI
Buprenorphine HCl	Higenamine	Pramiracetam
Bupropion HCI	Hydrocodone	Sertraline
Butyryl fentanyl carboxy metabolite salt	Isobutyryl norfentanyl	Sibutramine
Butyryl fentanyl HCI	Lidocaine	Sildenafil Citrate
Butyryl norfentanyl HCl	Lorazepam	Temazepam
Cardarine	Meclonazepam	Tetrahydrofuran fentanyl HCI
Cis-3-methyl Butyryl fentanyl HCl	Meta-Fluorofentanyl HCI	Thebaine
Cis-3-methyl fentanyl HCI	Meta-Fluoroisobutyryl fentanyl HCI	Thienyl fentanyl HCI
Cis-3-methyl norfentanyl salt	Meta-methylfentanyl HCI	Tramadol
Clenbuterol	Methoxyacetyl fentanyl HCI	Triazolam
Clonazepam	Metizolam	U-69593
Clonazolam	Midazolam	Valeryl fentanyl carboxy metabolite salt
Codeine	Modafinil	Valeryl fentanyl HCI
Cyclobuytl fentanyl HCI	Morphine	Vardenafil
Cyclohexyl fentanyl HCl	Norfentanyl salt	Zolpidem
Cyclopentyl fentanyl HCI	Noroxymorphone salt	α-methyl acetyl fentanyl HCI
Cyclopropyl fentanyl HCI	Noscapine	α-methyl butyryl fentanyl HCI
Cyproheptadine HCI	Ortho-fluoro Acrylfentanyl HCI	α-methyl thiofentany I HCI

(Continued)

TABLE 8.1 (*Continued*)
Current Target Analytes for SERS Method Development at FCC

Despropionyl meta-fluorofentanyl	Ortho-Fluorobutyryl fentanyl HCl	β-hydroxythiofentanyl HCI
Despropionyl ortho-fluorofentanyl	Ortho-fluorofentanyl HCI	β-methyl fentanyl
Desprop ionyl para-fluorofentanyl	Ortho-fluoroisobutyrylfentanyl HCl	
Diazepam	Ortha-methylfentanyl HCI	

FT-IR often permits examination of the as-received state of the sample, which includes tablet coatings, cores, packaging materials, etc. Unlike other vibrational spectroscopic techniques based on dispersive configurations, peak positions using FT-IR spectrometers are highly reproducible due to the frequency precision of the interferometer. Furthermore, when used in an attenuated total reflection (ATR) modality, peak heights using ATR-FT-IR spectrometers are highly reproducible due to the volumetric resolution of the ATR sampling accessory. Consequently, this instrument provides an excellent means for comparing two different spectra (i.e., authentic and suspect counterfeit products) due to high-precision peak positions and intensities.

Several studies describe the use of FT-IR spectroscopy to detect counterfeit drug products. Sacre et al. (2010), Deconinck et al. (2012b), Custers et al. (2016b), Lanzarotta et al. (2011, 2013, 2015), and Witkowski and DeWitt (2020) employed FT-IR spectroscopy with and without multivariate analysis to successfully differentiate authentic and counterfeit Viagra and Cialis tablets, among others. Similarly, Dorlo et al. (2012), Mariotti et al. (2013) Lawson (2014), and Tobias et al. (2021) used this method to authenticate suspect counterfeit miltefosine, Desobesi-M (fenproporex), acetaminophen, and Xanax (alprazolam) tablets/capsules, respectively. Wang et al. employed FT-IR spectroscopy to differentiate simulated authentic and counterfeit tablets containing artemether, efavirenz, and isoniazid (Wang et al., 2020). Degardin et al. employed FT-IR spectroscopy to determine the number of unique profiles in a batch of counterfeit capsules (Been et al., 2011) and to examine suspect packaging materials, including ink and paper associated with folding boxes and leaflets (Degardin et al., 2018), as well as packaging/labeling associated with counterfeit vials (Degardin et al., 2019). Andria et al. employed FT-IR spectroscopy to differentiate suspect counterfeit and authentic blister package components, including the plastic, inner and outer part of the foil safety seal, and printing (paper/ink) (Andria et al., 2012).

While FT-IR spectroscopy is an excellent tool for authenticating suspect tablets and for identifying APIs at high concentrations, it is much less effective for identifying APIs in these tablets at low concentrations and in the presence of interfering matrices. Individual tablet/capsule particle sizes (μm) are often much smaller than the sampling aperture (mm), which often results in a complex mixture spectrum for multi-component samples. Although spectral subtractions can be performed to identify individual ingredients in multi-component samples, this approach is hampered by artifacts that make comparisons to a standard difficult and is typically only effective for identifying a limited number of components. One way to overcome these limitations is to use FT-IR spectroscopic imaging, which utilizes a multi-channel detector to collect an infrared spectrum at each spatial position in a two-dimensional region of interest. The multi-channel detector pixel size or the diffraction-limited focused beam diameter (whichever is larger) dictates the size of each spatial element at the sample plane; instrument optics can often be configured so that particles of interest are larger than the pixel size or the diffraction-limited focused beam diameter. Consequently, individual ingredients that are physically separated from each other will often yield spectra that are characteristic of nearly pure compounds, which makes it possible to identify several ingredients in a multi-component sample. Ricci et al. employed FT-IR spectroscopic imaging to differentiate authentic and counterfeit artensunate tablets based on absence of the API and presence of an unlabeled excipient in the counterfeit product (Ricci et al., 2007a, b). Lanzarotta et al. compared the performance of FT-IR spectroscopic imaging to macro-FT-IR spectroscopy for examining counterfeit tablets, and demonstrated that the former was more effective for counterfeit sourcing, and the latter was more effective for counterfeit detection (Lanzarotta et al., 2011). The same group later described an effective method for characterizing counterfeit finished dosage forms (e.g., tablet coatings and cores) and packaging materials (e.g., multi-layered foil safety seals and bottle labels) using FT-IR spectroscopic imaging (Winner et al., 2016). Counterfeit tablets were differentiated from authentic tablets based on different coating and core formulations, and counterfeit packaging materials were differentiated from authentic packaging based on different paper and adhesive components of the pharmaceutical labels.

8.3.11 Nuclear Magnetic Resonance (NMR) Spectroscopy

One- and two-dimensional nuclear magnetic resonance (NMR) spectroscopy has been utilized to examine suspect counterfeit pharmaceutical products. Using these techniques, suspect counterfeit products can be differentiated from authentic products based on spectral differences and based on the presence of ingredients not found in the authentic formulation. Trefi et al. (2008) employed ^{1}H NMR and two-dimensional diffusion-ordered ^{1}H NMR spectroscopy (2D DOSY^{1}H NMR) to authenticate suspect counterfeit Cialis tablets and Nyadong et al. (2009) later employed the latter technique to examine 14 different suspect counterfeit antimalarial tables. In this study, 2D DOSY^{1}H NMR determined that only five of the 14 tablets contained the correct active ingredient, and the results were used to separate counterfeits into groups based on common excipients that were detected. Using ^{19}F, ^{1}H and ^{13}C NMR, McEwen et al. examined ten counterfeit corticosteroid-containing creams and ointments that contained various forms of hydrocortisone, dexamethasone, flumethasone, betamethasone, and triamcinolone (McEwen et al., 2012). Using^{13}C NMR, Silvestre et al. generated isotopic fingerprints of aspirin and acetaminophen tablets and capsules (Silvestre et al., 2009). Shortly thereafter, Bussy et al. used a similar approach to fingerprint ibuprofen tablets. These authors demonstrated feasibility for using this technique to compare suspect counterfeit tablets with authentic tablets and to compare suspect counterfeit tablets with each other for sourcing purposes (Bussy et al., 2011). Guadiano et al. employed ^{1}H and ^{13}C NMR to identify and quantitate an adulterant (aceclofenac) in counterfeit Adderall tablets (authentic Aderall tablets contain amphetamine and dextroamphetamine) (Gaudiano et al., 2016). Akhunzada et al. employed ^{1}H time-domain NMR (TD-NMR) to detect counterfeit biologic pharmaceuticals based on slight antibody concentration differences compared to known authentic products (Akhunzada et al., 2021).

8.3.12 X-Ray Spectroscopy

Ortiz et al. (2012), Anzanello et al. (2014), and later Soares et al. (2019) employed X-ray fluorescence (XRF) and multivariate analysis to differentiate elemental fingerprints of authentic and counterfeit sildenafil- and tadalafil-containing pharmaceutical tablets and to compare counterfeit tablets with each other. These studies even yielded semiquantitative results by using signals for sulfur, phosphorus, calcium, titanium, and iron, as surrogates for sildenafil, calcium phosphate, titanium oxide, and iron oxide, respectively. Li-hui et al. (2013), Alsallal et al. (2018), Rebiere et al. (2019) used XRF with and without multivariate analyses to differentiate authentic and suspect counterfeit lamivudine-/zidovudine-/nevirapine-containing and atenolol- and clopidogrel-containing tablets, respectively. Degardin et al. employed XRF and scanning electron microscopy with energy dispersive X-ray spectroscopy (SEM-EDS) to examine components of drug packaging products (Dégardin and Roggo, 2017, Degardin et al., 2018, 2019). This group differentiated suspect counterfeit and authentic products based on elemental differences of ink and paper on cartons and leaflets as well as the cap and glass portions of vials that contained a protein-based medication. Andria et al. employed SEM-EDS to examine components of a foil blister package, paper packaging, and ink (Andria et al., 2012). In this study, all suspect products were differentiated from the authentic products based on contrasting elemental profiles.

Using X-ray diffraction (XRD) spectroscopy, Dewitt and Witkowski created a library that consisted of 13 authentic finished dosages and 34 APIs/excipients known to be present in these products (DeWitt, 2015, Witkowski and DeWitt, 2020). While the method proved to be successful for differentiating suspect counterfeit and authentic drug products based on XRD pattern profile differences, the authors concluded that this method was not effective for identifying APIs or excipients in these products. Beckers et al. used XRD to differentiate suspect counterfeit and authentic Viagra tablets (Beckers, 2021). Analysis of the XRD data using multivariate analysis allowed the authors to separate the counterfeit tablets into clusters that may indicate that the tablets shared a common source. Crews et al. employed XRD along with multivariate analysis to compare API concentration levels of suspect counterfeit and authentic ibuprofen- and acetaminophen-containing tablets (Crews, 2018). Partial least squares (PLS) and partial least-squares regression (PLSR) analysis of the XRD data yielded concentration levels consistent with what were reported with higher resolution instruments, and allowed the authors to differentiate counterfeit and authentic tablets.

8.3.13 Gas Chromatography with Mass Spectrometry (GC/MS)

GC/MS is an effective tool for separating and identifying volatile and semi-volatile compounds. The presence of an unlabeled API or absence of a labeled API in a suspect product may be an indicator that the product is counterfeit. Comparison of chromatographic profiles of authentic and suspect counterfeit products can also be used to differentiate the two. Alabdalla employed GC/MS to examine 124 batches of suspect counterfeit Captagon (phenethylline) tablets (Alabdalla, 2005). This study indicated that the suspect products did not contain the correct API but instead contained amphetamine, caffeine, and other unlabeled ingredients. The authors also used the results of this analysis to compare similarities between different products for potential sourcing purposes. Graham et al. utilized GC/MS to examine 38 vials/sachets for parenteral use, and 19 oral suspect counterfeit performance-enhancing and image-enhancing drug products (Graham et al., 2009). Using trimethyl silyl (TMS) derivatization, GC/MS analysis was able to identify labeled and unlabeled steroids, and determine that approximately half of the suspect products were counterfeit. Been et al. employed GC/MS along with other analytical devices to divide chromatographic profiles of suspect samples into groups for potential sourcing purposes (Been et al., 2011). Mariotti et al. utilized GC/MS to examine suspect counterfeit Desobesi-M (fenproporex) capsules; results indicated that the suspect capsules actually contained sibutramine and were therefore counterfeit (Mariotti Kde et al., 2013). Foroughi et al. employed GC/MS to examine 80 suspect counterfeit herbal medicines used for opioid suppression therapy (Foroughi et al., 2017). Their study indicated that more than 96% of the products were adulterated with an unlabeled API, and, of these samples, 90% contained diphenoxylate and 67% contained tramadol. Tobias et al. used GC/MS as a confirmation tool, following other rapid screening instruments for the analysis of 20 counterfeit alprazolam tablets (Tobias et al., 2021). GC/MS analysis indicated that two (10%) of the samples contained only alprazolam, and 18 (90%) were counterfeit. Four (20%) counterfeit tablets contained alprazolam along with other APIs, and 14 (70%) contained unlabeled APIs such as antihistamines, other benzodiazepines, synthetic cannabinoids, and fentanyl analogs.

Deconinck et al. employed headspace GC/MS (HS-GC/MS) with and without multivariate analysis to differentiate authentic and suspect counterfeit Viagra and Cialis tablets based on differences between the presence and amount of ten residual solvents (Deconinck et al., 2012a, Custers et al., 2014). Dos Santos et al. employed GC/MS along with flame ionization detection (GC-FID/MS) and chemometric analyses to examine 45 suspect counterfeit stimulants and antidepressants (Santos et al., 2020). This study was able to both differentiate authentic and counterfeit drug products, and identify similarities between the counterfeit products. The authors used data from this study to populate a database that will be continually updated so that chemical profiling can be conducted for traceability purposes.

8.3.14 High-Performance Liquid Chromatography (HPLC)

Counterfeit antimalarial tablets (Gaudiano et al., 2006, Hoellein and Holzgrabe, 2014), isometamidium-containing drug products (Schad et al., 2008), PDE-5-containing drug products (Fejős et al., 2014), Vasodarone (drondarone) and Cordarone (amiodarone) tablets (El-Bagary et al., 2015), dapoxetine-containing products (Toth et al., 2020), and Plavix (clopidogrel), Aspirin (acetylsalicylic acid), and Prasista (prasugrel) tablets(Mohammad et al. 2020) have been examined using HPLC with ultraviolet detection (HPLC/UV) and/or ultra-HPLC (UHPLC). Yang et al. (2010) employed HPLC with diode array detection (HPLC/DAD) to examine suspect counterfeit PDE-5-containing drug products, and McCord et al. describe a partnership between Purdue University and the Kilimanjaro School of Pharmacy to develop standardized HPLC methods to examine counterfeit drug products (McCord et al., 2015). McCord et al. demonstrate feasibility for their proposed standardized HPLC method by examining amoxicillin-, hydrochlorothiazide-, lumefantrine-, mefloquine-, and quinine-containing counterfeit drug products and plan to continue developing methods for a much wider range of commonly encountered products. In each of these studies, HPLC was successfully employed to differentiate authentic and suspect counterfeit drug products based on differences between the presence and amounts of APIs and impurities in these products. HPLC/UV has also been employed along with multivariate analysis techniques to differentiate authentic and suspect counterfeit drug products based on contrasting impurity fingerprints. This approach has been successfully employed to detect counterfeit Viagra (Sacre et al., 2011a, Deconinck et al., 2012c), Cialis (Sacre et al., 2011a, Custers et al., 2016a, Deconinck et al., 2012c) and several different polypeptide drugs (Janvier et al., 2018). High-performance anion exchange chromatography was used to discriminate authentic and suspect counterfeit drug products based on different carbohydrate excipient profiles (Beasley et al., 2015). Even though excipients are often considered inert, they can affect the bioavailability and stability of the API so that any deviation from the FDA-approved form can potentially be harmful to the consumer.

Several studies also describe the use of HPLC with mass spectrometric detection (LC/MS) and tandem mass spectrometric detection (LC/MS/MS) for the analysis of suspect counterfeit drug products. These techniques are effective for separating, identifying, and quantifying a wide range of drug compounds, and are not limited to volatile and semi-volatile, low-molecular-weight compounds such as with GC/MS. However, these methods typically need to be tailored for specific analytes much more so than with GC/MS. Using LC/MS or UHPLC/MS, Wolff et al. (2003) identified sulfamethazine (an antibacterial) in counterfeit Halfan (halofantrine) antimalarial suspensions, Yao et al. (2007) examined suspect counterfeit anti-diabetic drugs for the presence and amount of the labeled API, Graham et al. (2009) identified several steroids and other APIs in counterfeit performance- and image-enhancing drug products, Dorlo et al. (2012) examined suspect counterfeit capsules labeled to contain miltefosine for the presence and amount of the correct API, Cho et al. (2015) examined suspect counterfeit drug products for the presence of 26 anabolic-androgenic steroids, and Gaudiano et al. (2016) identified aceclofenac (a non-steroidal anti-inflammatory drug) in counterfeit Adderall (amphetamine/dextroamphetamine) tablets. Lebel et al. (2014) and Lee et al. (2016, 2019) describe methods for identifying and quantifying up to 80 PDE-5 analogs in counterfeit drug products marketed for erectile dysfunction. In each of these cases, LC/MS was employed to identify and/or quantify labeled ingredients, unlabeled ingredients, and/or impurities as a means to differentiate suspect counterfeit drug products from authentic products. Jiang et al. employed LC/MS to differentiate suspect counterfeit and authentic human growth hormone (hGH) based on oxidation, deamidation, and chain cleavage differences (Jiang et al., 2009). Hall described a method using LC/MS followed by multivariate analysis techniques to differentiate suspect counterfeit and authentic antimalarial tablets, and then chemically fingerprint those found to be counterfeit based on the presence and amount of labeled and unlabeled ingredients (Hall, 2005). Using this approach, the authors were able to draw a correlation between sample origin and sample composition.

8.3.15 Ambient Ionization Mass Spectrometry

Ambient ionization mass spectrometers do not require a vacuum and are typically able to rapidly collect MS data on a wide variety of samples in their native state. The most significant limitation of these instruments is that they lack front-end separation stages, which can cause them to be less effective for multi-component samples. However, instruments can be configured to yield accurate mass information and can be equipped with imaging stages that allow ingredients in a multi-component sample to be separated without the use of chromatography. Ambient ionization mass spectrometers have been employed to differentiate counterfeit and authentic drug products based on the presence of unlabeled ingredients and absence of labeled ingredients. For example, Wolff et al. employed a quadrupole orthogonal acceleration time-of-flight (TOF) mass spectrometer equipped with an electrospray ion (ESI) source to examine suspect counterfeit Halfan (phenanthrene) suspensions (Wolff et al., 2003). Along with other analytical techniques, accurate mass infusion MS and MS/MS results generated using this instrument helped differentiate the suspect counterfeit tablets from authentic tablets based on the presence of an unlabeled API and/or absence of the labeled API.

Ricci et al. (2007b) employed desorption electrospray-ionization ion-trap mass spectrometry (DESI-MS), Nyadong et al. (2007) employed reactive DESI-MS, and Nyadong et al. (2009) utilized imaging DESI-MS to differentiate suspect counterfeit and authentic antimalarial tablets. Using DESI-MS, ionization is enhanced by chemical reactions between the charged spray microdroplets and surface molecules of a sample, thereby increasing instrument sensitivity. Reactive DESI-MS further improves sensitivity by forming stable noncovalent complexes between compounds dissolved in the DESI spray solution and API molecules in a sample. Using imaging DESI-MS, mass spectra are collected at each defined spatial location in a two-dimensional region of interest. While this approach does not use a chemical mechanism or high-resolution mass spectrometer as a means to separate compounds in a multi-component sample, formulation ingredients spatially separated from each other using this technique typically exhibit spectra as "pure" as those separated chromatographically.

Fernandez et al. (2006), Nyadong et al. (2009), and Samms et al. (2011) utilized direct analysis in real-time TOF mass spectrometry (DART-MS), and Twohig et al. (2010) employed an atmospheric solids analysis probe (ASAP) along with TOF-MS to examine antimalarial, alprazolam, and PDE-5 tablets, respectively. These instruments were able to discriminate suspect counterfeit products from authentic products based on the presence of unlabeled ingredients and/or absence of labeled ingredients. The high resolution of TOF instruments provides the ability to separate compounds in a multi-component sample based on exact mass. However, high resolution is not necessarily required to authenticate suspect counterfeit tablets, especially for products that do not contain the labeled API. Consequently, several studies have focused on developing/applying methods to simplify and reduce the cost of these instruments. For example, Bernier et al. employed a DART-MS instrument equipped with a single quadrupole mass spectrometer as a lower cost and more portable alternative to TOF-based instruments for the analysis of antimalarials; this instrument detected the same compounds as a high-resolution TOF instrument (Bernier et al., 2016). The same group later utilized a triboelectric nanogenerator (TENG) mass spectrometric instrument to examine antimalarials; this instrument uses lower voltages to produce gas-phase ions, which significantly reduces instrument complexity and cost. (Bernier et al. 2018) This study indicated that the TENG device was able to yield comparable detection capabilities as other more complex ambient ionization devices at a fraction of the cost.

8.3.16 Less Commonly Employed Techniques/Methods

A few studies have described feasibility of novel approaches for counterfeit drug detection. For example, using a low-cost capillary electrophoresis instrument to quantitate APIs, Marini et al. determined that suspect counterfeit quinine tablets were sub-potent and possibly counterfeit (Marini et al., 2010). Velpandian et al. utilized sodium dodecyl sulfate polyacrylamide gel electrophoresis (SDS-PAGE) and Bradford assay to determine protein weight and concentration for authentic and suspect counterfeit bevacizumab ophthalmic formulations (Velpandian et al., 2016). Wilczynski et al. utilized dynamic thermal analysis to quickly differentiate authentic and suspect counterfeit drug products based on thermokinetic parameter differences; authentic tablets had a lower time constant (temperature decrease as a function of time) compared to that of the suspect counterfeit tablet, which is attributed to the products having different ingredients (Wilczynski, 2015). Dos Santos et al. employed a low-cost differential scanning calorimeter along with hierarchical cluster analysis to examine 25 counterfeit Viagra and Cialis tablets. Counterfeit products were differentiated from authentic products based on different thermal profiles (Dos Santos et al., 2019). Using laser ablation multicollector inductively coupled plasma mass spectrometry (LA-MC-ICP-MS), followed by high-performance liquid chromatography (LC-MC-ICP-MS), Santamaria-Fernandez et al. differentiated authentic and suspect counterfeit tablets based on sulfur isotopic ($\delta^{34}S$) signature differences related to the sulfur-containing API (Santamaria-Fernandez et al., 2008).

8.4 GUIDANCE FOR EXAMINING SUSPECT COUNTERFEIT DRUG PRODUCTS

The number of analytical instruments that have been used to examine suspect counterfeit drug products is staggering, which demonstrates that there is no single tool that is universally effective for all products. Since it can be overwhelming to compare the relative capabilities of these techniques to determine the best course of action for examining counterfeit drug products, the Asia-Pacific Economic Cooperation (APEC) Life Sciences Innovation Forum (LSIF) Regulatory Harmonization Steering Committee (RHSC) prepared a "Guidance Document on the Use of Detection Technologies and Overview of Detection Technologies for Drug Safety" (Witkowski, 2015). In this document, the group discusses the relative capabilities of the most commonly employed techniques used to examine counterfeit drug products. Consider Figure 8.25, which was taken from this RHSC guidance document. The *x*- and *y*-axes represent the sensitivity for detecting APIs and ability to differentiate counterfeits, respectively, for level I (poor quality),

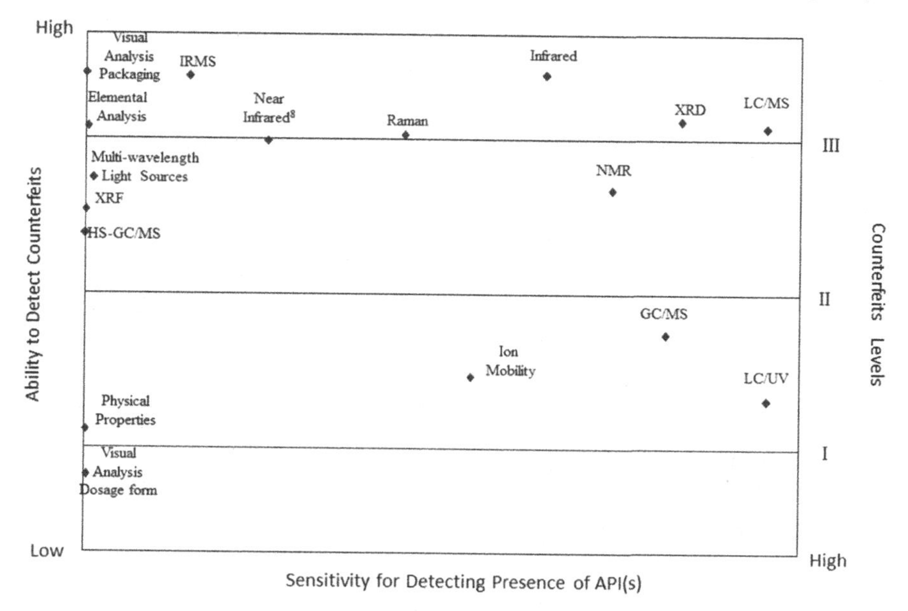

FIGURE 8.25 Relative strength for detecting counterfeit drug products and relative sensitivity of analytical devices for detecting APIs in suspect counterfeit drug products. Figure taken from the RHSC guidance document (used with permission).

level II (moderate quality), and level III (high-quality) counterfeits. While this document demonstrates that there is no single course of action that is effective for all counterfeit drug products, it also demonstrates that each counterfeit drug product requires a multi-disciplinary approach. For example, physical properties differentiate suspect counterfeit drug products, but they will not determine which APIs are present. Or, LC/UV may be able to determine that the labeled API is present but will not be able to determine if the product is counterfeit, unless an unlabeled API is detected or if the suspect product exhibits a chromatographic profile that is different from the authentic. In this case, additional techniques that have a strong ability to differentiate counterfeits is required (e.g., visual analysis, NIR, Raman, IR, etc.). Overall, a multi-faceted approach is recommended for counterfeit drug analysis that includes one or more techniques that are effective for counterfeit drug detection and one or more techniques that are effective for API detection. In fact, many studies described throughout this chapter employed two or more techniques for this exact purpose.

The APEC LSIF RHSC guidance document also contains a recommended flow chart for the analysis of suspect counterfeit drug products (Witkowski, 2015), which is shown in Figure 8.26. When an investigator encounters a suspect counterfeit drug product (Figure 8.26a), the packaging and finished dosage

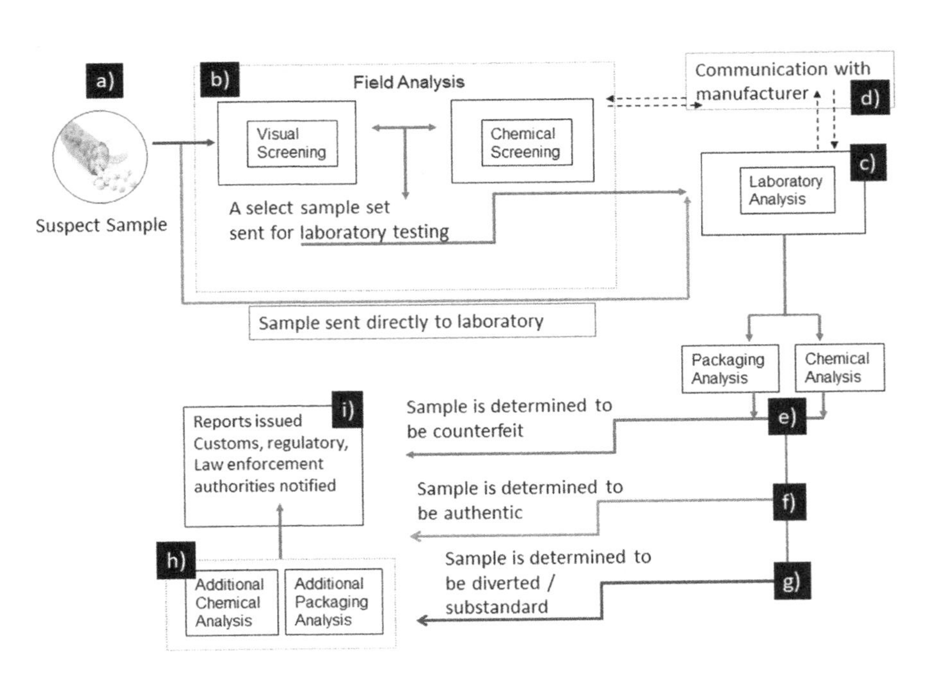

FIGURE 8.26 Recommended flow chart for the analysis of suspect counterfeit drug products by the RHSC. Figure taken from the RHSC guidance document (used with permission).

form can be visually screened using handheld ALS and other optical-enhancing devices, and can be chemically screened using a wide variety of handheld and field portable devices (Figure 8.26b). Results of these field examinations can be used to prioritize samples requiring confirmatory testing at a full-service laboratory (Figure 8.26c). Laboratory analysts receiving the sample and field analysts who sent the sample can communicate with the manufacturer to acquire authentic products for comparison, in addition to information regarding effective methods for differentiating the authentic product from counterfeit products (Figure 8.26d). Results of these analyses will help determine if the sample is counterfeit (Figure 8.26e), consistent with authentic (Figure 8.26f), or possibly diverted (Figure 8.26g). Diverted samples (those manufactured for a foreign market) may require additional testing (Figure 8.26h) before issuing reports to customs, regulatory, and/or law enforcement authorities (Figure 8.26i). Further discussion regarding how this approach is used in practice is described in the following section.

8.5 CURRENT COUNTERFEIT DRUG ANALYSIS TRENDS AT FDA'S FORENSIC CHEMISTRY CENTER

The FDA's FCC has used a similar approach to that described by the APEC LSIF RHSC for the analysis of suspect counterfeit products for the last 25+ years. This laboratory has seen a wide variety of suspect counterfeit products, including the non-controlled substance-containing pharmaceuticals and veterinary medications listed in Table 8.2.

TABLE 8.2
Various Non-Controlled Substance-Containing Suspect Counterfeit Drug Products Received by FDA's Forensic Chemistry Center

Solid Dosage Forms (Tablets And Capsules) Non-controlled Products		Veterinary Medications
Actonel	Levitra	Levamisole
Aricept	Lipitor	Nuflor
Casodex	Plavix/Iscover	
Celebrex	Propecia	
CellCept	Reductil/Meridia	
Cialis	Singulair	
Crestor	Tarniflu	
Evista	Viagra	
Ezetrol/Zetia	Xenical/Alli	
Hyzaar	Zyprexa	

FDA's involvement in counterfeit investigations usually begins when an FDA consumer safety officer (CSO) or FDA Office of Criminal Investigations (OCI) special agent seizes a suspect product or is referred a suspect product under FDA purview from other international, national, state, or local law enforcement agencies. ALS devices with over 70 authentic drug products and over 160 dosage forms (strengths) in the library are being used by CSOs at IMFs, and handheld Raman devices with over 120 authentic finished dosage forms in the library are being employed by OCI agents at ports of entry to screen drug products entering the United States. CSOs and OCI agents use the devices to prioritize which samples are sent to the laboratory for analysis. If FCC does not already have an authentic product for comparison, the laboratory communicates with the CSO or OCI agent and the manufacturer to obtain the appropriate packaging/finished product and any addition pertinent information on analytical methods that are useful for differentiating the authentic product from a counterfeit product. Currently, FCC has over 350 authentic products for comparison in-house, and this number increases as new suspect counterfeit products are received for analysis.

Peter Pitts, a former Associate Commissioner for the FDA, recently described how FDA's role regarding counterfeit drug products has changed significantly over the last two decades and discussed actions taken by FDA to create a multi-faceted response to protect the supply chain against these products (Pitts, 2020). He discussed how counterfeit drug products initially encountered by FDA primarily consisted of lifestyle and/or erectile dysfunction drugs, and recently evolved into much more lucrative (and much more dangerous) controlled substance-containing products (e.g., opioids, benzodiazepines) and lifesaving injectable medicines (e.g., oncology treatments, including biotherapeutics). Some of these products recently encountered by FCC are listed in Table 8.3.

Being able to authenticate and source potential counterfeit-controlled substance-containing products and lifesaving medicines is currently at the forefront of counterfeit drug analysis due to the significant health hazards faced by the consumers. Consequently, examples of multi-disciplinary approaches used by FCC for the analysis of these types of products are described below.

TABLE 8.3
Suspect Controlled Substance-Containing Tablets and Injectable Drug Products Recently Received by FDA's Forensic Chemistry Center

Suspect Product Declared API(S)	Injectable Product Declared API
Acetaminophen/hydrocodone bitartrate	Bevacizumab
Acetaminophen/oxycodone HCl	Epoetin alfa
Alprazolam	Somatropin
Clonazepam	Filgrastim
Oxycodone HCl	
Diazepam	

8.5.1 Example Analysis of Counterfeit Tablets Labeled to Contain Controlled Substances

FCC employs a multi-faceted analysis approach for authenticity testing. Suspect counterfeit tablets labeled to contain controlled substances are often received without any labeling, which precludes the ability to conduct a packaging examination. Therefore, all the tablets are first visually screened using white light and an ALS device to determine if there is any variation among the tablets. If more than one tablet type is observed using an ALS device, the product is subdivided into multiple groups (e.g. Item 1a, 1b, 1c, etc.). If all the tablets are visually consistent with each other, as shown in Figure 8.27, one or more representative tablets are visually compared to an authentic product, as shown in Figure 8.28. In this example, the suspect tablet is not visually consistent with the authentic tablet using monochromatic visible light (Figure 8.28a) and infrared light (Figure 8.28b).

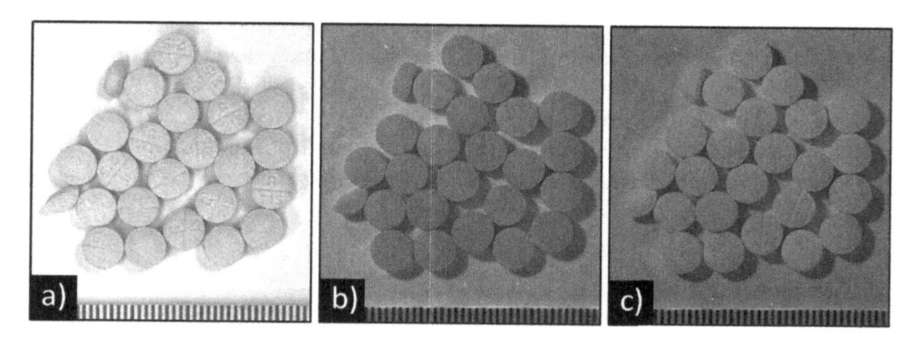

FIGURE 8.27 Benchtop ALS analysis of suspect generic oxycodone 30 mg tablets. White light/no filter (a), monochromatic light/yellow filter (b), and monochromatic light/orange filter (c). In this example, all the suspect tablets are consistent with each other using the same lighting and filter settings.

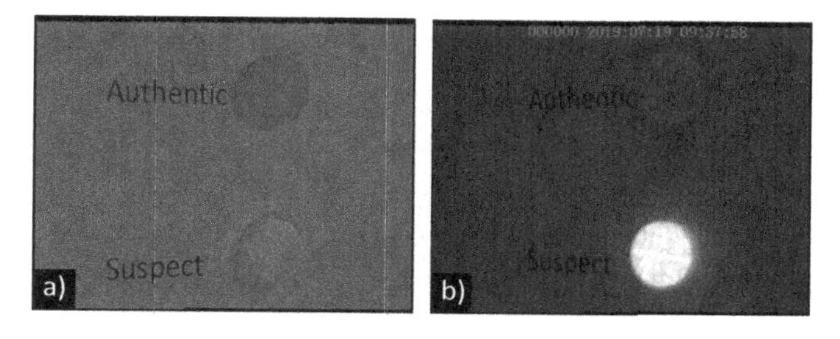

FIGURE 8.28 Handheld ALS analysis of suspect and authentic generic oxycodone 30 mg tablets. The suspect tablet was not visually consistent with the authentic tablet using monochromatic visible light with a CCD detector (a) and monochromatic visible light with an IR cut-on filter and/or visible cut-off filter (b).

Once ALS analyses are complete, representative tablets are examined using 3DIA profilometry to determine if suspect tablets are consistent with authentic tablets, consistent with each other, and consistent with tablets received from other investigations. Using this technique (example shown in Figure 8.29), the debossed characteristics of the surface feature "A" of one suspect tablet from this investigation and suspect tablets seized from eight other different investigations were found to be consistent with each other and not consistent with authentic tablets. Therefore, the possibility that the suspect tablets could have come from the same tablet punch tip and die could not be eliminated.

Following 3D IA, tablets are examined as it is using a Raman spectrometer to determine consistency with an authentic product. Using a handheld Raman device as described by Lanzarotta et al., the Raman spectrum of one suspect generic oxycodone 30 mg tablet (Figure 8.30a) was not consistent with that of the authentic tablet (Figure 8.30b) (Lanzarotta et al., 2020). SERS analysis is then performed on one or more suspect tablets to determine if any APIs are present, as described by Kimani et al. (2021a, b). For example, SERS analysis of one suspect generic oxycodone 30 mg tablet using a handheld Raman device (Figure 8.31a) did not yield a "match" to oxycodone (Figure 8.31b), but did yield "matches" to fentanyl (Figure 8.31c) and analogs, including valeryl fentanyl (Figure 8.31d), methoxy acetyl fentanyl (Figure 8.31e), and tetrahydrofuran fentanyl (Figure 8.31f). Although the device was unable to determine the exact analog due to

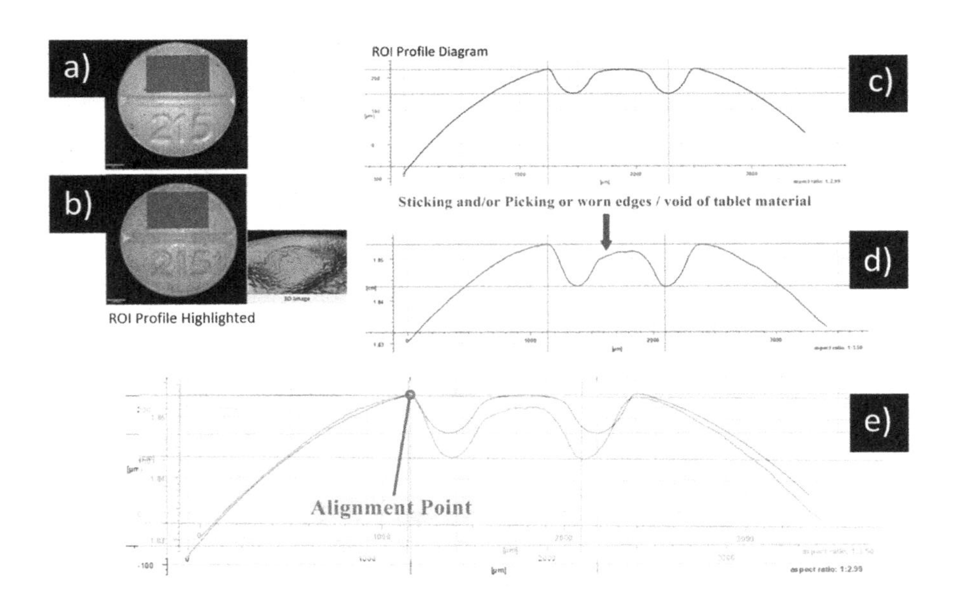

FIGURE 8.29 3DIA Profile Measurement analysis ROI of debossed characteristics of the surface feature "A" of a suspect tablet (a) and authentic tablet (b) along with the ROI diagrams of the suspect tablet (c) and authentic tablet (d). The comparison of diagrams, suspect vs. authentic (e), indicated that the suspect tablet was not consistent with the authentic tablet.

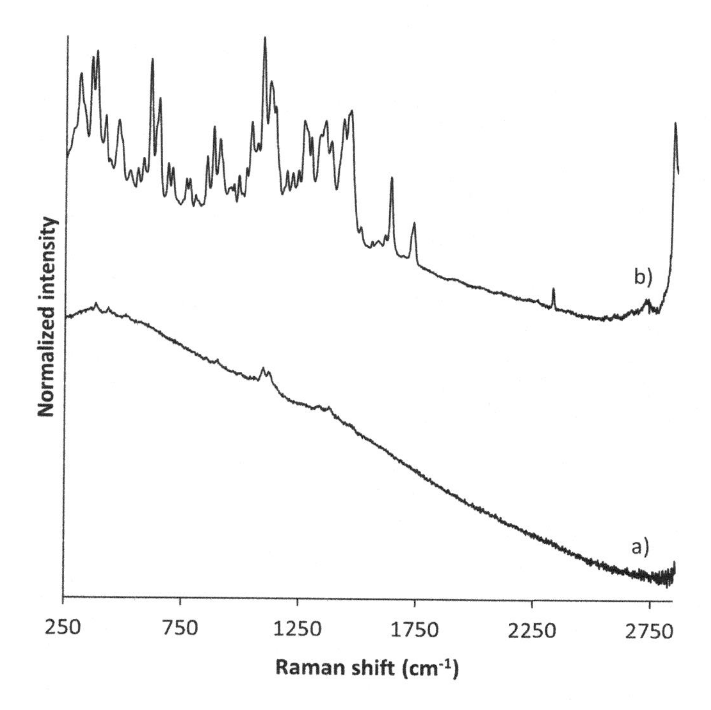

FIGURE 8.30 Handheld Raman spectra of a representative suspect counterfeit oxycodone 30 mg tablet (a) compared to that of an authentic tablet (b).

similarities between their SERS spectra, it was able to indicate that a compound with a structure similar to fentanyl was present.

FT-IR spectra of a tablet coating (or surface) and core were collected and are compared to those of an authentic tablet using procedures described by Lanzarotta et al. (2011, 2013, 2015). FT-IR spectra of a suspect oxycodone 30 mg tablet coating/surface (Figure 8.32a) and core (Figure 8.32c) are not consistent with those of an authentic tablet coating/surface (Figure 8.32b) and core (Figure 8.32d). Spectra of the suspect tablet core are examined for the presence of APIs using spectral searching tools and/or spectral subtractions (Lanzarotta et al. 2011, 2013). If no APIs are detected using these processes, such as the example in Figure 8.32, a liquid–liquid or solid–liquid extraction is performed followed by analysis of the dried residue using FT-IR spectroscopy (Lanzarotta et al., 2013). FT-IR spectra of a solid–liquid extraction residue from a suspect counterfeit oxycodone 30 mg tablet using ammonium hydroxide saturated hexanes (Figure 8.33a) is compared to that of a fentanyl standard (Figure 8.33b) and oxycodone hydrochloride standard (Figure 8.33c) prepared using the same procedure. In this example, the infrared spectrum of the extraction residue is consistent with fentanyl and not consistent with the labeled ingredient—oxycodone hydrochloride.

Once FT-IR analyses are complete, the tablets are analyzed for the presence of APIs using GC/MS. Fentanyl and despropionyl fentanyl (4-ANPP) were identified

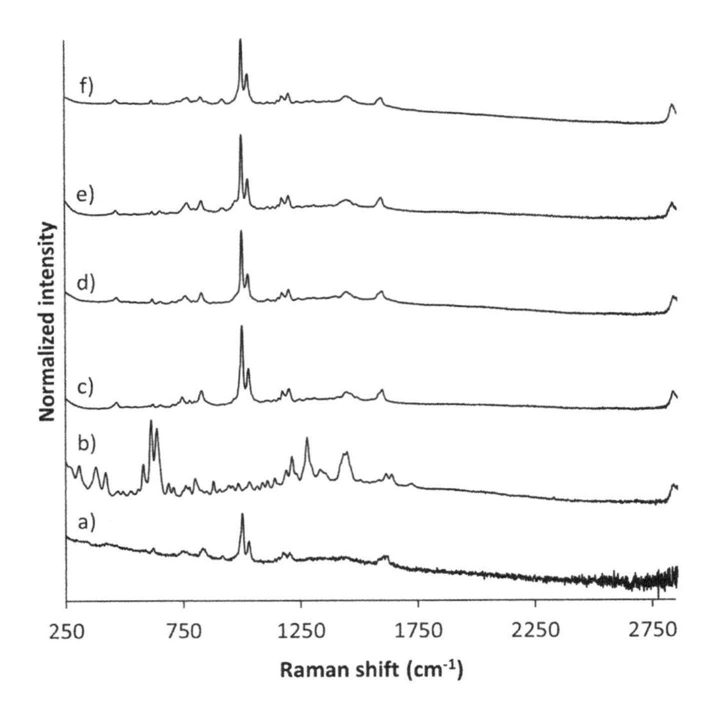

FIGURE 8.31 SERS results using a handheld Raman spectrometer for a representative suspect counterfeit oxycodone 30 mg tablet (a) oxycodone (b), fentanyl (c), valeryl fentanyl (d), methoxy acetyl fentanyl (e), and tetrahydrofuran fentanyl (f).

in the 30 mg oxycodone tablets based on retention time and mass spectral comparisons to reference standards analyzed using the same experimental conditions (Figure 8.34). Specifically, the suspect sample yielded total ion chromatogram (TIC) peaks with retention times of 21.35 and 23.39 min. These retention times and mass spectra corresponding to these peaks were consistent with those of 4-ANPP and fentanyl, respectively.

Following GC/MS, LC/MS can be conducted to screen for compounds not amenable to GC/MS, and to detect compounds that are below the detection limit of GC/MS. The tablets declared to contain 30 mg oxycodone yielded extracted ion chromatogram (EIC) peaks (measured from m/z 275–375) with retention times of 2.07, 2.10, and 2.42 min, as shown in Figure 8.35. The retention times, mass spectra, and MS/MS spectra (data not shown) corresponding to these peaks were consistent with those of 4-ANPP, fentanyl, and valeryl fentanyl, respectively. In some cases, APIs are quantitated using HPLC/UV, but this is typically only conducted at the request of the submitting official. The summary results for each technique and overall results for the analysis of the suspect counterfeit oxycodone 30 mg tablet discussed in this example are listed in Table 8.4.

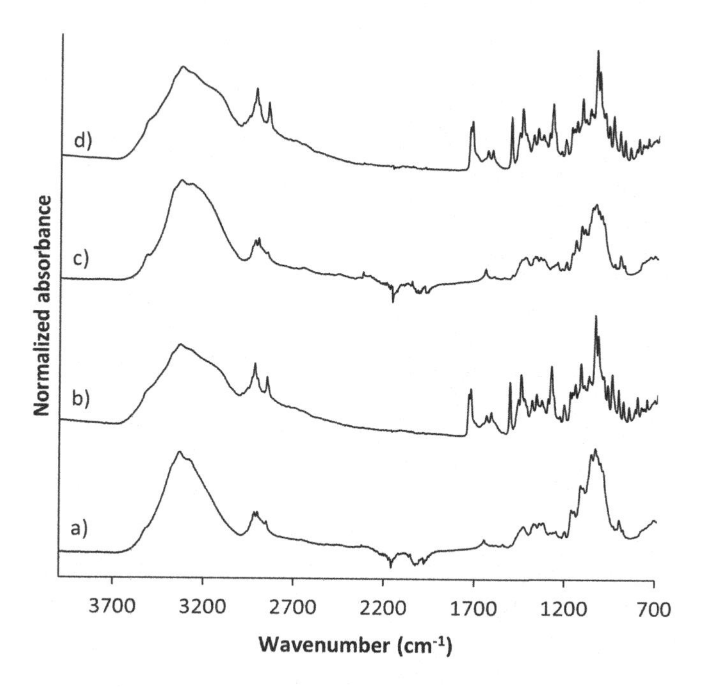

FIGURE 8.32 FT-IR spectra of a suspect and authentic oxycodone 30 mg tablet surface (a and b, respectively) and core (c and d, respectively).

8.5.2 Recent Trends Observed by FCC for Opioid-Containing Counterfeit Drug Products

Between February of 2017 and May of 2019, FCC received 66 unique suspect counterfeit samples declared to contain controlled substances. Most of these products were variations of innovator and generic tablets declared to contain oxycodone and alprazolam, as shown in Figure 8.36. Of the 66 tablets, 49 (74.2%) were found to be not consistent with authentic products via analytical testing (Figure 8.37). Many of these products were often found to contain fentanyl analogs and other unlabeled APIs, and are therefore at an exceptionally high risk for consumers. One product alone accounted for nearly one-third of these samples: 21 suspect counterfeit generic oxycodone 30 mg tablets manufactured to mimic authentic tablets manufactured by Actavis (blue, half-scored tablets debossed "A 215" on the scored side). Each of these 21 products were determined to be not consistent with the authentic product and were found to contain as many as six opioids. APIs detected in suspect counterfeit tablets labeled to contain oxycodone and alprazolam are listed in Table 8.5 (left) and (right), respectively. Since May of 2019 even more counterfeit tablets labeled to contain controlled substances have been received by this laboratory and the number of fentanyl analogs and other controlled substances found in these products continues to rise.

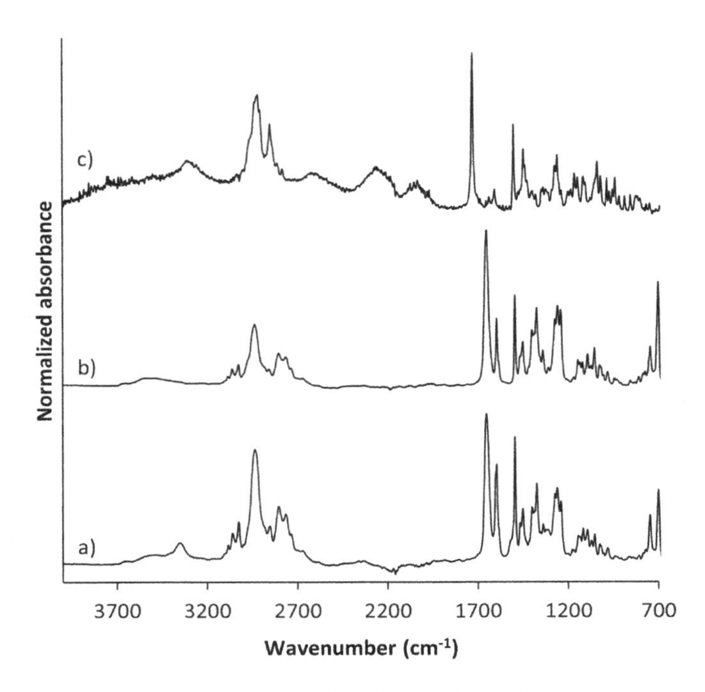

FIGURE 8.33 FT-IR spectra of a solid–liquid extraction residue from a suspect counterfeit oxycodone 30 mg tablet using ammonium hydroxide saturated hexanes (a) compared to that of fentanyl (b) and oxycodone HCl (c) recrystallized using ammonium hydroxide saturated hexanes.

8.5.3 Biotherapeutics

Most of the discussion to this point has been focused on solid dosage forms that contain small-molecule APIs. Biological-based (biotherapeutic) drug products represent a rapidly growing sector of the pharmaceutical industry. These types of drug products have been developed and prescribed for the treatment of many life-threatening and chronic health conditions. Biotherapeutic products are created using living cells and proprietary processes under regulated good manufacturing practices. Therapeutic protein molecules are hundreds of times larger than small-molecule APIs and often contain post-translational modifications (PTMs), typically glycosylation. Between 1997 and 2020, a total of 97 antibody therapeutics were granted first approval in either the US or the EU. In 2021, eight antibody therapeutics have been granted first approval, and 18 investigational antibody therapeutics are in regulatory review (AS, 2021). As the patent protection for the first generation of these products expires, biosimilar development has been accelerating among biopharmaceutical firms. In the past 5 years, 20 biosimilar drugs for five innovator monoclonal antibody (mAb) therapeutics, including adalizumab, trastuzumab, rituximab, infliximab, and bevacizumab, have gained marketing approval from the US FDA (FDA, 2022).

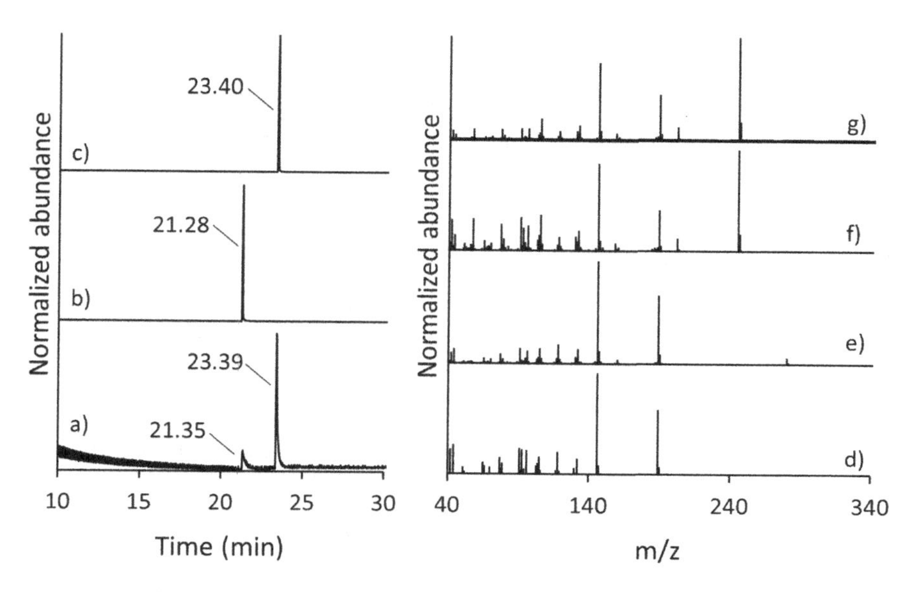

FIGURE 8.34 GC/MS analysis of a suspect counterfeit oxycodone 30 mg tablet. Total ion chromatograms (TICs) of a suspect counterfeit generic 30 mg oxycodone tablet (a) along with those of a 4-ANPP (b) and fentanyl (c). Mass spectra corresponding to the suspect TIC peak at 21.35 min (d), the 4-ANPP TIC peak at 21.28 min (e), the suspect TIC peak at 23.39 min (f), and the fentanyl TIC peak at 23.40 min (g).

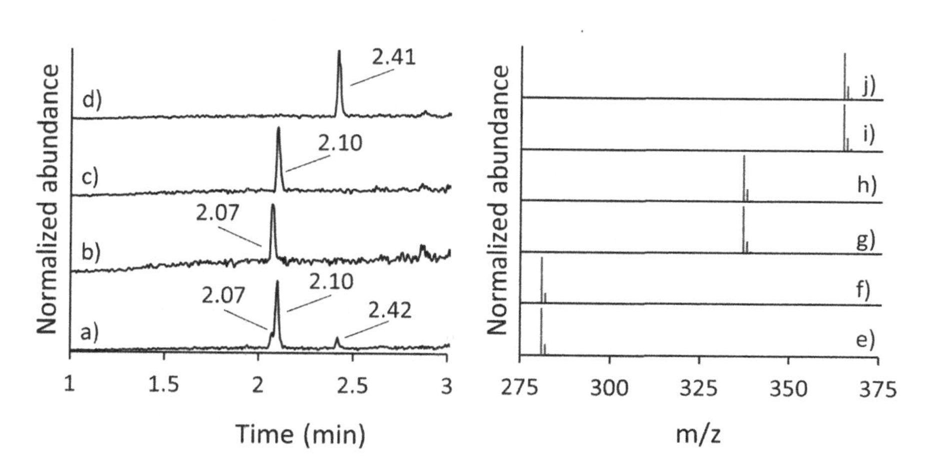

FIGURE 8.35 LC/MS analysis of a suspect counterfeit oxycodone 30 mg tablet. Extracted ion chromatogram (EIC) from m/z 275–375 of a suspect counterfeit generic 30 mg oxycodone tablet (a) along with those of a 4-ANPP (b), fentanyl (c), and valeryl fentanyl (d). Mass spectra corresponding to the suspect EIC peak at 2.07 min (e), the 4-ANPP EIC peak at 2.07 min (f), the suspect EIC peak at 2.10 min (g), the fentanyl EIC peak at 2.10 min (h), the suspect EIC peak at 2.42 min (i), and the valeryl fentanyl EIC peak at 2.41 min (j).

TABLE 8.4
Summary Results For Each Technique and Overall Results for the Analysis of the Suspect Counterfeit Oxycodone 30 mg Tablets Discussed in this Example

Technique	Summary Results
Benchtop ALS	Using the same lighting and filter settings, all tablets are visually consistent with each other
Handheld ALS	Using the same lighting and filter settings, one representative tablet was not consistent with an authentic tablet
3DIA profilometry	Debossed characteristics of the surface feature "A" of one suspect tablet was not consistent with that of authentic tablet This suspect tablet and tables seized from eight other investigations were consistent with each other
Handheld Raman	The Raman spectrum of one representative tablet was not consistent with that of an authentic tablet
SERS via handheld Raman	The suspect Raman spectrum indicated the presence of a compound with a structure similar to fentanyl
FT-IR	The infrared spectra of the suspect tablet coating and core were not consistent with those of an authentic tablet. The suspect tablet indicated the presence of fentanyl based on comparison with a standard but oxycodone HCI was not detected
GC/MS	Fentanyl and despropionyl fentanyl (4-ANPP) were identified in the suspect tablets based on retention time and mass spectral comparisons to reference standards analyzed using the same experimental conditions
LC/MS	Fentanyl, 4-ANPP, and valeryl fentanyl were identified in the suspect tables based on the retention time and mass spectral comparisons to reference standards analyzed using the same experimental conditions
Overall	The suspect tablets examined were not consistent with authentic oxycodone HCI 30 mg tablets. Fentanyl, 4-ANPP, and valeryl fentanyl were identified in the suspect tablets examined. The possibility that the suspect tablets (from this investigation and eight other investigations) could have come from the same tablet punch tip and die could not be eliminated

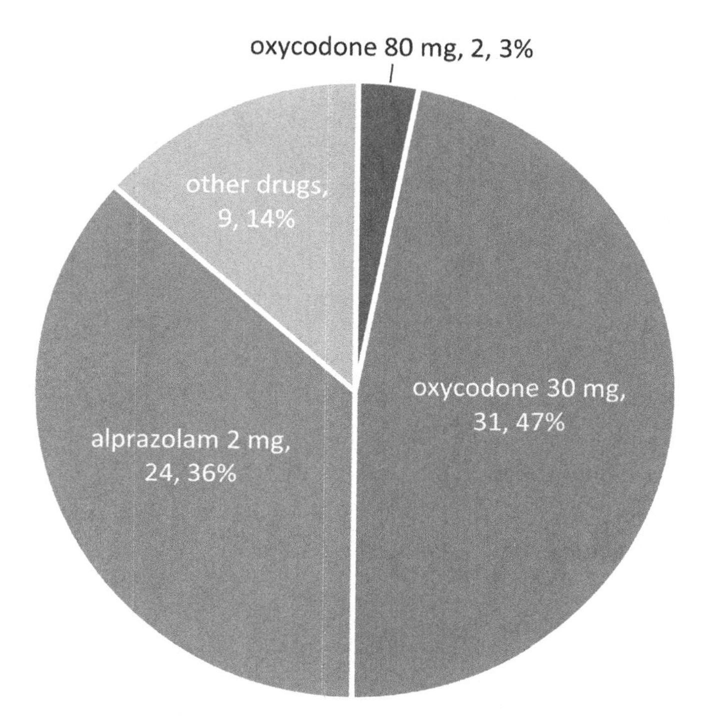

FIGURE 8.36 Controlled substance-containing suspect counterfeit drug products received by FDA's Forensic Chemistry Center between February 2017 and May 2019.

8.5.3.1 Example Analysis of Suspect Counterfeit Altuzan

Suspect counterfeit Altuzan (bevacizumab) products were received for analysis; the products are referred to as Items A and B, as shown in Figure 8.38. Each item consisted of a carton and a vial that contained a liquid. A summary of all analytical results is provided in Table 8.6.

8.5.3.2 Altuzan Example Printing Analysis

Using a digital light microscope (DLM), printing processes on the suspect cartons and vials were determined and are listed in Table 8.6. Carton graphics (artwork) for Items A (Figure 8.39a) and B (Figure 8.39b) were consistent with offset lithography. Carton variable data (lot/batch and expiration dates) for Items A (Figure 8.39c) and B (Figure 8.39d) were consistent with offset lithography and flexography, respectively.

Printing processes for the vial components of Items A (left column) and B (right column) are shown in Figure 8.40. Artwork on the vial crimp for Items A (Figure 8.40a) and B (Figure 8.40b) were consistent with screen printing and pad printing, respectively. Vial label graphics for Items A (Figure 8.40c) and B (Figure 8.40d) were both consistent with offset lithography, and the vial label variable data for Items A (Figure 8.40e) and B (Figure 8.40f) were consistent

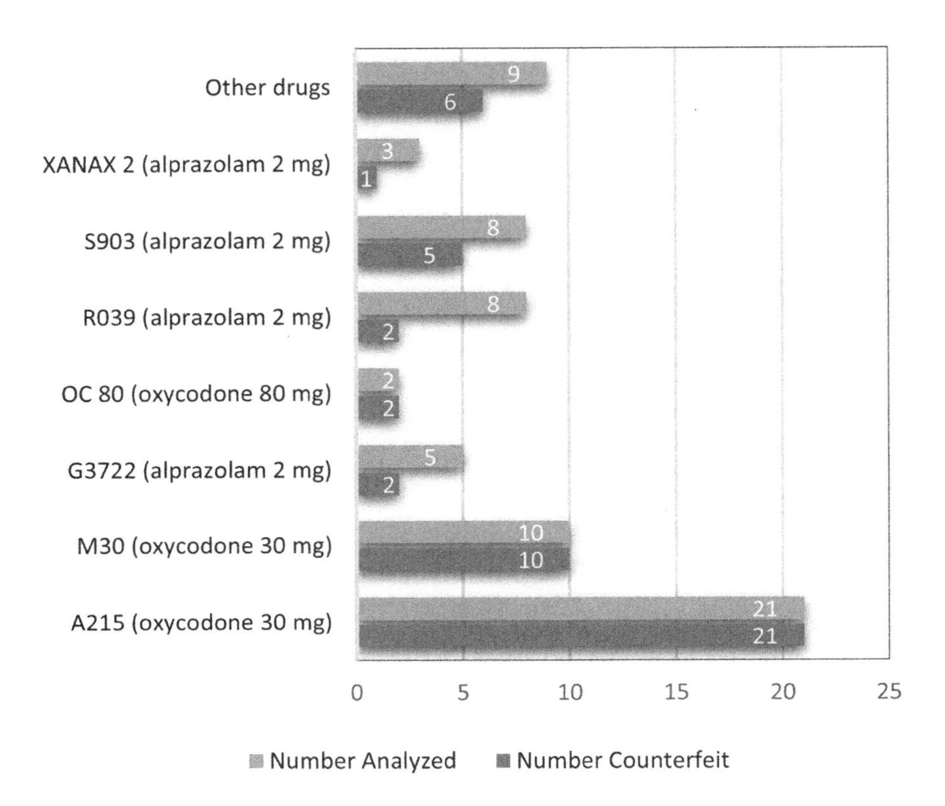

FIGURE 8.37 Number of tablets analyzed vs. number of tablets found to be not consistent with the authentic product for suspect controlled substance-containing products received by FDA's Forensic Chemistry Center between February 2017 and May 2019.

with offset lithography and xerography, respectively. Overall, the graphics printing and variable data printing on the cartons and vial labels from Items A and B were not consistent with authentic printing processes (authentic printing process are not shown or stated because they are proprietary). The printing process of the artwork on the Item A vial crimp was not consistent with that of the authentic but the printing process of the artwork on the Item B vial crimp was consistent with that of the authentic.

8.5.3.3 Altuzan Example Chemical Analysis

Using Raman spectroscopy, similar to what has been described previously, (Degardin et al., 2017, Lanzarotta et al., 2020) spectra of Items A (Figure 8.41a) and B (Figure 8.41b) were compared to a spectrum of an authentic product (Figure 8.41c); the Item A spectrum was not consistent, and the Item B spectrum was consistent with that of the authentic product, respectively. Utilizing FT-IR spectroscopy,

TABLE 8.5
APIs Detected in Suspect Counterfeit Products Labeled to Contain Oxycodone (Left) and Alprazolam (Right)

APIs Found in Tablets Labeled to Contain Oxycodone		APIs Found in Tablets Labeled to Contain Alprazolam
6-acetylcodeine	*n*-Methyl norfentanyl	Alprazolam
6-monoacetylmorphine	Norfentanyl	Ibuprofen
acetaminophen	Noscapine	Diphenhydramine
acetyl fentanyl	Papaverine	Noopept
alprazolam	Phenyl fentanyl	Etizolam
caffeine	Quinine	Phenazepam
carfentanil	Sildenafil	Lidocaine
cyclopropyl fentanyl	Tramadol	U-47700
despropionyl fentanyl	U-47700	Fentanyl
fentanyl	U-48800	
furanyl fentanyl	XLR-11	
heroin	Isonitazene	
methamphetamine	*n*-Pyrrolidino etonitazene	
methoxyacetyl fentanyl		

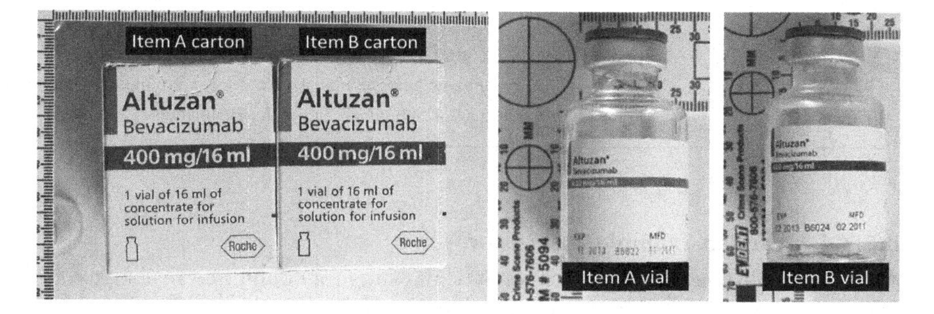

FIGURE 8.38 Photographs of Items A and B suspect Altuzan products.

spectra of dried residues from Items A (Figure 8.42a) and B (Figure 8.42b) were compared to the spectrum of an authentic product (Figure 8.42c); the Item A spectrum was not consistent, and the Item B spectrum was consistent with that of the authentic product, respectively. Using SDS-PAGE, a protein with a molecular weight consistent with that found in the authentic was not detected in Item A and was detected in Item B (Figure 8.43). Using GC/MS (data not shown), benzoic acid and albuterol were detected in Item A, and no additional drugs were identified in Item B. Using LC/MS (data not shown), Item 1A was found to contain 8 µg/mL albuterol (Item B was not examined using LC/MS.).

TABLE 8.6
Results of Analysis for Items A and B

Technique		Item A	Item B
Printing analysis	Carton graphics	Not consistent with authentic (offset lithography)	Not consistent with authentic (offset lithography)
	Carton variable data	Not consistent with authentic (offset lithography)	Not consistent with authentic (flexography)
	Vial graphics	Not consistent with authentic (offset lithography)	Not consistent with authentic (offset lithography)
	Vial variable data	Not consistent with authentic (offset lithography)	Not consistent with authentic (xerography)
	Vial crimp	Not consistent with authentic (screen printing)	Consistent with authentic
SDS-PAGE		Protein with molecular weight consistent with that of the authentic not detected	Protein with molecular weight consistent with that of the authentic detected
FT-IR		Not consistent with authentic	Consistent with authentic
Raman		Not consistent with authentic	Consistent with authentic
GC/MS		Benzoic acid, albuterol	No drugs were identified
LC/MS		8 μg/mL albuterol	Not performed

8.5.3.4 Altuzan Example Discussion

At the Forensic Chemistry Center, analytical testing for counterfeit investigations generally involves printing and chemical analysis. Based on the printing analysis described above for the suspect counterfeit Altuzan examples, it was determined that the cartons and vial labels from Items A and B were not consistent with the authentic, which perpetrators often do to conceal expired/spent authentic products and re-brand authentic products destined for a foreign market. Based on chemical analysis of the vial contents, it was determined that Item A was not consistent with the authentic product, because a protein with a molecular weight consistent with that of the authentic was not detected and because it was found to contain benzoic acid and albuterol. However, the vial contents of Item B could not be differentiated from the authentic, indicating they were possibly consistent with the authentic, counterfeit, or diverted (i.e., a product intended for a foreign market that was repackaged for domestic use). Reasons for not being able to make

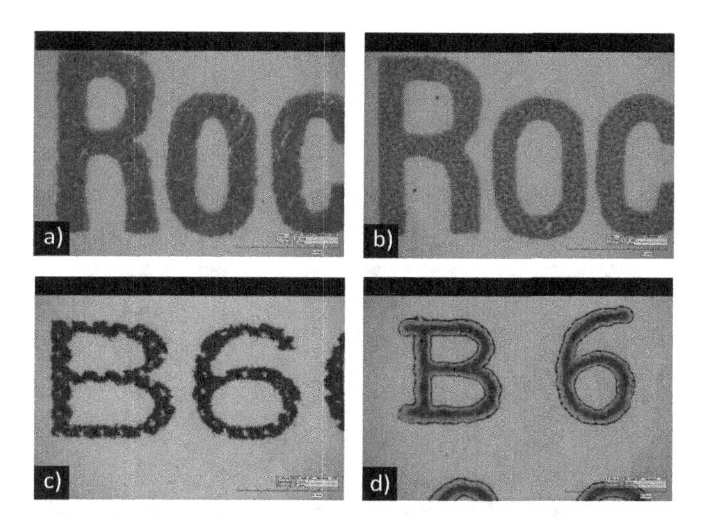

FIGURE 8.39 Digital light microscopy images of printing processes for carton components of Item A (left column) and Item B (right column). Carton graphics for Item A (a) and Item B (b) were consistent with offset lithography. Carton variable data for Item A (c) and Item B (d) were consistent with offset lithography and flexography, respectively.

a definitive conclusion regarding the authenticity of Item B and potential solutions for overcoming these challenges are discussed in more detail below.

8.5.3.5 Challenges and Potential Solutions for Authenticating Counterfeit Biotherapeutics

Definitive conclusions regarding the authenticity for Item B could not be drawn due to the complexity of the biotherapeutic molecules and the analytical testing techniques employed in this case work. For example, even authentic products, both innovator and biosimilar drugs, have batch-to-batch variability, since they are produced in cellular machinery through complex bioprocesses. Critical quality attributes must be monitored and controlled carefully (Guerra et al., 2019). Between innovator and biosimilar drugs, variations such as C-terminal lysine truncation can be present, resulting from unidentical manufacturing processes that do not necessarily impact biological function of the biomolecule (Jung et al., 2014). Just as with small-molecule drug counterfeiting, products in the biotherapeutics arena will become ever increasingly sophisticated. Analytical capabilities to discern innovators and biosimilars from counterfeit biotherapeutics are becoming critical in safeguarding the biopharmaceutical supply chain.

The FCC has been investigating the power of mass spectrometry techniques and developing mass spectrometry-based analytical methods that will permit the ability to rapidly differentiate and verify various mAb therapeutics (Jin, 2016, 2017, 2018, 2019). Mass spectrometry is a key technique used in the biopharmaceutical industry for the analysis and characterization of mAb therapeutics throughout the drug discovery, development, and manufacturing processes (Wei et al., 2013).

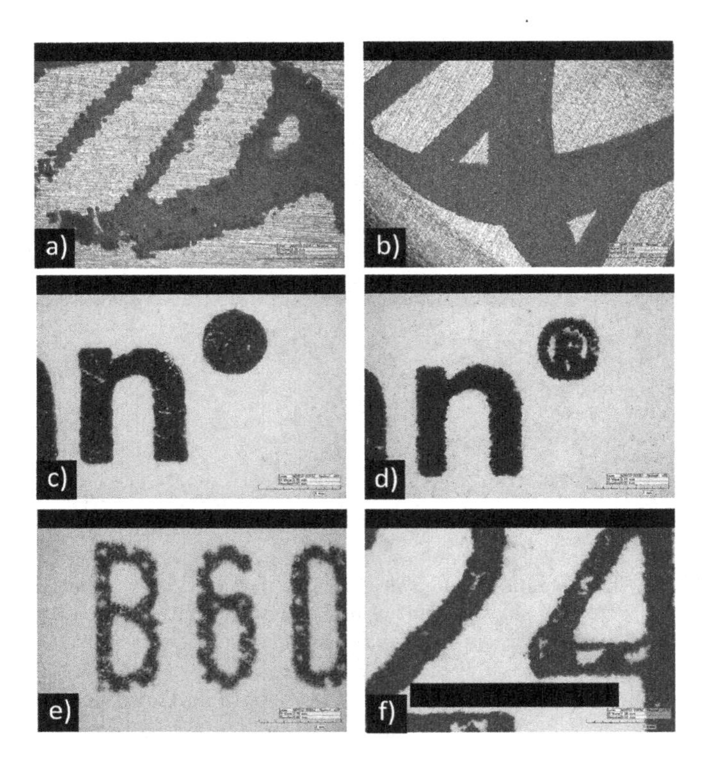

FIGURE 8.40 Digital light microscopy images of printing processes for vial components of Item A (left column) and Item B (right column). Artwork printed on the crimp for Item A (a) and Item B (b) were consistent with screen printing and pad printing, respectively. Vial graphics for Item A (c) and Item B (d) were consistent with offset lithography. Vial variable data for Item A (e) and Item B (f) were consistent with offset lithography and xerography, respectively.

The ability of mass spectrometry in protein primary sequence analysis is unsurpassed by any other analytical technique. Extensive sequence coverage is achievable at the peptide level, because the molecular size of smaller peptides falls in the optimal working range of mass spectrometry structural analysis. However, protein sequencing at the peptide level is tedious and time consuming and not amenable for use in the setting of rapid product screening. To explore alternative approaches, it was of interest to determine if it was possible to differentiate mAb molecules at intact, subunit, and subdomain levels without breaking up the molecules into smaller peptides. IgG-based mAbs are complex biopharmaceuticals with large molecular weight (~150 kDa) and glycosylated at the conserved Fc domain. A typical mAb molecule is composed of two heavy chains (HC) and two light chains (LC) that are linked through intra- and inter-molecular disulfide bonds at cysteine residues, forming a Y-shaped molecule. The disulfide bonds can be reduced to release the subunit LC (~25 kDa) and HC (~50 kDa). The HC can be further downsized using IdeS enzyme (Immunoglobulin-degrading enzyme from

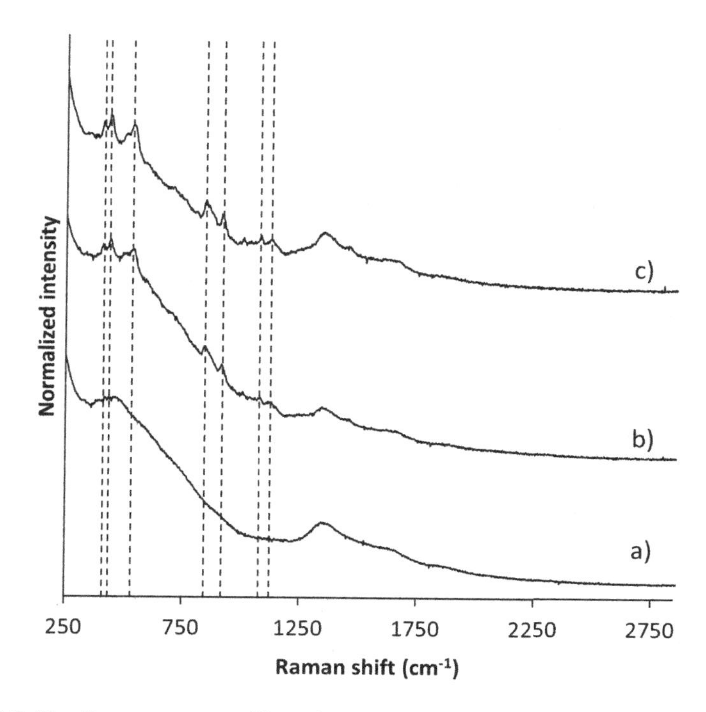

FIGURE 8.41 Raman spectra of Item A (a) and Item B (b) injectable liquids compared to a spectrum of an authentic product (c).

Streptococcus pyogenes), which cleaves the HC into Fc/2 and Fd subdomains similar in size to the LC (Xie et al., 2010, Janin-Bussat et al., 2013). The study integrating accurate mass analyses at the intact, subunit, and subdomain levels indicated that the two mAb therapeutics reference materials were readily differentiated based on the molecular masses and the distinct profiles resulting from the respective major glycoforms of the Fc domain. Note that the mass measurement accuracy at the subdomain level was much improved to less than 1 Da, which permitted the ability to discern different LCs, Fc/2, and Fd subdomains resulting from the two different mAb reference materials. Additionally, this integrated approach detected a 28 Da mass discrepancy between the measured mass and the calculated one using the sequence in public repository. A literature search helped pinpoint the mass discrepancy arising from A219V reporting error, demonstrating the power of the integrated approach corroborating information at intact, subunit, and subdomain levels.

The mAb Fc N-glycosylation profiles were then investigated in greater detail at the peptide level for possible use in molecular fingerprinting. IgG-based mAbs are N-glycosylated at the Fc domain through enzymatic processes during fermentation and is a carefully monitored quality attribute in product manufacturing (Reusch et al., 2015, Shah et al., 2014, Rogers et al., 2015). N-Glycopeptides are difficult to analyze using mass spectrometry due to their presence at low levels and

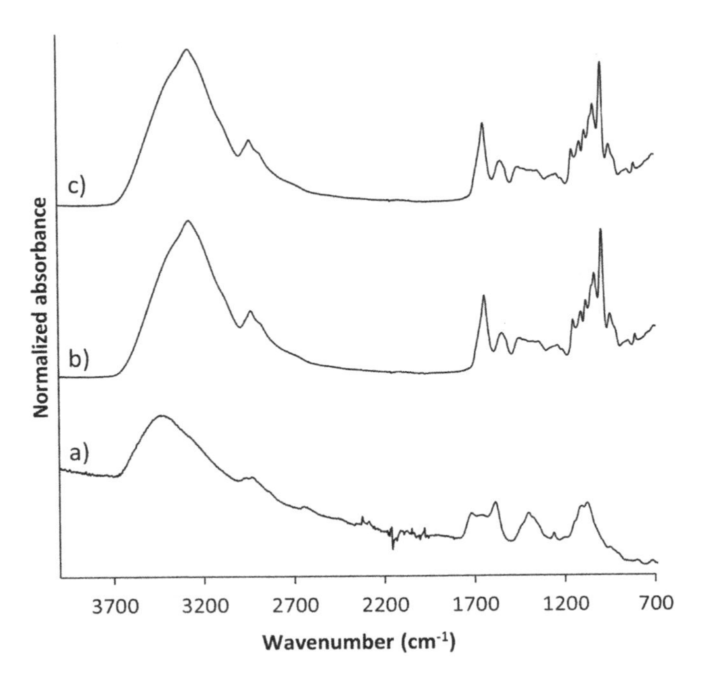

FIGURE 8.42 FT-IR spectra of Item A (a) and Item B (b) injectable liquid dried residues compared to a spectrum of an authentic product (c).

signal suppression by the normal peptides during mass spectrometry process. An enrichment procedure was employed using hydrophilic interaction chromatography (HILIC) beads where minor N-glycopeptide species not discerned at intact, subunit, and subdomain levels were now enriched and detected by mass spectrometry. Results showed that the Fc N-glycosylation profiling at the glycopeptide level was amenable as an in-depth molecular fingerprinting tool for differentiating and verifying mAb therapeutics. So far, the focus has been on developing rapid screening tools using mass spectrometry for mAb therapeutics products. It was also of interest to determine the extent to which it was possible to confirm the primary sequence at amino acid level. Consequently, the bottom-up approach was investigated, which is the gold standard in primary sequence analysis. An optimized low-pH digestion kit was used to minimize introduction of sample preparation artifacts, especially deamidation and oxidation (Hosfield, 2016). The Fc N-glycosylation profiling study showed that among the five mAb therapeutics (two mAb therapeutics reference materials and three mAb therapeutics products) high-sequence coverage was attainable using this approach, typically above 80% for the LC and over 90% for the HC. Of course, mass spectrometry exploration would not be complete without attempting top-down or middle-down sequencing (Fornelli et al., 2018). In particular, the middle-down sequencing allowed protein sequencing at the subdomain level that would simplify the sample preparation procedure, minimize introducing sample preparation associated artifact, and

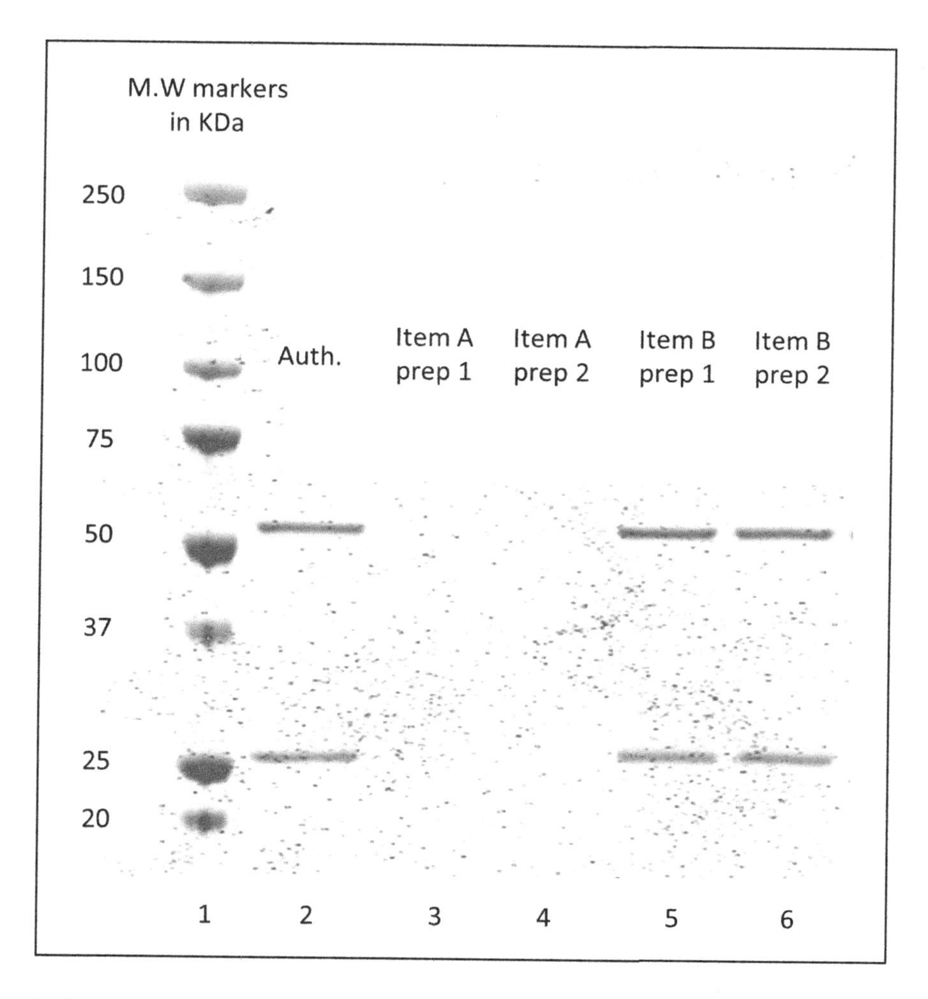

FIGURE 8.43 SDS-PAGE results in reduced condition for the authentic product that contains bevacizumab (lane 2), Item A preps 1 and 2 (lanes 3 and 4) and Item B preps 1 and 2 (lanes 5 and 6).

produce adequate size molecules for mass spectrometry. The middle-down sequencing study was able to obtain reasonable sequence coverage for all the subunits or subdomains investigated, specifically, 17%–27% for Fc/2-K (-K: C-terminus lysine truncation) or Fc/2, 23%–34% for LC or pLC (p: N-terminus pyroglutamylation), and 10%–15% for Fd or pFd. Although the sequence coverage in middle-down approach is not as extensive as those from the bottom-up approach, nonetheless, the sequence ions mostly clustered at the terminal regions add to the accurate molecular mass measurements at the subdomain level, further increasing the discerning power and thus confidence. In summary, it is anticipated that the knowledge accrued from these studies and the new methodologies developed will facilitate the rapid differentiation between different mAb therapeutic products, between innovators and biosimilars, and ultimately find their way into future counterfeit investigations to permit the ability to draw definitive conclusions at a molecular level.

8.6 CHAPTER CONCLUSIONS

To conclude this chapter, no one technology or analytical tool is versatile enough to detect all counterfeit drug products. Instead, a multi-disciplinary approach is required, and the choice of tools needed is highly dependent on the sample being analyzed. Analysts should consider establishing a target list of drug products to start with, and expand the list with time and experience. It is necessary to create a reference library of finished dosage and packaging information based on good relationships and communication with authentic product manufacturers. It is recommended to define a workflow or flow chart such as that described by RHSC, which includes sampling and testing strategies to help with decision making pertaining to screening, detecting, and sourcing counterfeit drug products. Since pharmaceutical counterfeiting is a global issue, working relationships and collaborations with foreign government scientists, law enforcement personnel, and global health organizations is critical. Counterfeit drug analysts are further encouraged to work with other laboratory scientists and experts to provide support and help regarding the development of screening and testing methods.

ACKNOWLEDGEMENTS

The authors would like to thank Dr. Colin G. Barry, JaCinta Batson, Kelsey M. Griffin, Dr. Skyler W. Smith, and Timothy Yi from the FCC for their contributions on this book chapter.

REFERENCES

Adar, F., P. Leary, and T. Kubic. 2014. "Raman microscopy for detecting counterfeit drugs-a study of the tablets versus the packaging." *Spectroscopy* 29 (6):10–17.

Akhunzada, Z., Y. Wu, T. Haby, D. Jayawickrama, G. McGeorge, M. La Colla, J. Bernstein, M. Semones, and A. Abraham. 2021. "Analysis of biopharmaceutical formulations by Time Domain Nuclear Magnetic Resonance (TD-NMR) spectroscopy: A potential method for detection of counterfeit biologic pharmaceuticals." *J Pharm Sci.* doi: 10.1016/j.xphs.2021.03.011.

Alabdalla, M.A. 2005. "Chemical characterization of counterfeit captagon tablets seized in Jordan." *Forensic Sci Int* 152 (2–3):185–8. doi: 10.1016/j.forsciint.2004.08.004.

Albright, D. 2016. "Detection and Link Analysis of Counterfeit Altuzan® Printing Defects Using Light Microscopy and Digital Imaging." *Microscopy and Microanalysis* 22 (S3):2038–2039. doi: 10.1017/s1431927616011028.

Alsallal, M., M.S. Sharif, B. Al-Ghzawi, and S. M. Mlkat al Mutoki. 2018. "A Machine Learning Technique to Detect Counterfeit Medicine Based on X-Ray Fluorescence Analyser." 2018 International Conference on Computing, Electronics & Communications Engineering (iCCECE), 118–122. doi: 10.1109/iCCECOME.2018.8659110.

Andria, S.E., S.F. Platek, M. Fulcher, and M.R. Witkowski. 2012. "COUNTERFEITING-The Use of SEM/EDS and FT-IR Analyses in the Identification of Counterfeit Pharmaceutical Packaging." *American Pharmaceutical Review* 15 (3):62.

Anzanello, M.J., R.S. Ortiz, R. Limberger, and K. Mariotti. 2014. "A framework for selecting analytical techniques in profiling authentic and counterfeit Viagra and Cialis." *Forensic Science International* 235:1–7 doi: 10.1016/j.forsciint.2013.12.005.

AS. 2021. "Antibody Society." accessed 10/13/2021. https://www.antibodysociety.org/resources/approved-antibodies.

Bakker-'t Hart, I.M.E., D. Ohana, and B.J. Venhuis. 2021. "Current challenges in the detection and analysis of falsified medicines." *J Pharm Biomed Anal* 197:113948. doi: 10.1016/j.jpba.2021.113948.

Bansal, D., S. Malla, K. Gudala, and P. Tiwari. 2013. "Anti-counterfeit technologies: a pharmaceutical industry perspective." *Scientia pharmaceutica* 81 (1):1–14.

Bate, R. 2012. "The deadly world of fake medicine." accessed 10/13/2021. https://www.aei.org/articles/the-deadly-world-of-fake-medicine/.

Batson, J.S., D.K. Bempong, P.H. Lukulay, N. Ranieri, R.D. Satzger, and L. Verbois. 2016. "Assessment of the effectiveness of the CD3+ tool to detect counterfeit and substandard anti-malarials." *Malar J* 15:119. doi: 10.1186/s12936-016-1180-2.

Bawuah, P., P. Pääkkönen, and K.-E. Peiponen. 2017. "Gloss measurement in detection of surface quality of pharmaceutical tablets: a case study of screening of genuine and counterfeit antimalaria tablets." *Journal of the European Optical Society-Rapid Publications* 13 (1). doi: 10.1186/s41476-017-0046-8.

Beasley, E., M. Suman, and T. Coleman. 2015. "The use of high performance anion exchange chromatography for the detection of counterfeit pharmaceutical products using the excipient content as a marker." *Fields: journal of Huddersfield student research* 1 (1):32–53.

Beckers, D. 2021. "Unambiguous identification of packaged counterfeit drugs by X-ray Powder Diffraction (XRPD)." accessed 10/13/2021. https://ec.europa.eu/health/sites/default/files/files/counterf_par_trade/doc_publ_consult_200803/22_panalytical_public_part_en.pdf.

Been, F., Y.K. Roggo, K. Degardin, P. Esseiva, and P. Margot. 2011. "Profiling of counterfeit medicines by vibrational spectroscopy." *Forensic Sci Int* 211 (1–3):83–100. doi: 10.1016/j.forsciint.2011.04.023.

Bernier, M.C., A. Li, L. Winalski, Y. Zi, Y. Li, C. Caillet, P. Newton, Z.L. Wang, and F.M. Fernandez. 2018. "Triboelectric Nanogenerator (TENG) Mass Spectrometry of Falsified Antimalarials." *Rapid Commun Mass Spectrom*. doi: 10.1002/rcm.8207.

Bernier, M.C., F. Li, B. Musselman, P.N. Newton, and F.M. Fernández. 2016. "Fingerprinting of falsified artemisinin combination therapies via direct analysis in real time coupled to a compact single quadrupole mass spectrometer." *Analytical Methods* 8 (36):6616–6624. doi: 10.1039/c6ay01418f.

Bliese, S.L., M. Maina, P. Were, and M. Lieberman. 2019. "Detection of degraded, adulterated, and falsified ceftriaxone using paper analytical devices." *Analytical Methods* 11 (37):4727–4732. doi: 10.1039/c9ay01489f.

Bonsu, D.O.M., C. Afoakwah, and M.d. Aguilar-Caballos. 2021. "Counterfeit formulations: analytical perspective on anorectics." *Forensic Toxicology* 39 (1):1–25. doi: 10.1007/s11419-020-00564-5.

Bottoni, P., and S. Caroli. 2019. "Fake pharmaceuticals: A review of current analytical approaches." *Microchemical Journal* 149. doi: 10.1016/j.microc.2019.104053.

Bureau International des Poids et Mesures. 2021. "Bureau International des Poids et Mesures." accessed 10/13/2021. https://www.bipm.org/en/home.

Bussy, U., C. Thibaudeau, F. Thomas, J.R. Desmurs, E. Jamin, G.S. Remaud, V. Silvestre, and S. Akoka. 2011. "Isotopic finger-printing of active pharmaceutical ingredients by 13C NMR and polarization transfer techniques as a tool to fight against counterfeiting." *Talanta* 85 (4):1909–14. doi: 10.1016/j.talanta.2011.07.022.

Caillet C., S. Vickers, P. Boupha, P. Newton. 2018. "An evaluation of portable screening devices to assess medicines quality for national medicines regulatory authorities." accessed 10/13/2021. https://www.iddo.org/external-publication/evaluation-portable-screening-devices-assess-medicines-quality-national.

Cho, S.H., H.J. Park, J.H. Lee, J.A. Do, S. Heo, J.H. Jo, and S. Cho. 2015. "Determination of anabolic-androgenic steroid adulterants in counterfeit drugs by UHPLC-MS/MS." *J Pharm Biomed Anal* 111:138–146 doi: 10.1016/j.jpba.2015.03.018.

Crews, C. 2018. "Non-destructive detection of counterfeit and substandard medicines using X-ray diffraction." Ph.D. dissertation, University College London.

Custers, D., M. Canfyn, P. Courselle, J.O. De Beer, S. Apers, and E. Deconinck. 2014. "Headspace–gas chromatographic fingerprints to discriminate and classify counterfeit medicines." *Talanta* 123:78–88 doi: 10.1016/j.talanta.2014.01.020.

Custers, D., B. Krakowska, J.O. De Beer, P. Courselle, M. Daszykowski, S. Apers, and E. Deconinck. 2016a. "Chromatographic impurity fingerprinting of genuine and counterfeit Cialis(R) as a means to compare the discriminating ability of PDA and MS detection." *Talanta* 146:540–8 doi: 10.1016/j.talanta.2015.09.029.

Custers, D., S. Vandemoortele, J.L. Bothy, J.O. De Beer, P. Courselle, S. Apers, and E. Deconinck. 2016b. "Physical profiling and IR spectroscopy: simple and effective methods to discriminate between genuine and counterfeit samples of Viagra(R) and Cialis(R)." *Drug Test Anal* 8 (3–4):378–387. doi: 10.1002/dta.1813.

de Veij, M., A. Deneckere, P. Vandenabeele, D. de Kaste, and L. Moens. 2008. "Detection of counterfeit Viagra® with Raman spectroscopy." *Journal of Pharmaceutical and Biomedical Analysis* 46 (2):303–309. doi: 10.1016/j.jpba.2007.10.021.

Deconinck, E., M. Canfyn, P.Y. Sacre, S. Baudewyns, P. Courselle, and J.O. De Beer. 2012a. "A validated GC-MS method for the determination and quantification of residual solvents in counterfeit tablets and capsules." *J Pharm Biomed Anal* 70:64–70 doi: 10.1016/j.jpba.2012.05.022.

Deconinck, E., P.Y. Sacre, D. Coomans, and J. De Beer. 2012b. "Classification trees based on infrared spectroscopic data to discriminate between genuine and counterfeit medicines." *J Pharm Biomed Anal* 57:68–75 doi: 10.1016/j.jpba.2011.08.036.

Deconinck, E., P.Y. Sacre, P. Courselle, and J.O. De Beer. 2012c. "Chemometrics and chromatographic fingerprints to discriminate and classify counterfeit medicines containing PDE-5 inhibitors." *Talanta* 100:123–33 doi: 10.1016/j.talanta.2012.08.029.

Deconinck, E., P.Y. Sacre, P. Courselle, and J.O. De Beer. 2013. "Chromatography in the detection and characterization of illegal pharmaceutical preparations." *J Chromatogr Sci* 51 (8):791–806. doi: 10.1093/chromsci/bmt006.

Dégardin, K., A. Desponds, and Y. Roggo. 2017. "Protein-based medicines analysis by Raman spectroscopy for the detection of counterfeits." *Forensic Sci Int* 278:313–325 doi: 10.1016/j.forsciint.2017.07.012.

Dégardin, K., A. Guillemain, N.V. Guerreiro, and Y. Roggo. 2016. "Near infrared spectroscopy for counterfeit detection using a large database of pharmaceutical tablets." *J Pharm Biomed Anal* 128:89–97 doi: 10.1016/j.jpba.2016.05.004.

Dégardin, K., A. Guillemain, P. Klespe, F. Hindelang, R. Zurbach, and Y. Roggo. 2018. "Packaging analysis of counterfeit medicines." *Forensic Sci Int* 291:144–157 doi: 10.1016/j.forsciint.2018.08.023.

Dégardin, K., M. Jamet, A. Guillemain, and T. Mohn. 2019. "Authentication of pharmaceutical vials." *Talanta* 198:487–500 doi: 10.1016/j.talanta.2019.01.121.

Dégardin, K., Y. Roggo, F. Been, and P. Margot. 2011. "Detection and chemical profiling of medicine counterfeits by Raman spectroscopy and chemometrics." *Anal Chim Acta* 705 (1–2):334–41. doi: 10.1016/j.aca.2011.07.043.

Dégardin, K., A. Guillemain, and Y. Roggo. 2017. "Comprehensive Study of a Handheld Raman Spectrometer for the Analysis of Counterfeits of Solid-Dosage Form Medicines." *Journal of Spectroscopy* 2017:1–13 doi: 10.1155/2017/3154035.

Dégardin, K., and Y. Roggo. 2017. "Innovative Strategy for Counterfeit Analysis." *Medicine Access @ Point of Care* 1. doi: 10.5301/maapoc.0000013.

Desai, D. 2014. "Pharmachk: robust device for counterfeit and substandard medicines screening on developing regions." Ph.D. dissertation, Boston University.

DeWitt, K. 2015. Internship report from the FDA's Forensic Chemistry Center: X-Ray Powder Diffraction Method Development and Validation for the Identification of Counterfeit Pharmaceuticals.

Dorlo, T.P., T.A. Eggelte, P.J. de Vries, and J.H. Beijnen. 2012. "Characterization and identification of suspected counterfeit miltefosine capsules." *Analyst* 137 (5):1265–74. doi: 10.1039/c2an15641e.

Dos Santos, M.K., K. de Cassia Mariotti, A. Kahmann, M.J. Anzanello, M.F. Ferrao, A. de Araujo Gomes, R.P. Limberger, and R.S. Ortiz. 2019. "Comparison between counterfeit and authentic medicines: A novel approach using differential scanning calorimetry and hierarchical cluster analysis." *J Pharm Biomed Anal* 166:304–309. doi: 10.1016/j.jpba.2019.01.029.

Dubois, J., J.-C. Wolff, J.K. Warrack, J. Schoppelrei, and E. Lewis. 2007. "NIR chemical imaging for counterfeit pharmaceutical products analysis." *Spectroscopy* 22 (2):40.

El-Bagary, R.I., E.F. Elkady, S. Mowaka, and M. Attallah. 2015. "Validated HPLC and Ultra-HPLC Methods for Determination of Dronedarone and Amiodarone Application for Counterfeit Drug Analysis." *J AOAC Int* 98 (6):1496–502. doi: 10.5740/jaoacint.15-054.

FDA. 2022. accessed January 21, 2022. https://www.accessdata.fda.gov/scripts/cder/daf.

FDA. 2013. "CD-3: A New Tool in FDA's Fight Against Counterfeit Products.", accessed 10/13/2021. http://www.youtube.com/watch?v=mfYUkiKAJvA.

Fejős, I., G. Neumajer, S. Béni, and P. Jankovics. 2014. "Qualitative and quantitative analysis of PDE-5 inhibitors in counterfeit medicines and dietary supplements by HPLC–UV using sildenafil as a sole reference." *Journal of Pharmaceutical and Biomedical Analysis* 98:327–333 doi: 10.1016/j.jpba.2014.06.010.

Fernandez, F.M., R.B. Cody, M.D. Green, C.Y. Hampton, R. McGready, S. Sengaloundeth, N.J. White, and P.N. Newton. 2006. "Characterization of solid counterfeit drug samples by desorption electrospray ionization and direct-analysis-in-real-time coupled to time-of-flight mass spectrometry." *Chem Med Chem* 1 (7):702–5. doi: 10.1002/cmdc.200600041.

Fernandez, F.M., D. Hostetler, K. Powell, H. Kaur, M.D. Green, D.C. Mildenhall, and P.N. Newton. 2011. "Poor quality drugs: grand challenges in high throughput detection, countrywide sampling, and forensics in developing countries." *Analyst* 136 (15):3073–82. doi: 10.1039/c0an00627k.

Fernandez, F.M., M.D. Green, and P.N. Newton. 2008. "Prevalence and Detection of Counterfeit Pharmaceuticals: A Mini Review." *Industrial & Engineering Chemistry Research* 47 (3):585–590. doi: 10.1021/ie0703787.

Flaherty, M.P., G.M. Gaul. 2003. "Miami Man Charged With Selling Counterfeit Lipitor." accessed 10/13/2021. https://www.washingtonpost.com/archive/business/2003/12/06/miami-man-charged-with-selling-counterfeit-lipitor/3c6033b8-91e5-4811-855f-def-3ca7a1b57/.

Fornelli, L., K. Srzentić, R. Huguet, C. Mullen, S. Sharma, V. Zabrouskov, R.T. Fellers, K.R. Durbin, P.D. Compton, and N.L. Kelleher. 2018. "Accurate sequence analysis of a monoclonal antibody by top-down and middle-down orbitrap mass spectrometry applying multiple ion activation techniques." *Analytical chemistry* 90 (14):8421–8429.

Foroughi, M.H., M. Akhgari, F. Jokar, and Z. Mousavi. 2017. "Identification of undeclared active pharmaceutical ingredients in counterfeit herbal medicines used as opioid substitution therapy." *Australian Journal of Forensic Sciences* 49 (6):720–729. doi: 10.1080/00450618.2016.1273387.

Frosch, T., E. Wyrwich, D. Yan, C. Domes, R. Domes, J. Popp, and T. Frosch. 2019. "Counterfeit and substandard test of the antimalarial tablet Riamet® by means of Raman hyperspectral multicomponent analysis." *Molecules* 24 (18):3229.

Gaudiano, M.C., A. Borioni, E. Antoniella, and L. Valvo. 2016. "Counterfeit Adderall Containing Aceclofenac from Internet Pharmacies." *J Forensic Sci* 61 (4):1126–1130. doi: 10.1111/1556-4029.13095.

Gaudiano, M.C., E. Antoniella, P. Bertocchi, and L. Valvo. 2006. "Development and validation of a reversed-phase LC method for analysing potentially counterfeit antimalarial medicines." *Journal of Pharmaceutical and Biomedical Analysis* 42 (1):132–135. doi: 10.1016/j.jpba.2006.01.059.

Graham, M.R., P. Ryan, J.S. Baker, B. Davies, N.-E. Thomas, S.-M. Cooper, P. Evans, S. Easmon, C.J. Walker, D. Cowan, and A.T. Kicman. 2009. "Counterfeiting in performance- and image-enhancing drugs." *Drug Testing and Analysis* 1 (3):135–142. doi: 10.1002/dta.30.

Green, M.D., D.M. Hostetler, H. Nettey, I. Swamidoss, N. Ranieri, and P.N. Newton. 2015. "Integration of novel low-cost colorimetric, laser photometric, and visual fluorescent techniques for rapid identification of falsified medicines in resource-poor areas: application to artemether-lumefantrine." *Am J Trop Med Hyg* 92 (6 Suppl):8–16. doi: 10.4269/ajtmh.14-0832.

Green, M.D., D.L. Mount, R.A. Wirtz, and N.J. White. 2000. "A colorimetric field method to assess the authenticity of drugs sold as the antimalarial artesunate." *J Pharm Biomed Anal* 24 (1):65–70. doi: 10.1016/s0731-7085(00)00360-5.

Green, M.D., D.L. Mount, and R.A. Wirtz. 2001. "Authentication of artemether, artesunate and dihydroartemisinin antimalarial tablets using a simple colorimetric method." *Tropical Medicine & International Health* 6 (12):980–982. doi: https://doi.org/10.1046/j.1365-3156.2001.00793.x.

Guerra, A., M. von Stosch, and J. Glassey. 2019. "Toward biotherapeutic product real-time quality monitoring." *Critical reviews in biotechnology* 39 (3):289–305.

Guillemain, A., K. Degardin, and Y. Roggo. 2017. "Performance of NIR handheld spectrometers for the detection of counterfeit tablets." *Talanta* 165:632–640 doi: 10.1016/j.talanta.2016.12.063.

Guo, S., L. He, D.J. Tisch, J. Kazura, S. Mharakurwa, J. Mahanta, S. Herrera, B. Wang, and L. Cui. 2016. "Pilot testing of dipsticks as point-of-care assays for rapid diagnosis of poor-quality artemisinin drugs in endemic settings." *Trop Med Health* 44:15. doi: 10.1186/s41182-016-0015-8.

Hajjou, M., Y. Qin, S. Bradby, D. Bempong, and P. Lukulay. 2013. "Assessment of the performance of a handheld Raman device for potential use as a screening tool in evaluating medicines quality." *J Pharm Biomed Anal* 74:47–55 doi: 10.1016/j.jpba.2012.09.016.

Hall, K.A. 2005. "Chemical'Fingerprinting'and Identification of Unknowns in Counterfeit Artesunate Antimalarial Tablets from Southeast Asia by Liquid Chromatography/ Time-of-flight Mass Spectrometry." M.S. dissertation, Georgia Institute of Technology.

He, L., T. Nan, Y. Cui, S. Guo, W. Zhang, R. Zhang, G. Tan, B. Wang, and L. Cui. 2014. "Development of a colloidal gold-based lateral flow dipstick immunoassay for rapid qualitative and semi-quantitative analysis of artesunate and dihydroartemisinin." *Malar J* 13:127. doi: 10.1186/1475-2875-13-127.

Higgins, T. 2015. "Preventing tablet tooling problems." *Pharm Technol Europe* 27 (2):47–48.

Hoellein, L., and U. Holzgrabe. 2014. "Development of simplified HPLC methods for the detection of counterfeit antimalarials in resource-restraint environments." *J Pharm Biomed Anal* 98:434–45 doi: 10.1016/j.jpba.2014.06.013.

Hosfield, C., E. Largy, A. Catrain, F. Cantais, G. Van Vyncht, M. Rosenblatt, S. Saveliev, M. Urh, and A. Delobel. Characterizing deamidation and oxidation in adalimumab with low pH peptide mapping and middle-up mass spec analysis. Poster presentation at the 64th ASMS Conference on Mass Spectrometry and Allied Topics, American Society for Mass Spectrometry, June 5–9, 2016, San Antonio, Texas, USA. 2016.

Hu, C.Q., W.B. Zou, W.S. Hu, X.K. Ma, M.Z. Yang, S.L. Zhou, J.F. Sheng, Y. Li, S.H. Cheng, and J. Xue. 2006. "Establishment of a Fast Chemical Identification System for screening of counterfeit drugs of macrolide antibiotics." *J Pharm Biomed Anal* 40 (1):68–74. doi: 10.1016/j.jpba.2005.06.026.

JÄHNKE, Richard WO. 2004. "Counterfeit medicines and the GPHF-Minilab for rapid drug quality verification." *Pharmazeutische Industrie* 66 (10):1187–1193.

Janin-Bussat, M.-C., L. Tonini, C. Huillet, O. Colas, C. Klinguer-Hamour, N. Corvaïa, and A. Beck. 2013. "Cetuximab Fab and Fc N-glycan fast characterization using IdeS digestion and liquid chromatography coupled to electrospray ionization mass spectrometry." In *Glycosylation Engineering of Biopharmaceuticals*, 93–113. Springer.

Janvier, S., K. Cheyns, M. Canfyn, S. Goscinny, B. De Spiegeleer, C. Vanhee, and E. Deconinck. 2018. "Impurity profiling of the most frequently encountered falsified polypeptide drugs on the Belgian market." *Talanta* 188:795–807 doi: 10.1016/j.talanta.2018.06.023.

Jiang, H., S.-L. Wu, B.L. Karger, and W.S. Hancock. 2009. "Mass spectrometric analysis of innovator, counterfeit, and follow-on recombinant human growth hormone." *Biotechnology Progress* 25 (1):207–218.

Jin, L. 2016. "Integrating intact, reduced and IdeS approaches for rapid differentiation and verification of monoclonal antibodies using liquid chromatography Q-Exactive mass spectrometry. Poster presentation at the 64th ASMS Conference on Mass Spectrometry and Allied Topics, American Society for Mass Spectrometry, June 5–9, 2016, San Antonio, Texas, USA.".

Jin, L. 2017. "Comparative Fc N-glycosylation profiling at glycopeptide level for differentiation and verification of mAb therapeutics using HILIC-SPE coupled to MALDI-TOF-MS and LC-HRAMMS. Poster presentation at the 65th ASMS Conference on Mass Spectrometry and Allied Topics, American Society for Mass Spectrometry, June 4–8, 2017, Indianapolis, Indiana, USA.".

Jin, L. 2018. "Bottom-up approach with low pH digestion for differentiation of monoclonal antibody therapeutics using high resolution accurate mass LC-MS/MS. Poster presentation at the 15th Symposium on the Practical Applications of Mass Spectrometry in the Biotechnology Industry (Mass Spec 2018), California Separation Science Society, September 9–12, 2018, San Francisco, California, USA."

Jin, L. 2019. "Middle-Down Analysis with Multiplexing Higher Energy Collisional Dissociation for Sequence Verification of Monoclonal Antibody Therapeutics Using Q-Exactive LC-MS/MS. Poster presentation at the 16th Symposium on the Practical Applications of Mass Spectrometry in the Biotechnology Industry (Mass Spec 2019), California Separation Science Society, September 17–20, 2019, Chicago, Illinois, USA."

Jung, C.R., R.S. Ortiz, R. Limberger, and P. Mayorga. 2012. "A new methodology for detection of counterfeit Viagra(R) and Cialis(R) tablets by image processing and statistical analysis." *Forensic Sci Int* 216 (1–3):92–96. doi: 10.1016/j.forsciint.2011.09.002.

Jung, S.K., K.H. Lee, J.W. Jeon, J.W. Lee, B. Oh Kwon, Y.J. Kim, J.S. Bae, D.-I. Kim, S.Y. Lee, and S.J. Chang. 2014. "Physicochemical characterization of Remsima®." *MAbs*. 6 (5):1163–1177. doi:10.4161/mabs.32221.

Kalyanaraman, R., G. Dobler, and M. Ribick. 2010. "Portable spectrometers for pharmaceutical counterfeit detection." *American Pharmaceutical Review* 13 (3):38–45.

Kimani, M.M., A. Lanzarotta, and J.S. Batson. 2021a. "Trace level detection of select opioids (fentanyl, hydrocodone, oxycodone, and tramadol) in suspect pharmaceutical tablets using surface-enhanced Raman scattering (SERS) with handheld devices." *J Forensic Sci* 66 (2):491–504. doi: 10.1111/1556-4029.14600.

Kimani, M.M, A. Lanzarotta, and J.S. Batson. 2021b. "Rapid determination of eight benzodiazepines in suspected counterfeit pharmaceuticals using surfac… enhanced Raman scattering with handheld Raman spectrometers." *Journal of Forensic Sciences* 66: 2167–2179.

Koesdjojo, M.T., Y. Wu, A. Boonloed, E.M. Dunfield, and V. T. Remcho. 2014. "Low-cost, high-speed identification of counterfeit antimalarial drugs on paper." *Talanta* 130:122–7 doi: 10.1016/j.talanta.2014.05.050.

Kumar, A.K., N. Vishal Gupta, P. Lalasa, and S. Sandhil. 2013. "A review on packaging materials with anti-counterfeit, tamper-evident features for pharmaceuticals." *Int. J. Drug Dev. Res* 5: 26–34.

Kumar, B., and A. Baldi. 2016. "The Challenge of Counterfeit Drugs: A Comprehensive Review on Prevalence, Detection and Preventive Measures." *Curr Drug Saf* 11 (2):112–20. doi: 10.2174/1574886310666151014114633.

Kwok, K., and L.S. Taylor. 2012. "Analysis of counterfeit Cialis(R) tablets using Raman microscopy and multivariate curve resolution." *J Pharm Biomed Anal* 66:126–135 doi: 10.1016/j.jpba.2012.03.026.

Lanzarotta, A., N. Ranieri, and D. Albright. 2015. "Analysis of counterfeit FDA-Regulated products at the forensic chemistry center: Rapid visual and chemical screening procedures inside and outside of the laboratory." *American Pharmaceutical Review* 18.

Lanzarotta, A., M.M. Kimani, M.D. Thatcher, J. Lynch, M. Fulcher, M.R. Witkowski, and J.S. Batson. 2020. "Evaluation of Suspected Counterfeit Pharmaceutical Tablets Declared to Contain Controlled Substances Using Handheld Raman Spectrometers." *J Forensic Sci* 65 (4):1274–1279. doi: 10.1111/1556-4029.14287.

Lanzarotta, A., K. Lakes, C.A. Marcott, M.R. Witkowski, and A.J. Sommer. 2011. "Analysis of counterfeit pharmaceutical tablet cores utilizing macroscopic infrared spectroscopy and infrared spectroscopic imaging." *Anal Chem* 83 (15):5972–8. doi: 10.1021/ac200957d.

Lanzarotta, A., L. Lorenz, J.S. Batson, and C. Flurer. 2017. "Development and implementation of a pass/fail field-friendly method for detecting sildenafil in suspect pharmaceutical tablets using a handheld Raman spectrometer and silver colloids." *J Pharm Biomed Anal* 146:420–425. doi: 10.1016/j.jpba.2017.09.005.

Lanzarotta, A., J. Crowe, and S. Andria. 2013. "A Systematic Procedure for Screening Counterfeit Pharmaceutical Tablet Coatings and Cores Utilizing Infrared Spectroscopy." *Journal of Regulatory Science* 5 (1):21–28.

Lawson, G. 2014. "Counterfeit Tablet Investigations: Can ATR FT/IR Provide Rapid Targeted Quantitative Analyses?" *Journal of Analytical & Bioanalytical Techniques* 5 (6). doi: 10.4172/2155-9872.1000214.

Lawson, L.S., and J.D. Rodriguez. 2016. "Raman Barcode for Counterfeit Drug Product Detection." *Anal Chem* 88 (9):4706–13. doi: 10.1021/acs.analchem.5b04636.

Leary, P.E., R.A. Crocombe, R. Kalyanaraman, Portable Spectroscopy and Spectrometry 2: Applications. 2021. "The Value of Portable Spectrometers for the Analysis of Counterfeit Pharmaceuticals," 85–123.

Lebel, P., J. Gagnon, A. Furtos, and K.C. Waldron. 2014. "A rapid, quantitative liquid chromatography-mass spectrometry screening method for 71 active and 11 natural erectile dysfunction ingredients present in potentially adulterated or counterfeit products." *J Chromatogr A* 1343:143–151 doi: 10.1016/j.chroma.2014.03.078.

Lee, J.H., H.N. Park, O.R. Park, N.S. Kim, S.K. Park, and H. Kang. 2019. "Screening of illegal sexual enhancement supplements and counterfeit drugs sold in the online and offline markets between 2014 and 2017." *Forensic Sci Int* 298:10–19 doi: 10.1016/j.forsciint.2019.02.014.

Lee, J.H., J.H. Jeong, H.-J. Park, J. Ah Do, S. Heo, S. Cho, and C.-Y. Yoon. 2016. "Analysis of erectile dysfunction drugs and their analogues in counterfeit drugs and herbal medicines by LC-ESI-MS/MS." *Analytical Science and Technology* 29 (4):155–161. doi: 10.5806/ast.2016.29.4.155.

Li-hui, Y., L. Jun-qing, C. Jin-quan, W. Jun, Z. Xue-bo, Y. Mei, Z. Li, Z. Yu, X. Xin-yue, and J. Shao-hong. 2013. "Falsified lamivudine/zidovudine/nevirapine tablets: rapid identification using X-ray fluorescence technique." *WHO Drug Information* 27 (3):213.

Lopes, M.B., and J.C. Wolff. 2009. "Investigation into classification/sourcing of suspect counterfeit Heptodintrade mark tablets by near infrared chemical imaging." *Anal Chim Acta* 633 (1):149–155. doi: 10.1016/j.aca.2008.11.036.

Mackey, T.K, and G. Nayyar. 2017. "A review of existing and emerging digital technologies to combat the global trade in fake medicines." *Expert opinion on drug safety* 16 (5):587–602.

Marini, R.D., E. Rozet, M.L. Montes, C. Rohrbasser, S. Roht, D. Rheme, P. Bonnabry, J. Schappler, J.L. Veuthey, P. Hubert, and S. Rudaz. 2010. "Reliable low-cost capillary electrophoresis device for drug quality control and counterfeit medicines." *J Pharm Biomed Anal* 53 (5):1278–87. doi: 10.1016/j.jpba.2010.07.026.

Mariotti Kde, C., R.S. Ortiz, D.Z. Souza, T.C. Mileski, P.E. Froehlich, and R.P. Limberger. 2013. "Trends in counterfeits amphetamine-type stimulants after its prohibition in Brazil." *Forensic Sci Int* 229 (1–3):23–26. doi: 10.1016/j.forsciint.2013.03.026.

Martino, R., M. Malet-Martino, V. Gilard, and S. Balayssac. 2010. "Counterfeit drugs: analytical techniques for their identification." *Analytical and Bioanalytical Chemistry* 398 (1):77–92. doi: 10.1007/s00216-010-3748-y.

McCord, J., M. Mavity, S. Damayo, and D. Wintczak. 2015. "Universal HPLC Analysis for Counterfeit Medication: A Partnership of Purdue University and the Kilimanjaro School of Pharmacy." *Purdue Journal of Service-Learning and International Engagement* 2 (1):23–26. doi: 10.5703/1288284315692.

McEwen, I., A. Elmsjo, A. Lehnstrom, B. Hakkarainen, and M. Johansson. 2012. "Screening of counterfeit corticosteroid in creams and ointments by NMR spectroscopy." *J Pharm Biomed Anal* 70:245–250 doi: 10.1016/j.jpba.2012.07.005.

Mohammad, M.A., E.F. Elkady, M.A. Fouad, and W.A. Salem. 2020. "Analysis of Aspirin, Prasugrel and Clopidogrel in Counterfeit Pharmaceutical and Herbal Products: Plackett-Burman Screening and Box-Behnken Optimization." *J Chromatogr Sci*. doi: 10.1093/chromsci/bmaa113.

Neuberger, S., and C. Neususs. 2015. "Determination of counterfeit medicines by Raman spectroscopy: Systematic study based on a large set of model tablets." *J Pharm Biomed Anal* 112:70–78. doi: 10.1016/j.jpba.2015.04.001.

Ning, X., G. Tan, X. Chen, M. Wang, B. Wang, and L. Cui. 2019. "Development of a lateral flow dipstick for simultaneous and semi-quantitative analysis of dihydroartemisinin and piperaquine in an artemisinin combination therapy." *Drug Test Anal* 11 (9):1444–1452. doi: 10.1002/dta.2656.

Nyadong, L., M.D. Green, V.R. De Jesus, P.N. Newton, and F.M. Fernández. 2007. "Reactive Desorption Electrospray Ionization Linear Ion Trap Mass Spectrometry of Latest-Generation Counterfeit Antimalarials via Noncovalent Complex Formation." *Analytical Chemistry* 79 (5):2150–2157. doi: 10.1021/ac062205h.

Nyadong, L., G.A. Harris, S. Balayssac, A.S. Galhena, M. Malet-Martino, R. Martino, R.M. Parry, M.D. Wang, F.M. Fernández, and V. Gilard. 2009. "Combining Two-Dimensional Diffusion-Ordered Nuclear Magnetic Resonance Spectroscopy, Imaging Desorption Electrospray Ionization Mass Spectrometry, and Direct Analysis in Real-Time Mass Spectrometry for the Integral Investigation of Counterfeit Pharmaceuticals." *Analytical Chemistry* 81 (12):4803–4812. doi: 10.1021/ac900384j.

Opuni, K.F., H. Nettey, M.A. Larbi, S.N.A. Amartey, G. Nti, A. Dzidonu, P. Owusu-Danso, N. A. Owusu, and A.K. Nyarko. 2019. "Usefulness of combined screening methods for rapid detection of falsified and/or substandard medicines in the absence of a confirmatory method." *Malar J* 18 (1):403. doi: 10.1186/s12936-019-3045-y.

Ortiz, R.S., K.C. Mariotti, N.V. Schwab, G.P. Sabin, W.F.C. Rocha, E.V.R. de Castro, R.P. Limberger, P. Mayorga, M. Izabel, M.S. Bueno, and W. Romão. 2012. "Fingerprinting of sildenafil citrate and tadalafil tablets in pharmaceutical formulations via X-ray fluorescence (XRF) spectrometry." *Journal of Pharmaceutical and Biomedical Analysis* 58:7–11 doi: 10.1016/j.jpba.2011.09.005.

Pharmaceutical Security Institute. 2020. "Incident trends." accessed 10/13/2021. https://www.psi-inc.org/incident-trends.

Pitts, P. 2020. "The Spreading Cancer of Counterfeit Drugs." *Journal of Commercial Biotechnology* 25 (3). doi: 10.5912/jcb940.

Puchert, T., D. Lochmann, J.C. Menezes, and G. Reich. 2010. "Near-infrared chemical imaging (NIR-CI) for counterfeit drug identification – a four-stage concept with a novel approach of data processing (Linear Image Signature)." *J Pharm Biomed Anal* 51 (1):138–45. doi: 10.1016/j.jpba.2009.08.022.

Ranieri, N., P. Tabernero, M.D. Green, L. Verbois, J. Herrington, E. Sampson, R.D. Satzger, C. Phonlavong, K. Thao, P.N. Newton, and M.R. Witkowski. 2014. "Evaluation of a new handheld instrument for the detection of counterfeit artesunate by visual fluorescence comparison." *Am J Trop Med Hyg* 91 (5):920–924. doi: 10.4269/ajtmh.13-0644.

Ranieri, N., M.R. Witkowski, W.G. Fateley, and R. Hammaker. 2016. Device and method for detection of counterfeit pharmaceuticals and/or drug packaging. U.S. Patent 9,476,839, issued October 25, 2016.

Ranieri, N., M.R. Witkowski, and M.D. Green. 2018. Device and method for detection of counterfeit pharmaceuticals and/or drug packaging. U.S. Patent 10,007,920, issued June 26, 2018.

Rasheed, H., L. Hollein, and U. Holzgrabe. 2018. "Future Information Technology Tools for Fighting Substandard and Falsified Medicines in Low- and Middle-Income Countries." *Front Pharmacol* 9:995. doi: 10.3389/fphar.2018.00995.

Rebiere, H., P. Guinot, D. Chauvey, and C. Brenier. 2017. "Fighting falsified medicines: The analytical approach." *J Pharm Biomed Anal* 142:286–306. doi: 10.1016/j.jpba.2017.05.010.

Rebiere, H., A. Kermaïdic, C. Ghyselinck, and C. Brenier. 2019. "Inorganic analysis of falsified medical products using X-ray fluorescence spectroscopy and chemometrics." *Talanta* 195:490–496 doi: 10.1016/j.talanta.2018.11.051.

Reusch, D., M. Haberger, B. Maier, M. Maier, R. Kloseck, B. Zimmermann, M. Hook, Z. Szabo, S. Tep, and J. Wegstein. 2015. "Comparison of methods for the analysis of therapeutic immunoglobulin G Fc-glycosylation profiles—Part 1: Separation-based methods." *MAbs*. 7(1): 167–179. doi: 10.4161/19420862.2014.986000.

Ricci, C., C. Eliasson, N.A. Macleod, P.N. Newton, P. Matousek, and S.G. Kazarian. 2007a. "Characterization of genuine and fake artesunate anti-malarial tablets using Fourier transform infrared imaging and spatially offset Raman spectroscopy through blister packs." *Analytical and Bioanalytical Chemistry* 389 (5):1525–1532. doi: 10.1007/s00216-007-1543-1.

Ricci, C., L. Nyadong, F.M. Fernandez, P.N. Newton, and S.G. Kazarian. 2007b. "Combined Fourier-transform infrared imaging and desorption electrospray-ionization linear ion-trap mass spectrometry for analysis of counterfeit antimalarial tablets." *Anal Bioanal Chem* 387 (2):551–9. doi: 10.1007/s00216-006-0950-z.

Ricci, C., L. Nyadong, F. Yang, F.M. Fernandez, C.D. Brown, P.N. Newton, and S.G. Kazarian. 2008. "Assessment of hand-held Raman instrumentation for in situ screening for potentially counterfeit artesunate antimalarial tablets by FT-Raman spectroscopy and direct ionization mass spectrometry." *Anal Chim Acta* 623 (2):178–186. doi: 10.1016/j.aca.2008.06.007.

Rodionova, O., A. Pomerantsev, L. Houmoller, A. Shpak, and O. Shpigun. 2010. "Noninvasive detection of counterfeited ampoules of dexamethasone using NIR with confirmation by HPLC-DAD-MS and CE-UV methods." *Anal Bioanal Chem* 397 (5):1927–35. doi: 10.1007/s00216-010-3711-y.

Rodionova, O.Y., A.V. Titova, K.S. Balyklova, and A.L. Pomerantsev. 2019a. "Detection of counterfeit and substandard tablets using non-invasive NIR and chemometrics – A conceptual framework for a big screening system." *Talanta* 205:120150. doi: 10.1016/j.talanta.2019.120150.

Rodionova, O.Y., A.V. Titova, N.A. Demkin, K.S. Balyklova, and A.L. Pomerantsev. 2019b. "Qualitative and quantitative analysis of counterfeit fluconazole capsules: A non-invasive approach using NIR spectroscopy and chemometrics." *Talanta* 195:662–667. doi: 10.1016/j.talanta.2018.11.088.

Rodionova, O.Y., and A.L. Pomerantsev. 2010. "NIR-based approach to counterfeit-drug detection." *TrAC Trends in Analytical Chemistry* 29 (8):795–803. doi: 10.1016/j.trac.2010.05.004.

Rodionova, O.Y., L.P. Houmøller, A.L. Pomerantsev, P. Geladi, J. Burger, V.L. Dorofeyev, and A.P. Arzamastsev. 2005. "NIR spectrometry for counterfeit drug detection." *Analytica Chimica Acta* 549 (1–2):151–158. doi: 10.1016/j.aca.2005.06.018.

Rodomonte, A.L., M.C. Gaudiano, E. Antoniella, D. Lucente, V. Crusco, M. Bartolomei, P. Bertocchi, L. Manna, L. Valvo, F. Alhaique, and N. Muleri. 2010. "Counterfeit drugs detection by measurement of tablets and secondary packaging colour." *J Pharm Biomed Anal* 53 (2):215–20. doi: 10.1016/j.jpba.2010.03.044.

Rogers, R.S., N.S. Nightlinger, B. Livingston, P. Campbell, R. Bailey, and A. Balland. 2015. "Development of a quantitative mass spectrometry multi-attribute method for characterization, quality control testing and disposition of biologics." *MAbs.* 7 (5):881–890. doi: 10.1080/19420862.2015.1069454.

Romano, F., and M. Riordan. 2003. "Pocket Pal." *History* :8–22.

Sacre, P.Y., E. Deconinck, T. De Beer, P. Courselle, R. Vancauwenberghe, P. Chiap, J. Crommen, and J.O. De Beer. 2010. "Comparison and combination of spectroscopic techniques for the detection of counterfeit medicines." *J Pharm Biomed Anal* 53 (3):445–453. doi: 10.1016/j.jpba.2010.05.012.

Sacre, P.Y., E. Deconinck, M. Daszykowski, P. Courselle, R. Vancauwenberghe, P. Chiap, J. Crommen, and J.O. De Beer. 2011a. "Impurity fingerprints for the identification of counterfeit medicines--a feasibility study." *Anal Chim Acta* 701 (2):224–231. doi: 10.1016/j.aca.2011.05.041.

Sacre, P.Y., E. Deconinck, L. Saerens, T. De Beer, P. Courselle, R. Vancauwenberghe, P. Chiap, J. Crommen, and J.O. De Beer. 2011b. "Detection of counterfeit Viagra(R) by Raman microspectroscopy imaging and multivariate analysis." *J Pharm Biomed Anal* 56 (2):454–461. doi: 10.1016/j.jpba.2011.05.042.

Samms, W.C., Y.J. Jiang, M.D. Dixon, S.S. Houck, and A. Mozayani. 2011. "Analysis of alprazolam by DART-TOF mass spectrometry in counterfeit and routine drug identification cases." *J Forensic Sci* 56 (4):993–8. doi: 10.1111/j.1556-4029.2011.01767.x.

Santamaria-Fernandez, R., R. Hearn, and J.-C. Wolff. 2008. "Detection of counterfeit tablets of an antiviral drug using δ34S measurements by MC-ICP-MS and confirmation by LA-MC-ICP-MS and HPLC-MC-ICP-MS." *Journal of Analytical Atomic Spectrometry* 23 (9). doi: 10.1039/b802890g.

Santos, M., A. Kahmann, L.M. Caffarate, L.R. Ucha, R.P. Limberger, and R.S. Ortiz. 2020. "Counterfeit medicines: a pilot study for chemical profiling employing a different proposal of an usual technique." *Drug Analytical Research* 4 (2):19–23. doi: 10.22456/2527-2616.107986.

Schad, G.J., A. Allanson, S.P. Mackay, A. Cannavan, and J.N. Tettey. 2008. "Development and validation of an improved HPLC method for the control of potentially counterfeit isometamidium products." *J Pharm Biomed Anal* 46 (1):45–51. doi: 10.1016/j.jpba.2007.08.026.

Schroettner, H., M. Schmied, and S. Scherer. 2006. "Comparison of 3D surface reconstruction data from certified depth standards obtained by SEM and an infinite focus measurement machine (IFM)." *Microchimica Acta* 155 (1–2):279–284.

Shah, B., X.G. Jiang, L. Chen, and Z. Zhang. 2014. "LC-MS/MS peptide mapping with automated data processing for routine profiling of N-glycans in immunoglobulins." *Journal of The American Society for Mass Spectrometry* 25 (6):999–1011.

Sharief, S.A., P. Chahal, and E. Alocilja. 2021. "Application of DNA sequences in anti-counterfeiting: Current progress and challenges." *Int J Pharm* 602:120580. doi: 10.1016/j.ijpharm.2021.120580.

Sherma, J., and F. Rabel. 2019. "Advances in the thin layer chromatographic analysis of counterfeit pharmaceutical products: 2008–2019." *Journal of Liquid Chromatography & Related Technologies* 42 (11–12):367–379. doi: 10.1080/10826076.2019.1610772.

Silvestre, V., V.M. Mboula, C. Jouitteau, S. Akoka, R.J. Robins, and G.S. Remaud. 2009. "Isotopic 13C NMR spectrometry to assess counterfeiting of active pharmaceutical ingredients: site-specific 13C content of aspirin and paracetamol." *J Pharm Biomed Anal* 50 (3):336–341. doi: 10.1016/j.jpba.2009.04.030.

Singh, S., B. Prasad, A. Savaliya, R. Shah, V. Gohil, and A. Kaur. 2009. "Strategies for characterizing sildenafil, vardenafil, tadalafil and their analogues in herbal dietary supplements, and detecting counterfeit products containing these drugs." *TrAC Trends in Analytical Chemistry* 28 (1):13–28. doi: 10.1016/j.trac.2008.09.004.

Soares, F., M.J. Anzanello, F.S. Fogliatto, R.S. Ortiz, K.C. Mariotti, and M.F. Ferrão. 2019. "Enhancing counterfeit and illicit medicines grouping via feature selection and X-ray fluorescence spectrometry." *Journal of Pharmaceutical and Biomedical Analysis* 174:198–205 doi: 10.1016/j.jpba.2019.05.064.

Talati, R., S. Parikh, and Y.K Agrawal. 2011. "Pharmaceutical counterfeiting and analytical authentication." 7 (1):54–61.

Tobias, S., A.M. Shapiro, C.J. Grant, P. Patel, M. Lysyshyn, and L. Ti. 2021. "Drug checking identifies counterfeit alprazolam tablets." *Drug Alcohol Depend* 218:108300. doi: 10.1016/j.drugalcdep.2020.108300.

Toth, G., E. Fogarasi, A. Bartalis-Fabian, M. Foroughbakhshfasaei, I. Boldizsar, A. Darcsi, S. Lohner, G.K.E. Scriba, and Z.I. Szabo. 2020. "Liquid chromatographic method for the simultaneous determination of achiral and chiral impurities of dapoxetine in approved and counterfeit products." *J Chromatogr A* 1626:461388. doi: 10.1016/j.chroma.2020.461388.

Trefi, S., C. Routaboul, S. Hamieh, V. Gilard, M. Malet-Martino, and R. Martino. 2008. "Analysis of illegally manufactured formulations of tadalafil (Cialis) by 1H NMR, 2D DOSY 1H NMR and Raman spectroscopy." *J Pharm Biomed Anal* 47 (1):103–113. doi: 10.1016/j.jpba.2007.12.033.

Twohig, M., S.J. Skilton, G. Fujimoto, N. Ellor, and R.S. Plumb. 2010. "Rapid detection and identification of counterfeit and [corrected] adulterated products of synthetic phosphodiesterase type-5 inhibitors with an atmospheric solids analysis probe." *Drug Test Anal* 2 (2):45–50. doi: 10.1002/dta.115.

Velpandian, T., M. Nath, M. Laxmi, and N. Halder. 2016. "Finger printing of counterfeit bevacizumab and identifying it before clinical use." 29 (6):326.

Vickers, S., M. Bernier, S. Zambrzycki, F.M. Fernandez, P.N. Newton, and C. Caillet. 2018. "Field detection devices for screening the quality of medicines: a systematic review." *BMJ Glob Health* 3 (4):e000725. doi: 10.1136/bmjgh-2018-000725.

Wang, W., M.D. Keller, T. Baughman, and B.K. Wilson. 2020. "Evaluating Low-Cost Optical Spectrometers for the Detection of Simulated Substandard and Falsified Medicines." *Appl Spectrosc* 74 (3):323–333. doi: 10.1177/0003702819877422.

Wei, H., A.A Tymiak, and G. Chen. 2013. "High-resolution MS for structural characterization of protein therapeutics: advances and future directions." *Bioanalysis* 5 (10):1299–1313.

WHO. 1999. Counterfeit drugs: guidelines for the development of measures to combat counterfeit drugs. World Health Organization. No. WHO/EDM/QSM/99.1. 1999.

WHO. 2018. "World Health Organization Global Surveillance and Monitoring System for substandard and falsified medical products. 2017." *WHO, Geneva, Switzerland.*

Wilczynski, S. 2015. "The use of dynamic thermal analysis to distinguish between genuine and counterfeit drugs." *Int J Pharm* 490 (1–2):16–21. doi: 10.1016/j.ijpharm.2015.04.077.

Wilczynski, S., R. Koprowski, and B. Blonska-Fajfrowska. 2016a. "Directional reflectance analysis for identifying counterfeit drugs: Preliminary study." *J Pharm Biomed Anal* 124:341–346. doi: 10.1016/j.jpba.2016.03.014.

Wilczynski, S., R. Koprowski, M. Marmion, P. Duda, and B. Blonska-Fajfrowska. 2016b. "The use of hyperspectral imaging in the VNIR (400–1000nm) and SWIR range (1,000–2,500 nm) for detecting counterfeit drugs with identical API composition." *Talanta* 160:1–8 doi: 10.1016/j.talanta.2016.06.057.

Wilczynski, S., R. Koprowski, A. Stolecka-Warzecha, P. Duda, A. Deda, D. Ivanova, Y. Kiselova-Kaneva, and B. Blonska-Fajfrowska. 2019. "The use of microtomographic imaging in the identification of counterfeit medicines." *Talanta* 195:870–875 doi: 10.1016/j.talanta.2018.12.009.

Winner, T.L., A. Lanzarotta, and A.J. Sommer. 2016. "Analysis of Counterfeit Coated Tablets and Multi-Layer Packaging Materials Using Infrared Microspectroscopic Imaging." *Microsc Microanal* 22 (3):649–655. doi: 10.1017/s143192761600060x.

Witkowski, M.R., Chin, L.C., Luhse, B., Bon L.T., Hui, L.Y. 2015. "Guidance Document on the Use of Detection Technologies and Overview of Detection Technologies for Drug Safety." A publication of the APEC Life Sciences Innovation Forum (LSIF) Regulatory Harmonization Steering Committee.

Witkowski, M.R. 2005. "The use of Raman spectroscopy in the detection of counterfeit and adulterated pharmaceutical products." *American Pharmaceutical Review* 8 (1):56–62.

Witkowski, M.R, and K. DeWitt. 2020. "Topics See All." *Spectroscopy* 35 (7):41–48.

Wolff, J.C., L.A. Thomson, and C. Eckers. 2003. "Identification of the 'wrong' active pharmaceutical ingredient in a counterfeit Halfan drug product using accurate mass electrospray ionisation mass spectrometry, accurate mass tandem mass spectrometry and liquid chromatography/mass spectrometry." *Rapid Commun Mass Spectrom* 17 (3):215–221. doi: 10.1002/rcm.893.

Xie, H., A. Chakraborty, J. Ahn, Y.Q. Yu, D.P. Dakshinamoorthy, M. Gilar, W. Chen, St J. Skilton, and J.R. Mazzeo. 2010. "Rapid comparison of a candidate biosimilar to an innovator monoclonal antibody with advanced liquid chromatography and mass spectrometry technologies." *MAbs* 2 (4):379–394. doi: 10.4161/mabs.11986.

Xie, S., and H.-Z. Tan. 2021. "An Anti-Counterfeiting Architecture for Traceability System Based on Modified Two-Level Quick Response Codes." *Electronics* 10 (3):320.

Yang, Y.J., D.M. Song, W.M. Jiang, and B.R. Xiang. 2010. "Rapid Resolution RP-HPLC-DAD Method for Simultaneous Determination of Sildenafil, Vardenafil, and Tadalafil in Pharmaceutical Preparations and Counterfeit Drugs." *Analytical Letters* 43 (3):373–380. doi: 10.1080/00032710903402283.

Yao, J., Y.-Q. Shi, Z.-R. Li, and S.-H. Jin. 2007. "Development of a RP-HPLC method for screening potentially counterfeit anti-diabetic drugs." *Journal of Chromatography B* 853 (1–2):254–259. doi: 10.1016/j.jchromb.2007.03.022.

Zhang, H., D. Hua, C. Huang, S.K. Samal, R. Xiong, F. Sauvage, K. Braeckmans, K. Remaut, and S.C. De Smedt. 2020. "Materials and Technologies to Combat Counterfeiting of Pharmaceuticals: Current and Future Problem Tackling." *Adv Mater* 32 (11):e1905486. doi: 10.1002/adma.201905486.

9 Coordinated Response to Poisonings and Fatalities from Counterfeit Tablets

Nikolas P. Lemos
Queen Mary University of London

CONTENTS

9.1 IN THE BEGINNING

On a fall October night, a group of friends decided to spice up their evening by seeking out drugs in the streets of San Francisco to party at home. They quickly came across some cocaine and Xanax® tablets for a handful of dollars. The group consisted of a 27-year-old man and two women (37 and 34 years old, respectively). After acquiring the drugs, they returned home ready to party the night, all excited about what was to be: a night not just fueled by alcohol but by alcohol combined with cocaine and alprazolam (the drug sold under the brand name Xanax®). A few hours after the combined ethanol and drug consumption, the partiers felt unwell and summoned help by dialing 9-1-1. When paramedics arrived at the scene, they found the 34-year-old woman dead at the scene and the other two friends in medical distress. The police and medical examiner took jurisdiction of the death scene and the decedent, while the two survivors were urgently transported to the local hospital.

DOI: 10.1201/9781003183327-9

9.2 AT THE HOSPITAL

The two friends who survived the drug-fueled party were urgently transported to the local hospital, where they reported that they had experienced unusually intense and prolonged calmness and sleepiness, and awoke with weakness on one side of their body as well as with a burning and prickling sensation in their upper and lower extremities (paresthesia). This paresthesia was described as tingling or numbness, skin crawling, or itching. Specifically, the 27-year-old man experienced paresthesia distal to his left elbow, while the 37-year-old woman reported it in her right lower extremity. The attending hospital physicians reported that both patients did not display any particular toxidrome, and that the reported weakness appeared to be consistent with compression neuropathy—a medical condition caused by direct pressure on a nerve. This is better known as a "trapped nerve."

Hospital laboratory test results for the 27-year-old man were significant for leukocytosis (white blood cell count, 30,000/µL) and acute renal insufficiency (creatinine level, 1.68 mg/dL). This man had rhabdomyolysis with a creatine kinase concentration of 1,012 IU/L, and higher than usual concentrations of aspartate aminotransferase (528 IU/L) and alanine aminotransferase (683 IU/L) with normal hepatic function. The troponin concentration of this patient was measured at 1.89 ng/mL, which is considered elevated and suggests a patient with unstable angina. Magnetic resonance imaging studies were unremarkable, further supporting the diagnosis of compression neuropathy.

The hospital laboratory test results for the 37-year-old woman were unremarkable for leukocytosis and for renal insufficiency. This woman had rhabdomyolysis with a creatine kinase concentration of 354 IU/L, and higher than usual concentrations of aspartate aminotransferase (120 IU/L) and alanine aminotransferase (64 IU/L) with normal hepatic function. The troponin concentration of this patient was measured at 0.9 ng/mL, which is also considered elevated and suggests that she suffered from unstable angina. Magnetic resonance imaging studies for the 37-year-old woman were unremarkable, also supporting the diagnosis of compression neuropathy.

The results of toxicology studies in serum specimens from these two patients are summarized in Table 9.1.

9.3 AT THE MORGUE

The third friend from the group was a 34-year-old woman who was found dead at the scene. On such occasions, the local office of the medical examiner/coroner assumes jurisdiction and undertakes medical, investigative, and laboratory studies to determine the cause and manner of death. That office's investigators typically visit the scene not only to remove the decedent back to the morgue but also to identify and collect physical evidence items that may prove useful in the investigation of the death. In this instance, the investigators collected drug paraphernalia and drugs including two tablets, as can be seen in Figure 9.1.

TABLE 9.1
Hospital Toxicology Studies Performed on Admission Serum Specimens of the Two Surviving Friends

	Patient 1	Patient 2
Sex	Man	Woman
Age (years)	27	37
Ethanol	None detected	0.068 mg/dL
Cocaine	Detected & confirmed (not quantified)	Detected & confirmed (not quantified)
Benzoylecgonine	Detected & confirmed (not quantified)	Detected & confirmed (not quantified)
Cocaethylene	Detected & confirmed (not quantified)	Detected & confirmed (Not quantified)
Levamisole	Detected	Detected
Fentanyl	1.6 ng/mL	0.61 ng/mL
Norfentanyl	Detected & confirmed (not quantified)	Detected & confirmed (not quantified)
Etizolam	0.60 ng/mL	< 0.24 ng/mL
Diphenhydramine	Detected	None detected
Other findings	None	Cotinine detected

Based on the physical evidence recovered at the scene of this death, the laboratory undertook forensic toxicologic tests that would detect any alcohol and any prescription medications and illicit drugs that were present in the postmortem biological specimens of this decedent. The postmortem forensic toxicology studies in this case detected no alcohol in any of the analyzed postmortem

FIGURE 9.1 Counterfeit tablets imprinted with "XANAX" recovered from the scene of the death, as photographed by the author.

specimens, but detected and confirmed cocaine (below the laboratory's limit of quantitation), benzoylecgonine (1.4 mg/L), and alprazolam (16 ng/mL) in her postmortem peripheral blood. Unexpectedly, the laboratory's toxicology studies also detected, confirmed, and quantified fentanyl in the decedent's postmortem peripheral blood at a concentration of 21 ng/mL. The results of the decedent's postmortem toxicology studies in peripheral blood and urine are shown in Table 9.2.

Upon the realization that fentanyl was present in significant and potentially toxic concentrations in the specimens of the decedent (a finding that contradicted the alleged drug use by the decedent and her friends), the laboratory expeditiously undertook the analysis of the recovered tablets from the scene of the death. Within hours, it was determined that the two tablets imprinted with "XANAX" contained no alprazolam, but that they, instead, contained fentanyl and etizolam. These results were urgently verified by a second reference laboratory in the interest of quality assurance.

The medical examiner also requested chemistry studies be performed in the vitreous humor of the decedent, the results of which are tabulated in Table 9.3.

TABLE 9.2
Postmortem Forensic Toxicology Findings in the Case of the 34-year-old Woman Found Dead at the Scene of a Group Gathering after Allegedly Having Consumed Ethanol, Cocaine, and Tablets Imprinted with "XANAX" Acquired in the Streets of San Francisco

Specimen Type	Compound	Result
Blood (Peripheral)	Fentanyl	21 ng/mL
Blood (Peripheral)	Alprazolam	16 ng/mL
Blood (Peripheral)	Cocaine	Confirmed present (below limit of quantitation)
Blood (Peripheral)	Benzoylecgonine	1.4 mg/L
Blood (Cardiac/Central)	Ecgonine methyl ester	Confirmed present
Urine	Fentanyl	Confirmed present
Urine	Norfentanyl	Detected
Urine	Alprazolam	Confirmed present
Urine	α-Hydroxy alprazolam	Confirmed present
Urine	Clonazepam	Confirmed present
Urine	Nordiazepam	Confirmed present
Urine	Cocaine	Confirmed present
Urine	Benzoylecgonine	Confirmed present
Urine	Ecgonine methyl ester	Confirmed present
Urine	Anhydroecgonine methyl ester	Detected
Urine	Doxylamine	Detected

TABLE 9.3
Postmortem Vitreous Humor Chemistry Findings in the Case of the 34-year-old Woman Found Dead at the Scene of a Group Gathering after Allegedly Having Consumed Ethanol, Cocaine, and Tablets Imprinted with "XANAX" Acquired in the Streets of San Francisco

Compound	Result
Sodium	145 mmol/L
Potassium	12.6 mmol/L
Chloride	128 mmol/L
Glucose	< 25 mg/dL
Urea Nitrogen	20 mg/dL
Creatinine	1.2 mg/dL
Ketones	Negative

9.4 IN THE PUBLIC EYE

Within hours of the discovery that the decedent had a significant amount of fentanyl in her biological specimens without having knowingly consumed any fentanyl products, and that the tablets acquired in the streets of San Francisco with the "XANAX" imprint on them contained no alprazolam but instead contained fentanyl and etizolam, a coordinated response team was formed consisting of senior members for the forensic toxicology laboratory, the local hospital and medical school, the state poison control system, and the local police. Efforts were immediately made to raise the public's awareness that counterfeit "XANAX" tablets containing potent opioids are poisoning and killing unsuspecting users in the area. Within just a few days from the two poisonings and fatality, several red alert public health warnings were issued on the radio, television, press, and over all social media alerting people of the identified threat and potential dangers. Health and Social Workers as well as Police Officers and Outreach Volunteers walked the streets of San Francisco and all surrounding counties talking to people about the fake tablets that may cause users to die.

Forensic laboratory personnel performed online searches and identified several popular websites offering Punches with "XANAX" stamps and "Single Punch Tablet Press Machines" for sale, like the ones shown in Figure 9.2.

At the request of the local police department, these "XANAX" press machines were immediately removed from the websites and online stores where they were available for purchase.

9.5 IN THE AFTERMATH

Unfortunately, and despite the significant outreach efforts undertaken by the public health and public safety teams, a whole group of cases similar to the ones

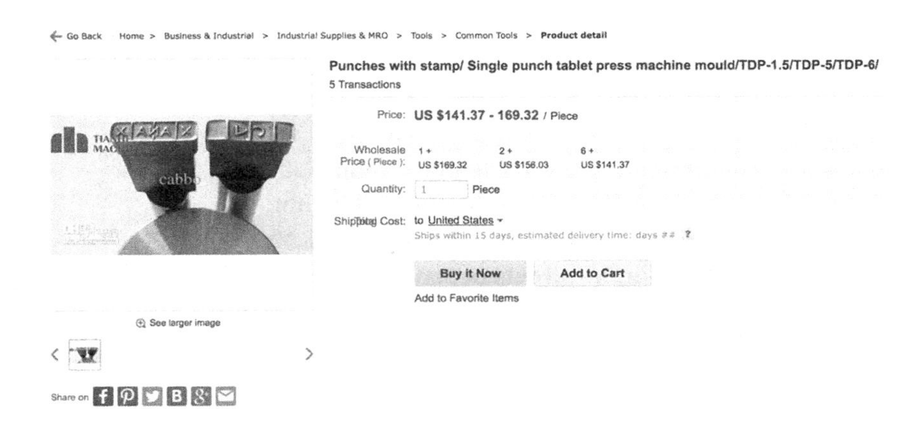

FIGURE 9.2 Screenshot of "XANAX" tablet press machine available for sale online, as captured by the author.

described in this chapter took place in the two months following the 34-year-old's fatality. These are tabulated in Table 9.4.

Most of these cases resulted in hospitalizations, with the patients eventually fully recovering. Three of the cases, however, resulted in fatalities with fentanyl being found after hospital and/or postmortem toxicology studies, even though the decedents had not knowingly acquired or consumed this potent opioid.

In the case series presented in this chapter, three drugs were the protagonists: alprazolam, fentanyl, and etizolam. Alprazolam is sold as "XANAX." It is a triazolo analog of the 1,4-benzodiazepine class of CNS-active compounds. Benzodiazepines produce sedation, induce sleep, relieve anxiety and muscle spasms, and prevent seizures. These drugs are legally available only through prescription, but abusers may acquire prescriptions from several doctors, forge prescriptions, or buy them illicitly. Alprazolam is the most frequently encountered benzodiazepine on the illicit market, and its abuse is particularly high among heroin and cocaine abusers. It is frequently associated with adolescents and young adults who take the drug orally or crush it and snort it. Fentanyl was first synthesized in Belgium in the late 1950s and has analgesic potency of many times that of morphine. It was introduced into medical practice in the 1960s as an intravenous anesthetic under the trade name of Sublimaze® and first appeared in the illicit drug scene in the mid-1970s in the medical community. This drug and its chemical analogues (often referred to as "fentalogues") continue to be a serious public health problem in the USA. Finally, etizolam is a benzodiazepine derivative currently legally available by prescription in Italy, India, and Japan. Etizolam does not, currently, have approved medical uses in the USA and is not currently controlled under the USA's Controlled Substances Act as of March 2020.

TABLE 9.4
Case Demographics, Histories, and Laboratory Findings in Seven Additional Cases that were Encountered in San Francisco in the Two Months Following the First Death with Exposure to Counterfeit Tablets Imprinted with "XANAX"

Case number	Age (Years)	Sex	Available History	Laboratory Findings ISV: Insufficient Specimen Volume for Confirmation or Quantitation. BL(P): Blood (Peripheral)
4	29	M	Found unresponsive by sister after ingesting 'Xanax.' Brain death declared three days later	**Hospital Toxicology Results** Serum cocaine: detected; ISV Serum fentanyl: detected; ISV Serum benzodiazepines: detected; ISV **Postmortem Toxicology Results** Liver cocaine: <1.25 mg/kg Liver midazolam: <25 ng/g Liver diphenhydramine: < 1000 ng/g Liver diazepam: confirmed present (below limit of quantitation)
5	21	M	Found down "in a wheelbarrow." Obtundation requiring intubation, cardiogenic pulmonary edema, and biventricular heart failure. Symptoms improved after 5 days	**Hospital Toxicology Results** Serum fentanyl: 1.4 ng/mL Serum etizolam: 0.26 ng/mL Serum norfentanyl: detected Serum benzoylecgonine: detected
6	25	M	Found deceased by girlfriend after ingesting "XANAX" bought on the street	**Postmortem Toxicology Results** Blood (P) fentanyl: 17 ng/mL Blood (P) alprazolam: 15 ng/mL Blood (P) cocaine: 0.02 mg/L Blood (P) benzoylecgonine: 0.18 mg/mL Blood (P) THC-OH: 1 ng/mL Blood (P) THC-COOH: 21 ng/mL
7	42	M	Found deceased by mother; Pills marked "XANAX" found in pocket of his denim jeans	**Postmortem Toxicology Results** Blood (P) fentanyl: 6 ng/mL Blood (P) methamphetamine: 1.49 mg/L Blood (P) amphetamine: 0.14 mg/L

(Continued)

TABLE 9.4 (*Continued*)
Case Demographics, Histories, and Laboratory Findings in Seven Additional Cases that were Encountered in San Francisco in the Two Months Following the First Death with Exposure to Counterfeit Tablets Imprinted with "XANAX"

Case number	Age (Years)	Sex	Available History	Laboratory Findings ISV: Insufficient Specimen Volume for Confirmation or Quantitation. BL(P): Blood (Peripheral)
8	45	M	Lethargy after ingesting chlordiazepoxide, ethanol and 3 "XANAX" bought from a friend	**Hospital Toxicology Results** Serum etizolam: 22 ng/mL Serum chlordiazepoxide: detected Serum demoxepam: detected
9	19	M	Lethargy and ataxia after ingesting "XANAX"	**Hospital Toxicology Results** Serum fentanyl: 0.15 ng/mL Serum norfentanyl: detected Serum etizolam: 55 ng/mL Serum alprazolam: 101 ng/mL
10	19	M	CNS and respiratory depression requiring naloxone after taking "XANAX"	**Hospital Toxicology Results** Serum fentanyl: 2.6 ng/mL Serum norfentanyl: detected Serum diphenhydramine: detected Serum naloxone: detected Serum sertraline: detected
11	25	M	Cardiac arrest after ingesting "XANAX" purchased on the street	**Hospital Toxicology Results** Serum fentanyl: 1.4 ng/mL Serum norfentanyl: detected Serum acetaminophen: detected Serum diazepam: detected Serum nordiazepam: detected Serum temazepam: detected Serum oxazepam: detected Serum naloxone: detected

9.6 IN THE END

The team that was urgently mobilized after the forensic toxicology laboratory and hospital staff alerted the wider community about the original fatality, and two hospitalizations included clinical, forensic and law enforcement specialists from within the local jurisdiction and many adjacent counties. The outbreak of fake "XANAX" tablets described in this chapter resulted in 11 cases: 7 survivors and 4 decedents who were exposed to counterfeit "XANAX" tablets containing potentially lethal concentrations of fentanyl and etizolam. Toxicology studies urgently undertaken by the hospital laboratory and the forensic toxicology laboratory suggest that the group of survivors had serum fentanyl equal or less than 2.6 ng/mL. The forensic toxicology laboratory's findings suggest that decedents' blood peripheral blood concentrations were greater or equal to 6 ng/mL. These concentrations may appear significantly different when considered in isolation, but are not necessarily so when we take into consideration the lipophilic properties of fentanyl and the well-established postmortem redistribution it undergoes. Consequently, the likelihood of one's survival or death when exposed to the fake tablets imprinted "XANAX" and containing fentanyl and etizolam cannot be estimated, making survival or death an equally likely outcome from such exposure.

Clusters of patients with unexpected symptoms and effects after ingestion of illicit or illegally obtained substances should raise red flags with public health care providers of a potential epidemic by a counterfeit or adulterated product.

Coordination of all public health agencies is key in the successful handling of such outbreaks, and could result in lower number of cases and harm reduction. Agencies that should be part of such a multidisciplinary team include the local law enforcement, social services, and outreach agencies, media outlets, as well as hospital personnel, the local poison control system, and the medical examiner/coroner who must act in a timely fashion to collect and analyze medical and case histories as well as biological specimens and physical evidence, which should be quickly processed using validated procedures by an accredited laboratory staffed by certified scientists. Delays or improper storage, handling, or analysis could result in higher numbers of poisoning cases and/or fatalities. The public must be informed as quickly as possible about the possible hazards from the discovery of counterfeit tablets through the issuance of Public Health Advisories using all possible communication avenues, including radio, television, newspapers, and all available social media.

Manufacturers of counterfeit drugs are relentless in their efforts to increase their profits with no regard to the social costs. As one counterfeit tablet press is taken off the market and becomes unavailable for purchase, another surfaces. Just three months after the cluster of fake "XANAX" cases described in this chapter, another group of counterfeit tablets appeared in the streets of San Francisco: counterfeit tablets imprinted with "ONAX" were discovered at another death scene in San Francisco (Figure 9.3). These tablets are the generic form of "XANAX" and should contain alprazolam. Analysis of the tablets recovered from this death scene showed that they contained the potent dissociative anesthetic ketamine.

It is, therefore, paramount that the lines of investigation, communication, and interagency cooperation not only must be established post-outbreak, but they must continually be maintained in the interest of public health and safety.

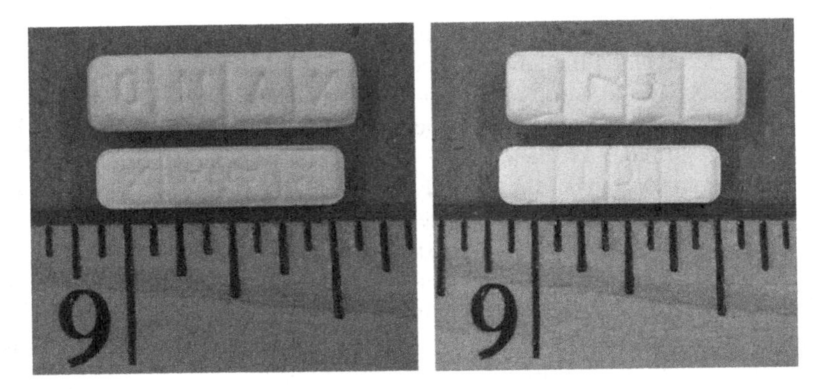

FIGURE 9.3 Counterfeit tablets imprinted with "ONAX," the generic form of "XANAX" (also shown for comparison purposes) recovered from the scene of a death, as photographed by the author.

ACKNOWLEDGMENTS

The author wishes to acknowledge the invaluable contributions and cooperation of the following:

- The California Poison Control System,
- The Zuckerberg San Francisco General Hospital and Trauma Center,
- The University of California—San Francisco,
- The Office of the Chief Medical Examiner of the City and County of San Francisco,
- The San Francisco Police Department, and
- The San Francisco Department of Public Health.

BIBLIOGRAPHY

Arens AM, KT Vo, C Smollin, X Van Wijk, K Lynch, AHB Wu, HS Narula, and NP Lemos. Poisoning outbreak in the streets of San Francisco: A case series involving counterfeit Xanax® Proceedings of the 54th Annual Meeting of The International Association of Forensic Toxicologists, Brisbane, Australia. September 2016. *Proceedings* 2016:150.

Arens AM, KT Vo, X Van Wijk, NP Lemos, KR Olson, and C Smollin. Adulterated Xanax: A Case Series from San Francisco. Proceedings of the 2016 American College of Medical Toxicology (ACMT) Annual Scientific Meeting. *Journal of Medical Toxicology*. (2016) 12:8. https://www.dea.gov/factsheets/fentanyl (accessed on 12 December 2021)

Lung DD and NP Lemos. Fentanyl: Cause of Death or Incidental Finding? Postmortem Peripheral Blood Concentrations with and without Documented Transdermal Patch Use. *Forensic Toxicology* 2014;32(1):118–125.

Lung DD, P Armenian, AM Gordon, TE Kearney and NP Lemos. Postmortem Peripheral Blood Fentanyl Concentrations in 20 Cases: Analysis and Comparison to Prior Studies. Presented at the Joint Meeting of the Society of Forensic Toxicologists (SOFT) and The International Association of Forensic Toxicologists (TIAFT), San Francisco, CA, USA, September 2011.

Lemos NP, SB Karch, E Lin, G Nazareno, V Azar, J Smith, J Melinek, AP Hart and BG Stephens. Fentanyl in Seven Medical Examiner's Cases in the City and County of San Francisco. *Proceedings of the Annual Meeting of the American Academy of Forensic Sciences* (2005),11: 376.

San NA and NP Lemos. A Survey of Human Performance and Postmortem Cases Involving Ketamine in San Francisco between 1997 and 2013. *Proceedings of the 67th Annual Meeting of the American Academy of Forensic Sciences*, Orlando, Florida, February 2015, 21: 1141.

10 Laws and Procedures to Counter the Problem of Counterfeit Drugs

Kirby B. Drake
Kirby Drake Law

CONTENTS

10.1 INTRODUCTION

There is no fool proof way for consumers, drug manufacturers/suppliers/distributors, medical personnel, or the government to counter the problem of counterfeit drugs. The good news is that only 1% of medicine in the United States is presumed to be counterfeit (Partnership for Safe Medicines, 2011). However, the World Health Organization (WHO) estimates that 50% of all medicines sold over the Internet from sites that conceal their physical address are fake (Partnership for Safe Medicines). This is a growing problem for United States consumers, especially given the increase in online purchases of drugs having international origins during the COVID-19 pandemic.

This chapter will address several mechanisms that have been employed to combat and address the problem of counterfeit drugs, including the Drug Quality and Security Act, the SHOP Safe Act, the Lanham Act, the Federal Food, Drug, and Cosmetic Act, Food and Drug Administration (FDA) action, and other actions or recommendations for consumers, medical personnel, and the government.

DOI 10.1201/9781003183327-10

10.2 DRUG QUALITY AND SECURITY ACT

Of the 35,000 online pharmacies worldwide, 95% operate illegally, including selling medicines without requiring a prescription and operating without a pharmacy license (Myshko). "Each year, as many as 19 million people in the United States purchase medicine outside the normal manufacturer-to-supplier direct supply chain, and use online pharmacies or other unconventional sources" (Magdun, 2021).

As a means for addressing the rise in online pharmacies and drug contamination outbreaks, Congress created the Drug Quality and Security Act ("DQSA") in 2013. The DSCSA outlines requirements for manufacturers, repackagers, wholesale distributors, dispensers, and third-party logistics providers (trading partners). The Office of Compliance in the Center for Drug Evaluation and Research (CDER) developed guidance on the best practices for these entities to exchange transaction information, transaction history, and a transaction statement (product tracing information) when engaging in transactions involving prescription drugs (DSCSA Standards). This involves all transactions that include "products" that are defined as a prescription drug in a finished dosage form for administration to a patient without substantial further manufacturing (such as capsules, tablets, and lyophilized products before reconstitution) (DSCSA Standards). If a trading partner does not provide the proper transaction documentation, proper documentation must be obtained to minimize disruption in the supply chain (ready for DSCSA).

The Office of Compliance in the Center for Drug Evaluation and Research (CDER) also developed guidance to aid trading partners in identifying a suspect product (DSCSA Implementation). If a product is identified as illegitimate, the trading partner must notify the Food and Drug Administration (FDA or Agency) and certain immediate trading partners under Section 582 of the FD&C Act (21 U.S.C. 360eee-1), as added by the Drug Supply Chain Security Act (DSCSA, 2013). "Manufacturers are additionally required under Section 582 to notify FDA and certain immediate trading partners, as applicable, after the manufacturer determines or is notified by FDA or a trading partner that there is a high risk that a suspect product is illegitimate" (DSCSA Implementation).

Systems must be put in place to enable a trading partner to quarantine suspect product and promptly investigate to determine whether a suspect product is illegitimate (FD&C Act, Section 582(b)(4)(A)(i), (c)(4)(A)(i), (d)(4)(A)(i), and (e)(4)(A)(i)). The FDA guidance provides specific scenarios that could significantly increase the risk of a suspect product entering the pharmaceutical distribution supply chain (DSCSA Implementation). More specifically, the FDA notes that diligence should be exercised with respect to trading partners and product sources when purchasing from a new source, receiving an unsolicited sales offer from an unknown source, purchasing on the Internet from an unknown source, and/or purchasing from a source that may have engaged in questionable business practices in the past (DSCSA Implementation). In addition, there may be considerations based on the supply, demand, history, and value of the product in question. For example, a product may be in "higher demand because of its potential or perceived relationship to a public health or other emergency (e.g., antiviral drugs)" (DSCSA Implementation). There also may be questions raised depending on the appearance of the product, such as the packaging missing information about the lot number or expiration date, the finished dosage form not matching with the color or shape of the FDA-approved product, or the packaging containing misspellings or other questionable items. (DSCSA Implementation).

In DSCSA Implementation, the FDA also provides strategies for how to identify suspect product including:

- Be alert for offers of product for sale at a very low price or one that is "too good to be true."
- Closely examine the package and the transport container (such as the case or tote):
  - To look for signs that it has been compromised (e.g., opened, broken seal, damaged, repaired, or otherwise altered). If a trading partner receives a product in a secured transport container or sealed homogenous case, trading partners should examine the appearance of that container to see if anything about that appearance seems questionable, such as shrink wrap that has unexpected markings or a seal that is broken, torn, or repaired. Such examinations may include:
    - Seeing if the package or the transport container has changed, since the last shipment of the same product type was received for an unexplained reason (e.g., a notification about the change from the manufacturer has not been received).
    - Seeing if product inserts are missing, do not correspond to the product, or are questionable in some way.
  - For shipping addresses, postmarks, or other materials indicating that the product came from an unexpected foreign entity or source.
- Closely examine the label on the package, and the label on the individual retail unit, if applicable, for:
  - Any missing information, such as the lot number or other lot identification, NDC, or strength of the drug.
  - Any altered product information, such as smudged print or print that is very difficult to read.
  - Misspelled words.
  - Bubbling in the surface of a label.
  - Lack of an "Rx only" symbol.
  - Foreign language with little or no English provided.
  - Foreign language that is used to describe the lot number.
  - A product name that differs from the name that appears on the FDA-approved drug label or labeling.
  - A product name that is the product name for a foreign version of the drug.
  - A product that is transported in a case or tote, when not expected under the circumstances.
  - Lot numbers and expiration dates on product that do not match the lot numbers and expiration dates of its outer container.

- While these strategies are intended for trading partners, they also can be helpful for consumers, especially when product is being purchased online.

10.3 FEDERAL FOOD, DRUG, AND COSMETIC ACT (FD&C ACT)

The United States also has employed the Federal Food, Drug, and Cosmetic Act (FD&C Act, 2022) to combat counterfeit drugs (FD&C Act). According to this Act, a counterfeit drug is "a drug which, or the container or labeling of which, without authorization, bears the trademark, trade name, or other identifying mark, imprint, or device, or any likeness thereof, of a drug manufacturer, processor, packer, or distributor other than the person or persons who in fact manufactured, processed, packed, or distributed such drug and which thereby falsely purports or is represented to be the product of, or to have been packed or distributed by such other drug manufacturer, processor, packer, or distributor" (FD&C Act, Section 201(2)). In general, the FD&C Act requires that drugs and devices be safe and effective for their intended uses, and that food, drugs, and devices be accurately labeled and handled in ways that prevent them from becoming contaminated. The FD&C Act prohibits "[m]aking, selling, disposing of, or keeping in possession, control, or custody, or concealing any punch, die, plate, stone, or other thing designed to print, imprint, or reproduce the trademark, trade name, or other identifying mark, imprint, or device of another or any likeness of any of the foregoing upon any drug or container or labeling thereof so as to render such drugs a counterfeit drug" (FD&C Act, Section 301(i)(2)). It also prohibits "[t]he doing of any act which causes a drug to be a counterfeit drug, or the sale or dispensing, or the holding for sale or dispensing, of a counterfeit drug" (FD&C Act, Section 301(i)(3)). A violation of Section 301(i)(3) can result in imprisonment for not more than 10 years, or a fine, or both (FDA, Section 303(b)(8)). If a violation is committed "with the intent to defraud or mislead," the violator is guilty of a felony punishable by 3 years imprisonment (FD&C Act, Section 303(a)(2)). The violation may be with the intent to defraud or mislead a consumer or a government authority, such the FDA.

Fraud on a consumer may occur when the supplier does not provide consumers the product that was purportedly sold. This can qualify as monetary fraud that satisfies the "intent to defraud" requirement (DOJ, Consumer Protection Branch). Fraud against the FDA can involve black market operations where the business is hidden from the FDA (DOJ, Consumer Protection Branch). There also are instances where a business sells products regulated by the FDA, but they attempt to avoid regulation. For example, in *United States v. Cambra*, 933 F.2d 752, 755 (9th Cir. 1991), the defendant was convicted of violating the FD&C Act by holding, causing to be held for sale, and selling counterfeit steroids with intent to defraud and mislead. He had sold products counterfeited to represent different products made by reputable manufacturers. The government and the defendant agreed at sentencing that the defendant "had at least the intent that the FDA not realize what he was doing, and he certainly was trying to hide his activities from the FDA, because he was worried that they certainly wouldn't approve of what he was doing." The defendant was sentenced to 24 months plus one year of supervised release, and the sentence was upheld on appeal, finding the court was justified in concluding that the offense involved fraud and deceit (Cambra). There are also misdemeanor offenses under the FD&C Act that do not require proof of fraudulent intent, or even knowing or willful conduct.

10.4 OTHER FDA ACTIONS

In addition to implementing provisions of the DSCSA, the FDA is taking various actions to protect consumers. Several of these efforts are described herein.

The Office of Criminal Investigations (OCI) conducts criminal investigations of illegal activities involving FDA-regulated products, arresting those responsible, and bringing them to the Department of Justice for prosecution. For example, a Florida man was sentenced to 3 years for selling counterfeit prescription drug pills through the Internet. He sold at least 249,700 counterfeit Xanax pills through online hidden marketplaces. He used a pill press at home to manufacture pills that did not contain just alprazolam but also flualprazolam, etizolam, adinazolam, and microcrystalline cellulose (Florida Man). The FDA Office of Criminal Investigations noted that "[s]elling counterfeit prescription drugs in the U.S. marketplace puts consumers' health at risk," and "[t]he FDA remains fully committed to disrupting and dismantling illegal prescription drug distribution networks that place profits ahead of public health and safety" (Florida Man). Also, in April 2021, a Pennsylvania man was sentenced to 70 months' imprisonment for trafficking counterfeit drugs, and he was ordered to pay over $3 million in restitution (Lebanon County Man). He admitted that he trafficked drugs, knowing them to contain counterfeit marks of Pfizer Pharmaceuticals, Bayer AG, Eli Lilly and Company, and Roche Holding AG. He also acknowledged that he trafficked counterfeit Viagra, Aurogra, Xanax, Levitra, Cialis, and Valium, all using counterfeit trademarks of the respective pharmaceutical companies (Lebanon County Man).

The FDA also issues warning letters to inform website operators believed to be engaged in illegal activity in violation of the FD&C Act (Internet Pharmacy Warning Letters). Internet warning letters have been issued for offering unapproved prescription drugs of unknown origin, safety, and effectiveness, offering prescription drugs without a prescription, offering prescription drugs without adequate directions for safe use, and offering prescription drugs without FDA-required warnings about health risks associated with the drugs. The FDA website allows Internet pharmacy complaints to be reported at https://www.fda.gov/safety/report-problem-fda/reporting-unlawful-sales-medical-products-internet. The FDA website also maintains a list of warning letters that have been sent, when the letters have been issued, and the public may review the actual warning letters as well (Internet Pharmacy Warning Letters).

In the science/forensic context, the FDA's Forensic Chemistry Center provides "rapid response and specialized analytical services in forensic chemistry and molecular/microbiology related to product tampering, counterfeiting, and adulteration/contamination" (Forensic Chemistry Center (FCC)).

The FDA has run a Counterfeit Alert Network, which is a coalition of health profession and consumer groups that disseminate alerts about specific counterfeit drug incidents and measures to minimize exposure, develop educational information about how to identify counterfeit drugs, report suspected counterfeit drugs, and prevent them from entering the United States (Counterfeit Alert Network). Groups that are part of the Counterfeit Alert Network include American College of Physicians, American Medical Association, National Association of Chain Drug Stores, and National Consumer League (Counterfeit Alert Network). When there is a confirmed counterfeit case in the United States, the FDA sends an alert to the groups. The FDA also will send groups a notice if a counterfeit incident happens outside the United States that could affect them (Counterfeit Alert Network).

The FDA also cooperates with organizations, such as the U.S. Customs and Border Protection (CBP), to maximize inspection and detection to prevent illegal and harmful products from entering the United States through the International Mail Facilities (IMFs) and ports of entry that may threaten public health (FDA and CBP). The efforts include expanded information sharing to identify trends to target future entries, such as general data points on frequent countries of origin as well as specific products and volumes of packages at each location (FDA and CBP).

10.5 SHOP SAFE ACT

The Stopping Harmful Offers on Platforms by Screening Against Fakes in E-Commerce (SHOP SAFE) Act was introduced in Congress in May 2021 to provide policing of online marketplaces and help prevent sales of counterfeit goods through online marketplaces. The House Judiciary Committee describes the SHOP SAFE Act as a way of "making platforms appropriately responsible for harmful counterfeits sold through their websites by others and encouraging them to go on the offensive in the fight against fakes" (House Committee on the Judiciary, 2021). "This legislation ensures that online platforms hosting third-party sellers are incentivized to engage in best practices for screening and vetting sellers and products, addressing repeat counterfeiter sellers, and ensuring that consumers have access to valid information when they make their purchases online" (House Committee on the Judiciary, 2021).

More specifically, Congress believes that the SHOP SAFE Act will:

- Establish trademark liability for online marketplace platforms when a third party sells a counterfeit product that poses a risk to consumer health or safety and that platform does not follow certain best practices;
- Incentivize online platforms to establish best practices, such as vetting sellers to ensure their legitimacy, removing counterfeit listings, and removing sellers who repeatedly sell counterfeits; and
- Call for online marketplaces to take steps necessary to prevent the continued sale of counterfeits by the third-party seller or face contributory liability for their actions.

(SHOP SAFE Act Section-by-Section)

Under *Tiffany (NJ), Inc. v. eBay*, 600 F.3d 93 (2d Cir. 2010), a platform is only liable for the activities of a third party if it intentionally induced that party to infringe, or it continued to supply its services to a party that it had reason to know was engaging in trademark infringement. Congress asserts that the SHOP SAFE Act provides a statutory scheme for assessing contributory liability that replaces the *Tiffany v. eBay* framework in covered circumstances (SHOP SAFE Act Section-by-Section). The Act requires that the platform undertake certain best-practice efforts for the platform to avoid liability under the commonly accepted contributory infringement theory articulated in Tiffany. The "best practices" the platforms must adopt are:

- Confirm the seller has a designated agent for service of process in the United States or a verified U.S. address for service of process in the United States (§ 1114(4)(A)(i)).
- Verify the seller's identity, location, and contact information (§ 1114(4)(A)(ii)).
- Require the seller to verify and attest that its goods are authentic (§ 1114(4)(A)(iii)).
- Condition the seller's use of the platform on agreeing not to sell counterfeits, consenting to being sued in U.S. court, and designating an agent or verified address for service of process in the United States (§ 1114(4)(A)(iv)).
- Display in listings the seller's identity, location, and contact information, and the country from which the goods will be shipped, with exceptions for personal information (§ 1114(4)(A)(v)).
- Display in listings the country of origin of the goods (§ 1114(4)(A)(vi)).
- Require sellers to use images that accurately depict the actual goods offered for sale, and that the seller owns or has permission to use (§ 1114(4)(A)(vii)).
- Use technology to screen for counterfeits before a seller's goods appear on the platform (§ 1114(4)(A)(viii)).
- Expeditiously remove listings selling counterfeit goods (§ 1114(4)(A)(ix)).
- Terminate sellers that have repeatedly listed or sold counterfeit goods on the platform (§ 1114(4)(A)(x)).
- Screen sellers to prevent terminated sellers from rejoining or remaining on the platform under a different alias or storefront (§ 1114(4)(A)(xi)).
- Provide a means to contact an allegedly infringing seller upon a registrant's request (§ 1114(4)(A)(xii)).

(SHOP SAFE Act Section-by-Section)

While much of what is described in the SHOP SAFE Act is generally directed to online sales, it is believed that the framework of this Act may provide steps to reduce the number of counterfeit drugs circulating in the marketplace. The Alliance for Safe Online Pharmacies (ASOP Global, 2022) has endorsed the

Act noting "The SHOP Safe Act reflects ASOP Global's core policy priorities of ensuring transparency, accountability, and consumer safety online. While more must be done at the federal level to protect Americans from criminals peddling dangerous drugs online through marketplaces, social media, and websites, ASOP Global is proud to support the bipartisan SHOP SAFE Act's policy goals. We believe, it is time to take a comprehensive approach to ensure consumers' safety and well-being in this online ecosystem, and the SHOP SAFE Act is a substantial leap in the right direction" (ASOP Global).

10.6 LANHAM ACT—COMBATTING PRICE-GOUGING AND COUNTERFEIT ISSUES THROUGH TRADEMARKS

Federal trademark law is governed by the Lanham Act. The Lanham Act provides for federal trademark registration and protects the owner of a federally registered mark against the use of similar marks if such use is likely to result in consumer confusion. Thus, trademarks can be important in the pharmaceutical field and/or with medical-oriented products, such as N95 respirators, as will be discussed herein, to ensure that consumers are getting what they intend to purchase.

Since the onset of the COVID-19 pandemic, there has been a significant need for personal protective equipment, particularly N95 respirators. 3M Company ("3M") manufactures N95 respirators, and there have been several bad actors that have manufactured counterfeit respirators and engaged in price gouging to improperly profit on the need for these products. These price-gouging efforts hurt those who desperately need these products and also damage 3M's brand image and goodwill, through no fault of its own. To that end, 3M has had to respond to the price-gouging, fraud, and counterfeiting issues that have arisen with respect to its 3M-branded N95 respirators.

In January 2020, 3M had zero trademark lawsuits pending relating to its 3M-branded N95 respirators. In the next six months alone, 3M received more than 4,000 complaints of fraud in connection with the sale or distribution of N95 respirators, reviewed each of these complaints, referred hundreds of matters to law enforcement for possible investigation, and filed numerous trademark-infringement lawsuits against bad actors claiming to be affiliated with 3M. 3M has brought numerous trademark suits, claiming its brand is being irreparably harmed by companies: (1) that are falsely claiming that they are authorized distributors of 3M products and creating consumer confusion; and (2) pricing 3M-brand N95 masks at approximately four to five times 3M's list price. As of January 11, 2022, 3M has filed 41 lawsuits and sent over 220 cease-and desist letters (3M Multimedia Infographic, 2022).

3M's lawsuits have sought relief under a variety of theories, including trademark theories of liability, including deceptive business practices, unfair competition, false advertising, false association, trademark dilution, and trademark infringement. While resale at an inflated price is not necessarily trademark infringement, the defendants have sought (or obtained) large-scale contractual sales with public and private entities by implying an authorization or relationship with 3M, when there is no such authorization or endorsement, and the defendants are not authorized distributors of the N95 respirators.

Early results have been promising, including obtaining multiple preliminary injunctions (3M/Performance Supply, 2020). Further, 3M's lawsuits have publicly shown that 3M distances itself from the price gouging that is occurring. The lawsuits also demonstrate how 3M protects the goodwill associated with its brands.

Efforts undertaken by 3M have included:

- 3M delivered a letter to the Attorney General of the United States, Chair of the National Governors Association, and President of the National Association of Attorneys General highlighting 3M's actions to curb counterfeiting and price gouging of PPE during the COVID-19 pandemic, and urging the U.S. federal and state governments and law enforcement officials to continue to lead the fight, with 3M's support and assistance (3M Supports Efforts, 2022).
- 3M launched a hotline and website (https://engage.3m.com/covidfraud) to help identify and fight fraud, price gouging, and counterfeiting of 3M respirators. This allows fraud to be reported on online marketplaces, company websites, social media, and other means of contact (3M Reporting, 2022).

In addition to partnering with state attorneys general and state and federal law enforcement to stop fraud, price gouging, and counterfeiting of 3M respirators, 3M also filed multiple lawsuits against alleged price gougers in the United States and Canada. Some of these lawsuits include:

- In federal court in Tallahassee, Florida, 3M sued Atlanta, Georgia-based 1 Ignite Capital LLC, Institutional Financial Sales LLC, and Auta Lopes for attempting to sell 10 million N95 respirators to the Florida Division of Emergency Management at nearly 460% percent over list prices, falsely claiming that they were working with 3M. This lawsuit was resolved (3M Settles, 2022).
- In federal court in Tampa, Florida, 3M sued St. Petersburg, Florida-based TAC2 Global LLC for claiming to be a 3M distributor and for trying to sell the Florida Department of Management Services State Emergency Operations Center 5–10 million N95 respirators and hand sanitizer at highly inflated prices. TAC2 falsely claimed to be a 3M supplier (3M Has Sued, 2020).
- In federal court in Orlando, Florida 3M sued King Law Center, Chartered for twice pretending to be affiliated with 3M as a vendor and escrow agent, and for trying to sell the Florida Department of Management Services State Emergency Operation Center 5 million N95 respirators at 460% over list prices (3M Has Sued).

- In federal court in Indianapolis, Indiana, 3M sued Zachary Puznak and two related entities, Zenger LLC and ZeroAqua, after Puznak claimed to be working with 3M and purported to be able to sell up to 5 billion 3M respirators to the state of Indiana at more than double the list price. Puznak accused Indiana's state employees of "paranoid irrationality" for asking for confirmation of any connection to 3M and falsely claimed that 3M executives had told him to abandon the deal, according to 3M's complaint. In fact, Puznak has no connection whatsoever to 3M. 3M resolved this case, including the grant of a consent judgment ensuring no additional infringement and a payment that will be donated to Direct Relief, as well as assistance identifying other bad actors and an apology to the state of Indiana (3M Has Sued).
- On April 28, 2020, 3M filed a lawsuit in federal court in Madison, Wisconsin, against Hulomil LLC for trying to sell 250,000 N95 respirators to state officials at inflated prices, while trying to force Wisconsin to sign a nondisclosure agreement about the deal and falsely claiming to have "direct access from 3M." A consent judgment and permanent injunction was entered in June 2020. The defendant also agreed to make a payment that will be donated to Direct Relief and agreed to provide assistance identifying other bad actors (3M Has Sued).
- 3M filed a lawsuit in federal court in Minnesota against Legacy Medical Supplies, LLC, Mark Eckhardt, Carol Ann Korpi, Joseph Nelson, and Jeremy Reboulet for claiming affiliation with 3M, including misleading potential buyers that they had a direct relationship with 3M's Chief Financial Officer to get special access to 3M products. This was false. Two defendants quickly settled the claims against them, and the court entered a temporary restraining order against the remaining defendants on July 14, 2020 (Hughlett).
- 3M brought suit and won a temporary restraining order in the case against Matthew Starsiak and AMK Energy in federal court in Minnesota (Civil Action No. 0:20-cv-01314); the defendants falsely claimed to be affiliated with (among others) 3M, the Gates Foundation, and the law firm Dentons to attempt to deceive buyers into purchasing billions of fictitious 3M N95 respirators. A temporary restraining order was entered (Hughlett, 2020).
- 3M resolved a case in federal court in Ohio (Civil Action No. 2:20-cv-02932) against Preventative Wellness Consultants LLC d/b/a Preventative Wellness Solutions, which was identified by pending litigation with Rx2Live, Inc. and Rx2Live, LLC. Preventative Wellness falsely claimed to provide "direct" access to 3M respiratory products when it is not an authorized distributor. In resolving the case, Preventative Wellness has agreed to a consent judgment, preventing further infringement as well as a payment that will be donated by 3M to Direct Relief (3M Updates).

Intellectual property rights can be used in unique ways to stop illegal sales, price gouging, or other distribution problems that may cause harm to a business' reputation. It also is critical to monitor distributors and other "gray market" sales to prevent damage and unauthorized sales. The bullet points above provide but a few examples of the many actions 3M is taking to stop and deter price gouging, fraud, and counterfeiting to protect people around the world. However, these examples reflect the importance of having numerous avenues of attack open to address issues related to critical goods, whether in a pandemic or at other times of crisis.

10.7 OTHER ACTIONS OR RECOMMENDATIONS

In "The Health and Economic Effects of Counterfeit Drugs," Erwin A. Blackstone and co-authors outline several actions/recommendations that could combat the counterfeit drug problem (Blackstone et al., 2014). First, public awareness must be raised, especially as the number of Internet pharmacies is on the rise. There should be more voluntary cooperation from companies participating in the Internet chain of commerce including, but not limited to, credit card companies, domain registrars, Internet service providers, and couriers. This might make counterfeiting of drugs less profitable. Internet search algorithms may be constructed so that legitimate online pharmacies appear first. These would be pharmacies that meet the standards of recognized industry organizations or licensing authorities. Other recommendations include more due diligence on the part of physicians who are purchasing drugs and stiffer penalties (even loss of medical license) for physicians who knowingly provide counterfeit drugs to their patients. Similarly, stiffer fines and even jail sentences should be employed for those convicted of selling counterfeit drugs.

The supply chain for drug manufacturing and distribution must be managed better. This includes improved controls of secondary drug markets. Drug supplies should only be sold to licensed manufacturers. Technology may be used better to track and trace counterfeit drugs. Further, with respect to international distribution of drugs, cooperation with foreign governments should be improved (i.e., raising awareness when problems arise). Overall, Blackstone et al. (2014) assert that "technological approaches should be utilized when appropriate and feasible to help ameliorate the counterfeit problem."

Consumers should be encouraged to know where the online pharmacy is actually located and the actual source of their drug supply. Online pharmacies may be verified through NABP and/or confirmed by Verified Internet Pharmacy Practice Sites (VIPPS) program certification. An online pharmacy's license may be checked through the state board of pharmacy to evaluate whether it is a "safe" pharmacy. If the online pharmacy is not listed, it should not be used (Locate a State-Licensed Online Pharmacy, 2022). In addition to scrutinizing the location and/or licensing of an online pharmacy, consumers should also be leery of any company that will distribute prescription drugs without a prescription. Even if the pharmacy is listed, consumers should confirm that the pharmacy requires a doctor's prescription, provides a physical address and telephone number in the United States, and has a licensed pharmacist to answer questions. Consumers should also be informed about the extent of counterfeit drugs and the harm they cause. Consumers should be aware of possible signs of a counterfeit drug and confirm whether a drug has been purchased from an online seller, whether the drug or packaging looks different from what has been normally received, and whether any new or unusual side effects are experienced when using the drug. Finally, if a consumer has an experience with an unsafe online pharmacy, has adverse effects caused by a medicine, or suspects criminal counterfeit activity, a consumer should be prepared to report as follows:

- Report sales of medicine on the Internet by unsafe online pharmacies to the FDA (Reporting Unlawful Sales).
- Report adverse effects caused by any medicine to FDA's MedWatch program (MedWatch).
- Report suspected criminal counterfeit activity to FDA's Office of Criminal Investigations (OCI Field Office Contact Information).

10.8 CONCLUSION

Counterfeit drugs present a problem for consumers, drug manufacturers and suppliers, medical personnel, and the government to combat. Prior and current actions by the United States Congress as well as governmental agencies such as the FDA provide good practices and rules of thumb to follow both to reduce to the likelihood of counterfeit drugs entering the consumer marketplace and get them out of the consumer marketplace. Despite these efforts, consumers, drug manufacturers and suppliers, medical personnel, and the government will need to continue to devise new ways to address the problem.

REFERENCES

3M Infographic on Fighting Respirator Fraud. Accessed January 24, 2022. https://multimedia.3m.com/mws/media/1862180O/3m-covid-19-infographic-print-version.pdf.

3M Co. v. Performance Supply, LLC, Case No. 1:20-cv-02949 (S.D.N.Y. May 4, 2020).

"3M COVID-19 Anti-Fraud, Anti-Price Gouging, and Anti-Counterfeiting Reporting." Accessed January 24, 2022. https://engage.3m.com/covidfraud.

"3M Has Sued 5 Vendors Who Targeted Emergency Officials in 3 States Offering Billions of Nonexistent N95 Respirators." (May 1, 2020). https://investors.3m.com/news/news-details/2020/3M-Has-Sued-5-Vendors-Who-Targeted-Emergency-Officials-in-3-States-Offering-Billions-of-Nonexistent-N95-Respirators/default.aspx.

"3M Settles Two Lawsuits in Florida." Accessed January 24, 2022. https://news.3m.com/2020-05-22-3M-Settles-Two-Lawsuits-in-Florida.

"3M Supports Efforts to Curb Pandemic Profiteers." Accessed January 24, 2022. https://news.3m.com/2020-03-24-3M-Supports-Efforts-to-Curb-Pandemic-Profiteers.

"3M Updates Ongoing Actions to Combat COVID-related Fraud." Accessed January 24, 2022. https://news.3m.com/2020-07-16-3M-Updates-Ongoing-Actions-to-Combat-COVID-related-Fraud.

"ASOP Global Expresses Support for SHOP SAFE Act." (May 26, 2021). Accessed January 23, 2022. https://buysaferx.pharmacy/shop-safe-act-support/.

Blackstone, Erwin A. et al. "The Health and Economic Effects of Counterfeit Drugs." *Am. Health Drug Benefits*, 2014 Jun; 7(4): 216–224.

Department of Justice, Consumer Protection Branch, "The Federal Food, Drug, and Cosmetic Act (FDCA)." (November 17, 2021). Accessed January 23, 2022. https://www.justice.gov/civil/consumer-protection-branch-29.

Department of Justice, "Florida Man Sentenced for Selling Counterfeit Drugs on the Dark Net." (August 24, 2021). Accessed January 24, 2022. https://www.fda.gov/inspections-compliance-enforcement-and-criminal-investigations/press-releases/florida-man-sentenced-selling-counterfeit-drugs-dark-net.

Department of Justice, "Lebanon County Man Sentenced to Seventy Months' Imprisonment for Trafficking Counterfeit Drugs." (April 15, 2021). Accessed January 24, 2022. https://www.fda.gov/inspections-compliance-enforcement-and-criminal-investigations/press-releases/lebanon-county-man-sentenced-seventy-months-imprisonment-trafficking-counterfeit-drugs.

Drug Supply Chain Security Act (DSCSA), FDA, http://www.fda.gov/Drugs/DrugSafety/DrugIntegrityandSupplyChainSecurity/DrugSupplyChainSecurityAct/default.htm [hereinafter Drug Quality and Security Act of 2013] (last visited Jan. 10, 2022).

Drug Supply Chain Security Act Implementation: Identification of Suspect Product and Notification Guidance for Industry. Accessed January 24, 2022. https://www.fda.gov/media/88790/download.

DSCSA Standards for the Interoperable Exchange of Information for Tracing of Certain Human, Finished, Prescription Drugs: How to Exchange Product Tracing Information, https://www.fda.gov/media/90548/download.

Federal Food, Drug, and Cosmetic Act, 21 U.S.C. 301. "About OCI". December 3, 2018. https://www.fda.gov/inspections-compliance-enforcement-and-criminal-investigations/criminal-investigations/about-oci

House Committee on the Judiciary, "Nadler, Johnson, Issa & Cline Introduce Bipartisan SHOP SAFE ACT." (May 20, 2021). https://judiciary.house.gov/news/documentsingle.aspx?DocumentID=4566.

Hughlett, Mike, "3M has investigated 4,000 reports of N95 fraud, filed 18 lawsuits." July 16, 2020. https://www.startribune.com/3m-has-investigated-4-000-reports-of-n95-fraud-filed-18-lawsuits/571790002/.

Locate a State-Licensed Online Pharmacy. (n.d.). Accessed January 24, 2022. https://www.fda.gov/drugs/besaferx-your-source-online-pharmacy-information/locate-state-licensed-online-pharmacy.

Magdun, Melanie, "NOTE: Trademark Enforcement of Counterfeit Drugs: A Guardian of the Rich and Poor Alike." (2021). 9 Ind. J.L. & Soc. Equality 281, 288.

Myshko, Denise, Online Pharmacies that Sell Counterfeit Drugs Grew During the Pandemic. (December 15, 2021). https://www.formularywatch.com/view/online-pharmacies-that-sell-counterfeit-drugs-grew-during-the-pandemic

Partnership for Safe Medicines, "Hard to Believe, But True." (August 15, 2011). Accessed January 10, 2022. https://www.safemedicines.org/2011/08/hard-to-believe-but-true-4018.html.

The SHOP SAFE Act of 2021 Section-by-Section. Accessed January 24, 2022. https://judiciary.house.gov/uploadedfiles/shop_safe_2021_-_section-by-section.pdf.

United States v. Cambra, 933 F.2d 752, 755 (9th Cir. 1991).

Index

Note: Page numbers in *italics* refer to figures and those in **bold** to tables.